D1176048

9 · TROPICAL SOILS AND SOIL SURVEY

CAMBRIDGE GEOGRAPHICAL STUDIES

TROPICAL SOILS AND SOIL SURVEY

ANTHONY YOUNG

Professor of Environmental Sciences, University of East Anglia

CAMBRIDGE UNIVERSITY PRESS

CAMBRIDGE

LONDON · NEW YORK · MELBOURNE

Published by the Syndics of the Cambridge University Press
The Pitt Building, Trumpington Street, Cambridge CB2 1RP
Bentley House, 200 Euston Road, London NW1 2DB
32 East 57th Street, New York, NY 10022, USA
296 Beaconsfield Parade, Middle Park, Melbourne 3206, Australia

Library of Congress catalogue card number: 75–19573

ISBN: 0 521 21054 2

First published 1976

Photoset and printed in Malta by Interprint (Malta) Ltd

Library of Congress Cataloging in Publication Data

Young, Anthony.

Tropical soils and soil survey.

(Cambridge geographical studies; 9)

Bibliography: p.

Includes index.

1. Soils—Tropics.   2. Soil-surveys—Tropics.
I. Title. II. Series.

S599.9.T76Y68   631.4'913   75–19573
ISBN 0 521 21054 2

CONTENTS

ACKNOWLEDGEMENTS

Parts of Chapters 18, 19 and 20 incorporate material from articles by the author in the following sources, and are reproduced by kind permission of the publishers: *Evaluating the human environment* (edited J. A. Dawson and J. C. Doornkamp), Edward Arnold, 1973; *Spatial aspects of development* (edited by B. S. Hoyle), John Wiley and Sons, 1974; and *Geographical Journal*, vol. 139, (1973), pp. 53–64. Figures 3, 4A, 5, 10, 12, 14, 15C, 18C, 22, 23 and 24 are based on or modified from the references cited in the captions and listed in full in the Bibliography; grateful acknowledgement is made of permission for this granted by the respective authors and publishers.

PREFACE

This is a book about tropical soils written from the point of view of the field soil scientist. I have had in mind particularly those who are concerned with, or training for, field studies of soils in tropical and subtropical regions, whether as soil surveyors or in other capacities. Whilst directed primarily at the specialist, I hope that the greater part of the text will also be both comprehensible and useful to non-pedologists, including agriculturalists, economists and planners, who wish to understand the significance of soils for questions of agricultural and economic development.

The orientation towards field study has involved differences of emphasis as compared with most books on soils. The tropical environment is treated at some length, since it is impossible to understand soils and, in particular, their resource potential, in isolation from other environmental factors. Conversely the sections on soil chemistry are shorter than is usual; they do not contain any attempts at new interpretations or syntheses, but rather seek to interpret laboratory data in terms of their consequences for the morphological properties of soils and their productive capacity. In discussing the identification and classification of soil types, greater emphasis is placed on properties which can be observed in the field than on those which require chemical or mineralogical analysis. There are other texts available, some of which are noted on p. 303, which treat the chemistry and chemical aspects of fertility of tropical soils in greater detail.

Aspects of both pure and applied soil science have been included. The chapters on processes, soil properties, classification, and problems of soil evolution may be regarded as pure soil science, the study of the soil as a natural body, whilst Part Four, on soil fertility, soil survey and land evaluation is concerned with applications in agricultural development. In Part Two, in which the main soil types of the tropics are described, both pure and applied aspects are included and, where relevant, discussed in relation to each other. It is a fundamental precept of the environmental sciences that the natural processes which have brought soils, as landforms, vegetation and other factors, to their present state have important consequences for their potential and hazards of development.

I have throughout preferred to depend on my own observations wherever possible, and to have recourse to published accounts only for certain soil types of which I have limited experience. This must inevitably result in some lack of balance in treatment or emphasis; but my intention is to avoid what has been all too common in written accounts of tropical soils, the copying of misconceptions from one source to another. Hence this is not a fully comprehensive and impartial description of the present state of knowledge, but more a personal interpretation based on fairly extensive field experience. Some of my firmly-held opinions, or prejudices, will be apparent, notably a dislike for complicated artificial systems of soil classification which, in the words of C. F. Charter, 'are unpleasantly reminiscent of the filing systems of Government Departments'.

An adequate treatment of soil erosion and conservation in the tropics appeared to require more space than was available, and there are recent books on the subject (Hudson, 1971). I have discussed the related, and in my view more serious, matter of soil degradation (p. 120), and erosion is referred to in the sections on individual soil types and on soil survey. The omission of a systematic treatment should not be taken to imply that I do not regard erosion as an important problem.

In addition to the publications cited in the text, short lists of selected references are given following chapters 4–20. These indicate what in my view are some of the most important books and articles on each subject, and some also contain bibliographies.

I owe a debt of thanks to those from whom I learnt about soils in the field: Alan Crompton, George Jackson and Ron Paton. I will not name individually the many field assistants who have accompanied me in Africa, Asia and South America, but thank all for their cheerful acceptance of an often arduous job. It is also a pleasure to acknowledge the almost invariably courteous and friendly way in which farmers in the tropics greet the unannounced appearance on their land of that strange being, the soil surveyor. I am grateful to Drs A. van der Wal, Librarian of the Royal Tropical Institute, Amsterdam, for his kind help with the history of soil survey in the tropics. Finally may I record my gratitude to my wife, who for nearly twenty years has accompanied me to far-flung lands, has been willing to act as a field assistant when circumstances required, and has been a constant source of help and encouragement.

Norwich 1976   Anthony Young

PART ONE

SOIL FORMATION IN THE TROPICS

CHAPTER 1
ENVIRONMENT

Tamino: Nun sag' du mir, in welcher Gegend wir sind.
Papageno: In welcher Gegend? (Sieht um sich.) Zwischen Tälern und Bergen.
Die Zauberflöte

The natural limits for the study of the tropical environment, including tropical soils, lie for the most part close to the limits of the Tropics defined in the astronomical sense, as latitudes $23\frac{1}{2}°$ North and South. On the western margins of the continental land masses these parallels pass through sparsely inhabited deserts – Sahara, Kalahari, Lower California, Atacama and Western Australia – which separate the Mediterranean lands on the poleward side from the humid lands of summer rainfall toward the Equator. On the eastern continental margins there is no such natural division, but instead a continuous zone of humid, well-populated lands with only a gradual and progressive change from cooler to warmer temperatures. Any limit in these regions must necessarily be arbitrary (cf. Fosberg *et al.*, 1961). The $23\frac{1}{2}°$ parallels are, however, as good a boundary as any. On such a basis all or most of the West Indies, Mexico, Bolivia, Brazil, Sudan, Rhodesia, Moçambique and Malagasy are tropical, whilst Florida, Argentina, South Africa and the inhabited parts of Algeria, Libya and Egypt are not. The cut-off through Canton is arbitrary, Formosa may be included in the tropics, and in eastern Australia it is convenient to bend the boundary southwards to 29° latitude to include Brisbane and the whole of Queensland.

In monsoonal South Asia, however, the natural limit lies not along a parallel of latitude but at the barrier formed by the Himalayan chain. Winter in the Punjab apart, the Indo-Gangetic Plains are unquestionably tropical in character. One may therefore allow the boundary of the natural tropics to follow the course of the Himalayas from Assam to the Punjab, reaching to some 34° North at the Tarbela Dam on the River Indus. Finally, the limit must again be arbitrary in the Middle East, for Iran, Iraq and Saudi Arabia form a natural unit which spans the tropical and warm temperate zones.

There are those who argue that there are no such things as tropical soils, only soils of the tropics; that is, that the soils lying within the inter-tropical zone do not differ in essentials from those of temperate latitudes. Insofar as it is possible to speak meaningfully at such a generalized level I would take issue with this view. It is true that the soils of semi-arid lands poleward and equatorward of the subtropical desert belt do not differ greatly. But where

3

humid environments are concerned, there is a group of distinctive soil characteristics normal to the tropics and exceptional, or less clearly developed, in the temperate zone: among these are the processes of rapid weathering and strong leaching, the properties of a deep and highly-weathered regolith with a predominance of kaolinitic clay minerals, and the importance of organic matter in soil fertility and soil management.

For soils as such it is the natural, climatically determined, inter-tropical zone which forms the convenient unit of study. When considering methods of soil survey and land evaluation, the area within which there are common approaches and problems is the economic unit formed by the less developed countries.

Defined as above, the natural unit of the tropics is to a considerable extent coincident with the less developed world. Allowing that China is *sui generis*, the major anomalies are the extension of high-income Australia to nearly 10° South and, in the opposite sense, the inclusion in the less developed world of the states of the north coast of Africa and the Middle East. Lesotho and Swaziland are minor exceptions. Apart from these, the inter-tropical zone with its northward extension in the Indian subcontinent corresponds well with the less developed, or low-income, countries.

There is thus a dual focus to the topics considered in this book: for the soils as such, the climatically determined tropics; for their survey and evaluation, the less developed countries. These two foci come together in the discussion of soil management, in which the agricultural techniques originating from traditional societies, onto which are being grafted the processes of development, are applied to soils with distinctively tropical characteristics and responses.

THE FACTORS OF SOIL FORMATION

There are eight main factors of the physical environment: climate, geology, relief, hydrology, soils, vegetation, fauna and man. In considering soil formation, the soils themselves become the dependent variable and the other factors independent and causative. In Dokuchaev's original formulation the factors of soil formation were climate, organisms, substrata and age; a factor of relief was subsequently added. These five factors, with substrata termed parent material, were given in the major re-statement of the theory of soil-forming factors by Jenny (1941, 1946), reproduced in many textbooks. Jenny noted that relief 'also includes certain hydrologic features'. The separation of hydrology as an independent factor was due to Stephens (1947), who distinguished factors R = relief, the shape of the land surface, and W = water, the height and other features of the water table. The fact that the water table is partly influenced by relief does not prevent it from being taken as an independent variable, for there are many other cases of inter-relations between factors. Man was formerly regarded as a passive factor, influenced by the environment but not acting upon it; the recognition that human activity can substantially alter the physical environment is now widespread, and man should therefore be included as a factor affecting the state of soils as observed at the present day. Hence the *factors of soil formation* are as follows:

1. Climate
2. Parent material
3. Relief: the shape of the land surface
4. Hydrology: drainage
5. Organisms: vegetation and soil fauna
6. Time: soil age
7. Man

There have been several attempts to formulate theoretical relations between soil-forming factors and soil properties (Jenny, 1941, 1946; Stephens, 1947; Crocker, 1952). The equations of Jenny, modified by the addition of hydrology and man, and using the initial letters of the factors as listed above, are as follows, where s = any soil property, S = the totality of the soil, and f = 'a function of' or 'dependent on':

$$s = f(C, P, R, H, O, T, M \ldots) \tag{1}$$

$$S = f(C, P, R, H, O, T, M \ldots) \tag{2}$$

$$s = f(C)_{P, R, H, O, T, M \ldots} \tag{3}$$

$$s = f(P)_{C, R, H, O, T, M \ldots} \tag{4}$$

$$\text{and similarly for } R, H, O, T \text{ and } M \tag{5–9}$$

$$S = f(C)_{P, R, H, O, T, M \ldots} \tag{10}$$

$$\text{and similarly for } P, R, H, O, T \text{ and } M \tag{11–16}$$

Equation (1) states that any single soil property, such as colour, clay content or reaction, is dependent on each of the seven factors, the dots indicating that there may be other factors operating (e.g. past climate). Equation (2) makes the same statement for the soil as an entity. Equation (3) states that any soil property will be a function of climate if the other six factors are held constant. Equation (4) makes the same statement for the case in which only parent material is variable. Five other equations of similar form to (3) and (4) follow for the remaining factors. Finally there are seven equations of the form of (10) in which the soil as a whole is related to changes in individual factors.

To place the complex relations between environment and soils in the form of mathematical equations is a gross over-simplification of the real situation, and one which it is unlikely that Dokuchaev would have sanctioned. Few of these 'functions' have been numerically quantified, although Jenny made a number of pioneer attempts to do so, mainly with organic properties as the dependent variables (Jenny *et al.*, 1948, 1949, 1960). It is rarely possible to find groups of soils in which all factors but one are sufficiently similar to be regarded as constant, and attempts to correlate soil properties with single environmental variables without restricting the range of other factors usually produce a wide scatter of points. Most of the factors are themselves multiple variables; for example climate includes temperature and rainfall, whilst rainfall may be considered as the mean annual total or the length of the dry season. It is difficult to assign numerical values to some variables, e.g. vegetation. The factors are far from being independent. Finally, this approach tacitly assumes that the soil results from present-day factors and processes, neglecting relict soils.

There are thus many objections to treating factor/soil relations as equations. Nevertheless, analysis of the effects of environment requires that it be separated into individual factors, each made up of several properties. The approach embodied in the above equations is valid and necessary if they are regarded not as strict mathematical relationships but as a way of codifying a series of questions of the general form: 'If all other environmental conditions were held constant, what would be the effect on a soil property, or the soil profile as a whole, of changes in one given variable?'

The relations between soil-forming factors and soil properties, or *environment* and *form*, is indirect. The two are causally linked by the *processes* of soil formation. The soil-forming factors lead to the action of a particular set of processes, e.g. weathering, leaching; the processes, acting over time, bring about changes in soil properties. The change in form over time is the *evolution* of the soil.

Relations between factors and soils only come to life when considered in relation to the processes which have brought them about. Without this process link the relations remain statistical correlations, and it is well known that causal relationships cannot be established by correlations alone. So the study of environment/soil relations needs to be one of environment → process → soil.

The factors are not all of the same nature. Climate, relief, hydrology and man affect the processes of soil formation. The time factor refers to the period over which these processes have acted. Parent material supplies the mineral substance from which the soil profile is developed, and its effects on processes are indirect. Vegetation both supplies organic substances and affects processes.

In the following sections each of the seven factors is examined with reference to its effects on tropical soils. The aspects considered include the effects of each factor, or its component variables, upon individual soil properties, soil types and their distribution; classification of the factor, considered with reference to the classes relevant to soil formation; and the description of the factor in the course of soil survey.

It is assumed throughout this chapter that soil properties result from the processes currently operating, as determined by the present-day soil-forming factors. The question of relict features, those inherited from environmental conditions differing substantially from those of the present, is discussed in chapter 14.

CLIMATE

Climate and soil formation

The factor which directly influences soil-forming processes is soil climate rather than air climate. Rates of weathering are related to soil temperature and the period of the year over which the soil remains moist, rates of leaching to the volume of water passing downwards through the profile and to a lesser extent to its temperature. Moreover soil climate cannot be treated as wholly dependent on air climate; in at least one important instance, that of the clearance of rainforest for agriculture, substantial changes to the soil moisture regime can occur with unaltered air climate. Hence the influence of air climate on soil properties is an indirect one and in specialized studies, whether of pedogenesis or for agricultural purposes, the conditions of soil temperature, soil moisture content and soil water movement should be determined directly.

For more general studies this is impracticable. Soil temperature and moisture are only recorded at a few special agro-climatic stations. Most work on climate/soil relations must be based on data recorded at standard meteorological stations. To use properties of soil climate as differentiating criteria in soil classification systems, as has occasionally been done, is unrealistic.

Temperature and rainfall are the two pedogenetically significant properties of climate. The temperature variable is adequately represented by mean annual air temperature, which does not differ greatly from the nearly constant temperature in the lower part of the regolith. In studies concerned with processes in the topsoil, such as organic matter decomposition, there are arguments for taking the mean temperature of the hottest month as the more significant parameter. There are no known relations between soils and the range of temperature as such, and frost frequency, whilst pedogenetically significant, is not of importance in the tropics.

The two rainfall parameters most widely available are the mean annual total and the length of the dry season; the latter may be defined in the tropics, following the Köppen classification (p. 10), as the number of months with less than 60 mm rainfall. In plant ecology the latter is the more significant, and in soils the period and depth to which the profile dries out to wilting point is certainly of importance. But the main effects of rainfall concern leaching, and the agent directly causing this is the volume of water passing through the profile. The best measure of this, in the absence of lysimeter records, is excess of rainfall over evapotranspiration for the period of moisture surplus. This can be estimated from monthly totals of rainfall and potential evapotranspiration, the latter calculated either by one of the standard methods (Penman, 1948, or Thornthwaite, 1948) or by multiplying measured open-water evaporation by a constant (commonly taken as 0.8). For special studies of leaching intensity, the estimated wet-season moisture surplus is the best parameter.

Difficulties in the use of moisture surplus for more general, especially comparative, purposes are first, the limited availability of data, and secondly, the fact that potential evapotranspiration estimates calculated by different methods not infrequently differ by substantial amounts. The simpler and more readily obtainable value of mean annual rainfall is here preferred as the major index for classifying tropical climates. In comparing annual rainfall with soil types some allowance must be made for the effects of altitude; for example, soils which occur at any given rainfall total on the plateaux of Central Africa, at altitudes of 900–1200 m, are found in the lower altitudes of West Africa under rainfall amounts some 200 mm higher.

Thus, in specialized studies of soil-forming processes, records of soil climate should be taken. But as a basis for comparisons of climate with soil properties, in which a wide measure of generalization is in any case inevitable, the simpler and readily available figures for mean annual temperature and mean annual rainfall are satisfactory.

Climate is the most important soil-forming factor affecting organic matter content and associated properties (notably nitrogen), reaction and base saturation. It has a substantial influence on profile depth and texture and is one of several factors influencing the type of clay mineral synthesized.

Topsoil organic matter and nitrogen content increase with a fall in temperature and an increase in rainfall; the former slows the rate of loss of humus by oxidation, the latter the supply from plant growth. The effects of temperature and rainfall are hard to separate (see p. 103). There are inverse relations between rainfall and both pH and base saturation, since both properties are related to intensity of leaching. The link between rainfall and total exchangeable bases is much weaker as the latter is strongly affected by cation exchange capacity and therefore indirectly by texture. The shorter the dry season the higher the percentages of clay, although this effect is frequently masked by the stronger influence of parent material. The silt : clay ratio tends to decrease with increase in rainfall. Thus soils of rainforest climates are frequently high in clay and very low in silt, whilst at the opposite extreme,

Literature cited

Hallsworth E.G., & GG Beckmann
1969. Gilgai in the Quaternary
Soil Sci 107: 409-420

2 1955 J Soil Scie 6: 1-31

PA Deposystems and Paleoclimate of
the Appalachian. Don Aubson Abstract

Redbeds, calcium carbonate concretions
and gilgai paleosols suggest
seasonally dry tropical climate.

gilgai — microrelief of heavy
clay soils w/ high coefficients
of expansion and contraction
according to changes in
moisture typical of
vertisols.

Soil Genesis & Classificat.
631.4
B887s

under the slower chemical weathering of semi-desert and desert soils, much material remains as sand or silt particles.

Clay mineral synthesis involves interactions between climate, parent material and drainage, and is discussed further below (p. 73). The principal relations with climate are that gibbsite is only found where the soil is subject to leaching throughout the year, whilst montmorillonite is absent in such conditions and only found (on freely-drained sites) where the profile dries out seasonally. Arising through the properties of clay minerals there are indirect relations of rainfall with soil structure and consistence. A distinctive type of consistence sometimes described as 'floury', or in standard terms as a moderately-developed crumb structure and very friable, is found in soils of the rainforest zone, where it probably results from an absence of 2:1 lattice clay minerals coupled with the presence of free iron oxides.

In high mountain ranges of tropical and subtropical latitudes, such as the Andes and Himalayas, there is a well-established altitudinal succession of vegetation zones. It may be presumed that a corresponding altitudinal succession of soil changes exists, although this has not been so thoroughly described. With rainfall throughout the year, leached ferrallitic soils of the lowlands pass into humic ferrallitic soils at altitudes of about 1600 m, ultimately giving place to montane podzols and peaty soils at 2000 m or more (Askew, 1964). By keeping parent material constant, the changes in soil properties with altitude may be traced. Thus, on soils derived from volcanic ash in Colombia, as the mean annual temperature falls from 24 °C (1000 m altitude) to 8 °C (3700 m) there is an increase in organic matter content and decreases in clay percentage, kaolinite and halloysite content, degree of weathering of primary minerals and degree of development of the B horizon (Cortes and Franzmeier, 1972). For someone seeking research in a challenging environment, the zonation of soils on tropical mountains offers an attractive and open field.

Classification

A classification of one factor for use in analyzing another needs to be simple, with few and broad classes. Of the standard systems of climatic classification, that of Köppen (1931, 1936) is preferable to the two more complex systems of Thornthwaite (1933, 1948) for this reason. The Köppen classification is the most widely known (sometimes in the modified form given by Trewartha, 1943), and the class of any station can be rapidly determined from monthly figures for average temperature and rainfall. The parameters used in Thornthwaite's 1933 system, namely temperature efficiency, precipitation effectiveness and seasonal distribution of precipitation, are all of significance with respect to pedogenetic processes, and it would certainly be possible to

9

analyze climate/soil relations in terms of this classification. However, the system depends on an estimate of potential evapotranspiration, and as such estimates are known to differ widely according to the method of calculation, no absolute significance can be attached to the estimate based on Thornthwaite's formula. Moreover, it is frequently impossible to determine the Thornthwaite class from the information given in published soil descriptions. Every climatic classification has some unsatisfactory features, but no other system has proved to be demonstrably superior to that of Köppen. It is regrettable that the FAO, in the first (South America) volume of the *Soil map of the world*, used the more complex and less widely known system of Papadakis (1966).

The main Köppen classes for tropical and subtropical latitudes are given in table 1, together with an identification key. Except at very high altitudes, tropical regions fall into the groups of A (hot), C (warm) and B (dry) climates. Excluding some subscripts which are not of pedological significance (e.g. for equable temperature regimes), the system gives nine classes of tropical and subtropical climates. It should be noted that although the '*m*' subscript is derived from monsoon, 'A*m*' climates are not confined to the area of the Asiatic monsoon but occur in other rainforest–savanna transition zones.

The limiting values for Köppen classes were chosen for their significance with respect to vegetation communities, and it is not to be expected that they will correspond to soil types; in particular, the limit of dry, B, climates occurs at annual rainfall totals of 680–780 mm (for mean annual temperatures of 20 and 25 °C respectively), which is appreciably wetter than the pedalfer–pedocal transition. An unsatisfactory feature is that the boundary between A and C climates occurs in tropical latitudes at an altitude of about 1400 m, which means that the higher parts of the Central and East African plateau are arbitrarily separated. Trewartha (1943) preferred to retain these tropical medium-altitude areas within the A*w* class, as a 'highland' modification, confining the C*wa* class to subtropical latitudes. The main altitudinal soil change, a substantial increase in organic matter, takes place at higher altitudes, nearer to the C*wa*/C*wb* boundary. Köppen's A*w* (savanna) class covers a wide range of annual rainfall and length of dry season, and can usefully be divided into moist and dry types.

A climatic classification based on pedogenetic significance is given in table 2. The definitions are largely in terms of the value most often given in soil site descriptions, that of mean annual rainfall. Length of dry season, defined on the Köppen basis as months with less than 60 mm rainfall, is given as an associated rather than a defining property. The climates of the Asian monsoonal region do not appear to be distinctive with respect to pedogenesis from climates with similar rainfall regimes in other continents, unless it be that a higher seasonal concentration of rainfall in the monsoon region leads to

TABLE 1   *Main tropical divisions of the Köppen (1931, 1936) classification of climates. For English-language accounts of the system see Trewartha (1943, Appendix A) or James (1966, Appendix B)*

Climatic class	Temperature of coldest month (°C)	Temperature of warmest month (°C)	Mean annual temperature (°C)	Rainfall of the driest month (mm)	Mean annual rainfall (mm)
'A' climates	> 18	–	–	–	$> 20(t + 14)$
A*f*	> 18	–	–	> 60	$> 20(t + 14)$
A*m*	> 18	–	–	$> (100 - r/25)$	$> 20(t + 14)$
A*w''*	> 18	–	–	< 60 [a]	$> 20(t + 14)$
A*w*	> 18	–	–	< 60	$> 20(t + 14)$
'B' climates	–	–	–	–	$< 20(t + 14)$ [b]
B*Sh*	–	–	–	–	$< 20(t + 14)$
B*Wh*	–	–	> 18	–	$< 10(t + 14)$ [b]
'C' climates	> − 3	–	–	–	$> 20(t + 14)$
C*fa*	> − 3	> 22	–	> 30	$> 20(t + 14)$
C*wa*	> − 3	> 22	–	< 30	$> 20(t + 14)$
C*wb*	> − 3	< 22	–	< 30	$> 20(t + 14)$

Symbols in last two columns: t = mean annual temperature, °C; r = mean annual rainfall, mm.

Derivation of symbols: f = *feucht*, moist; m = monsoonal; w = dry winter; a and b, no meaning; S = steppe; W = *Wüste*, desert; A (hot climates), B (dry climates), C (warm climates) are used conventionally.

[a] A*w''* has two seasonal rainfall maxima (not precisely defined).

[b] If less than 70 percent of the rainfall falls in the 6 summer months, $(t + 14)$ is replaced by $(t + 7)$; this is not usually the case in the tropics.

Identification key (tropical latitudes only)

1. Is mean annual rainfall less than $20(t + 14)$?	YES	2
	NO	3
2. Is mean annual rainfall less than $10(t + 14)$?	YES	B*Wh*
	NO	B*Sh*
3. Is temperature of coldest month over 18 °C?	YES	4
	NO	7
4. Is rainfall of driest month more than 60 mm?	YES	A*f*
	NO	5
5. Is rainfall of driest month more than $(100 - r/25)$?	YES	A*m*
	NO	6
6. Are there two distinct rainfall maxima?	YES	A*w''*
	NO	A*w*
7. Is rainfall of driest month more than 30 mm?	YES	C*fa*
	NO	8
8. Is temperature of warmest month over 22 °C?	YES	C*wa*
	NO	C*wb*

a somewhat lower effectiveness of a given annual rainfall total; these climates are therefore not classified separately. There is a broad correspondence with Köppen classes, with the modifications noted above. The pedological characteristics of these climatic regions are as follows:

11

TABLE 2 *Climatic classes of pedogenetic significance in the tropics*

Climatic region	Mean annual rainfall (mm)	Dry season: approximate number of months with < 60 mm rainfall	Other features	Approximate Köppen equivalent	West African zonal nomenclature equivalent
Rainforest	> 1800	0–2		*Af, Am* p.p.	Forest
Rainforest–savanna transition	1200–1800	2–6		*Am* p.p.	Forest–savanna mosaic
Moist savanna with two wet seasons	900–1200	3–5		*Aw"*	
Moist savanna	900–1200	5–7		*Aw* p.p.	Guinea
Dry savanna	600–900	6–8		*Cwa* p.p.	Sudan
Semi-arid	250–600	8–10		*BSh*	Sahel
Arid	< 250	10–12		*BWh*	Saharan
Tropical high-altitude	> 600	0–6	Altitude > 1600 m	*Cwb, Cwa* p.p.	
Subtropical humid	> 900	0–3	Subtropical latitudes	*Cfa*	

Note. The term *savanna* is used in a climatic sense, as a shorter alternative to 'wet-and-dry tropics', and refers to humid climates with a dry season; it does not refer, in this context, to a vegetation type. Asian monsoonal climates are not classified separately; each of the classes shown is represented in the Asian monsoon region.

Rainforest. These are lowland tropical rainforest areas either with no dry months or with a dry season insufficiently long for the soil moisture to fall to wilting point to any substantial depth. The permanently warm and moist conditions give rapid and intense weathering, with complete breakdown of most primary minerals other than quartz. The large moisture surplus for all or most months leads to strong leaching throughout the year, giving strongly acid soils with a low base saturation. Kaolinite dominates the clay fraction of soils, gibbsite is commonly present, but leaching is too intense for the synthesis of montmorillonite. Leached ferrallitic soils are the zonal soil type. There is no clear pedological differentiation between climates with no dry season and monsoonal climates with up to three dry months. The main agricultural systems are shifting cultivation (often of root crops), the growth of perennial crops and padi cultivation. Cultivation of annual crops without bush fallowing is difficult or impossible.

Rainforest–savanna transition. These areas have a natural vegetation either of semi-deciduous rainforest or a mosaic of forest and savanna. They include the wetter parts of the Asian monsoon region. Leaching is appreciably less intense than in the rainforest zone, resulting in somewhat higher values of acidity, base saturation and hence total exchangeable bases. It has been held that ferrisolic properties, a strongly-developed structure with clay skins, are particularly likely to develop in this zone, although parent rock of at least intermediate composition is probably necessary (p. 148). The profile dries out seasonally in the upper horizons but remains moist in depth. This is a favourable climate for perennial crops which permit or benefit from a dry season and which do not tolerate strong acidity and desaturation, e.g. cocoa. Annual crops requiring a long growing season, e.g. yams, are also cultivated.

Moist savanna with two wet seasons. This climatic type, well represented in Uganda, is particularly favourable for agriculture. Weathering and leaching are both of moderate intensity. The upper soil horizons dry to wilting point twice annually (or once, since the shorter dry season is not always apparent in individual years), whilst the soil remains moist within the reach of deeper-rooting crops. This is another climate favouring the development of ferrisols on rocks of basic and intermediate composition. The climate permits the cultivation of perennials that require a dry season, notably coffee, or alternatively double-cropping of annuals, e.g. cotton and a cereal crop.

Moist savanna. In the more widespread type of savanna climate there is one wet and one dry season. Soils are leached in the former and dry to wilting point to a depth of over one metre in the latter. The seasonal alternation of intense weathering under moist conditions with drying out and consequent oxidation of iron compounds makes rubefication a characteristic process. The clay fraction is still dominated by kaolinite, but small amounts of 2:1 lattice clay minerals are often present, and gibbsite absent. Acidity and base saturation values are moderate, e.g. pH 5.0–6.0, saturation 40–60 percent. Ferruginous soils are the zonal type, and weathered ferrallitic soils are also widely developed. The moist savannas have an adequate growing season for most annual crops and a low drought hazard.

Dry savanna. In the drier division of the savannas rainfall is concentrated into approximately five months (June–October or December–April). There is a long and intense dry season, during which the depth to which soil moisture falls to wilting point may reach 2 m. With the less intense leaching, base saturation values rise to 60–90 percent, but free calcium carbonate does not occur. Weathered ferrallitic soils of fairly low agricultural potential are widespread, together with more fertile ferruginous soils. This is another zone of annual cropping, but suited to crops and cultivars with a short growing season. There is a fairly high rainfall variability, and annual totals of less than 400 mm occur at intervals of the order of one in ten or twenty years causing total or partial crop failure. Both the moist and dry savanna climatic are represented in the Asian monsoon region, particularly in the Indian Deccan.

Semi-arid. This is sometimes called the steppe zone, although it should not be confused with the steppes of warm temperate latitudes characterized pedologically by chernozems. In the tropics, as the mean annual rainfall becomes less than about 600 mm (or about 500 mm at moderate altitudes) there is a fundamental change in pedogenetic processes. Calcium carbonate is no longer leached from the profile but accumulates in the lower horizons, giving the zone of pedocals. There is sufficient plant growth to give a clearly-developed humic topsoil. Brown calcimorphic soils and sierozems are the zonal soils of semi-arid climates, although very sandy soils (arenosols) are possibly more extensive; lithosols occur on slopes of even moderate steepness. One form of agricultural activity is pastoralism, often semi-nomadic. Another is the cultivation of drought-resistant cereals, e.g. barley and bulrush millet. There is a severe drought hazard.

Arid. Below about 250 mm rainfall plant growth is sparse or nil, and a humic topsoil only weakly-developed or absent. The soil remains continuously dry in depth, summer rainfall only penetrating the upper horizons. Grey and red desert soils are the zonal type, but are largely confined to alluvium. Most areas of solid rock carry lithosols or desert detritus. Saline soils are common in low-lying sites. Agriculture is dependent on irrigation, and measures must be taken to prevent the salinization of irrigated land. Both semi-arid and arid climates cover substantial parts of the Asian monsoon region, particularly in Pakistan and adjacent parts of India.

Tropical high-altitude. Altitude is a continuous variable and any class limit must necessarily be arbitrary. The Köppen boundary between A and C climates, designed for sub-tropical latitudes, is inconvenient. Latin American Andean nomenclature recognizes three zones, *tierra caliente, tierra tem-*

plada and *tierra fria*, separated at approximately 600 m and 1800 m altitude; tropical crops are grown in the two lower zones, being replaced by subtropical and temperate crops in the *tierra fria*. In Central Africa, land from 900 m to at least 1400 m is essentially a single pedological and environmental zone, whereas above 1700 m soil organic matter increases substantially. An altitude of 1600 m is here taken as the limit of the tropical high-altitude zone. Humic latosols are the zonal soil type. Farming is still based on tropical annual crops. Pasture grasses have a higher nutritive value than in the lowland tropics, and can be incorporated into rotations as leys, including for dairying.

Subtropical humid. This climatic region is peripheral to the tropics, occupying the same latitudes as the arid zone. It is humid throughout the year but has a cold winter. The leached ferruginous, or 'red yellow podzolic', soil type is common. As this region is not known personally to the author, the soils are not discussed in detail; its particular pedological interest is that the transition between temperate brown earths and tropical latosols can be examined on the eastern margins of the continents, whereas they are elsewhere separated by broad Mediterranean and desert belts.

It is convenient to be able to refer to all parts of the tropics in which arable farming is climatically possible and which are characterized pedologically by pedalfers. The term *humid tropics* is used in this sense, covering both rainforest and savanna climates, i.e. the first five classes in table 2. The *dry tropics* is used to refer to the semi-arid and arid zones.

PARENT MATERIAL

The identification of soil parent material

By definition, parent material means the material from which the present soil profile was derived. This can be identified only by inference. At some depth beneath a soil, rock unaltered by weathering is found, and the normal assumption is that rock of the same type formerly existed in the point in space now occupied by the soil profile, i.e. the soil parent material is similar to the underlying rock.

More often than not this is a correct inference, but there are also circumstances in which it is not. First, the soil may be derived from thin, superficial drift deposits; this is more common in temperate latitudes than in the tropics. Secondly, the soil may incorporate a proportion of material from a former cover of a different rock formation; it is not uncommon for mineralogical studies of tropical soils to identify resistant heavy minerals not present in the underlying rock, and a former cover is usually adduced to explain these. Thirdly, the underlying rock may be non-homogeneous, as in alternating sedimentary beds or banded metamorphic rocks, in which case it is usually impossible to extrapolate the succession up into the soil profile. Fourthly, the parent material of a soil on a slope may be derived from rocks higher on the slope, either transported by soil creep or as a colluvial deposit from surface wash. Soils on pediments sometimes show a clear relation to the rock of the inselberg above them.

Another problem is particular to the tropics. In many tropical soils, particularly ferrallitic soils, the unaltered parent rock is separated from the soil profile by many metres of highly weathered material. In such cases the soil parent material is sometimes regarded not as the original rock but as the altered regolith; this distinction was first made in the Belgian Congo, where the thick regolith covers were termed *nappes de recouvrements*. Rather than dispute the semantic point as to which should be termed parent material, it is useful in such cases to call the weathered regolith the *proximate parent material* and the unweathered rock the *original parent material* or *parent rock* (using 'rock' in the geological sense to include unlithified material). The recognition of proximate parent material implies distinguishing between pedogenesis *sensu stricto*, involving biological processes and internal translocation of material, and the weathering *in situ* of rock. Where the presence of successive layers of depositional origin is not suspected, a working definition of the proximate parent material is the material found at 2 m depth.

A further difficulty is that in tropical soil survey it is frequently impossible to find any unweathered rock at all. Weathered rock fragments can sometimes be used to give an indication, but in many instances even these lie well below the depth of profile pits. If there is a rock outcrop at the surface nearby, it is almost axiomatic that this is not typical of the main body of underlying rock, otherwise it would not project. Geological survey has to make use of such isolated outcrops, together with extrapolation from river beds, and may not be free of bias arising from the greater accessibility of less easily weatherable strata.

Finally there is the problem that even where geological maps are available, the basic unit of mapping is rock age and not lithology. Lithological information is given in qualitative form, e.g. 'hornblende–biotite gneiss', whereas for comparison with soils, quantitative chemical and mineralogical rock analyses are desirable.

It would be wrong to suppose from this catalogue of difficulties that soil parent material can rarely be identified. In a majority of cases some reasonable estimate can be made of the parent rock and of the proximate parent material if this is different. There is, however, often a danger of circular reasoning, for example of assuming that the parent material must be of basic composition because the soil possesses properties associated with such material.

Studies of the relations between parent material and soils should ideally be based on quantitative analyses of rock composition and grain size, and in work directed specifically to such purpose such data must be obtained. But *post hoc* comparisons of soil distribution with rock type inevitably rest upon a qualitative basis so far as the properties of the rocks are concerned.

Parent material and soil formation

There are three main variables in parent material as it affects soils: degree of consolidation, grain size and composition. From a pedogenetic point of view, non-lithified 'solid' geological strata, e.g. beds of Tertiary sands, have similar properties to superficial deposits, and the two may be grouped as unconsolidated materials. The bulk of the latter, however, are drift deposits, alluvial, marine and lacustrine. Their distinctive property is that soil profile development can proceed without the need for prior weathering of rock into fine material. Hence in arid regions, with slow weathering, it is often only in areas of superficial deposits that soils other than lithosols are found. The unconsolidated materials contrast with consolidated (lithified) rocks in which rock weathering must precede soil formation.

Grain size is the main determinant of soil texture, having a far greater effect than climate. Thus a material consisting mainly of sand-sized quartz grains, whether it be a sandstone or alluvial or coastal sands, will give a sandy soil even under conditions of intense weathering. The significance of this effect is that texture influences many other soil properties, including organic matter content, cation exchange capacity, profile drainage and moisture retaining capacity, and all these are properties of agricultural importance.

The dominance of parent material over climate as a determinant of soil texture is well demonstrated by soils on shales in a rainforest climate in Malaya. As a result of intense weathering, most rainforest soils have only 2–5 percent silt, but I have sampled soils with 30–40 percent silt (the dispersion checked by an independent laboratory), caused by a high insoluble residue of silt-sized grains in the parent rock.

The composition of parent material is of fundamental importance for the properties of tropical soils and their agricultural potential. The main variable is the proportion of silica. *Felsic** rocks are pale in colour and contain free silica in the form of quartz. The lower limit is 66 percent total, including combined, silica (this being the proportion present in orthoclase feldspar). Besides quartz, the most common minerals in felsic rocks are orthoclase and plagioclase feldspars, muscovite and biotite. Rocks of

* The term *felsic* (rich in felspars and silica) is preferred to 'acidic'; the latter is in common use in geology, but its similarity to acid, as applied to soil reaction, makes it confusing when used in soil science.

intermediate composition have 55–66 percent total silica. The ferromagnesian mineral hornblende is often present, and sometimes augite; plagioclase feldspars are common, orthoclase less so, and subsidiary amounts of quartz may still be present. *Basic** rocks are dark coloured or black and have less than 55 percent total silica, all of it in combined forms. The ferromagnesian minerals hornblende, olivine and augite are common; plagioclase feldspars may be present but not orthoclase. *Ultra-basic* rocks, which are uncommon, contain little or no feldspar and less than 45 percent combined silica.

Rock composition affects the rates of supply, to the soil solution, of the products of weathering. Felsic rocks provide a poor supply of calcium, magnesium, potassium, iron and manganese and a large insoluble residue of quartz. With a soil solution low in bases, kaolinite is the main secondary mineral produced. The soil acquires large quantities of residual quartz, and few or no weatherable minerals remain within the profile. Basic rocks provide a base-rich weathering environment, favouring the synthesis of 2:1 lattice clay minerals. There is less accumulation of insoluble products as a result of which weatherable minerals (feldspars and ferromagnesian minerals) remain within the lower parts of the profile. These minerals provide a continuous source of renewal of weathering products to the soil solution. Soils derived from basic rocks possess a strongly developed blocky structure with prominent clay skins covering the peds; this is probably the result of the presence of expanding lattice clay minerals together with the continuing supply of fine clay by weathering.

Zonal soil types are developed only on rocks of felsic to intermediate composition. This convention is justified on the grounds of the much greater extent of such rocks. Within each climatic zone, basic rocks are associated with distinctive soils. It is convenient to adopt the term *basisol* (taken from the Ghanaian classification, p. 252) to refer to soils derived from basic rocks in the humid tropics. The basisols thus include ferrisols, humic ferrisols and eutrophic brown soils. In the semi-arid zone, and also in the savannas if drainage is imperfect, basic rocks commonly give rise to vertisols.

Most basisols have a moderate to high fertility. They are exceptional in having the capacity to sustain continuous dry cropping (i.e. excluding padi) without the need for fallows or other measures to restore organic matter and nutrient levels. The reasons for this are not fully explained. Many nutrients, including calcium, magnesium, potassium and phosphorous (from apatite) are renewed by the weathering of primary minerals, and thus there is a continuous input to replace those removed in harvested crops. It may be that with an initially more fertile soil, crop residues that are hoed or ploughed into the soil also have a higher nutrient content, giving a higher

---

* *Basic* (rich in bases) is retained in preference to the more correct but less familiar 'mafic' (rich in ferromagnesian minerals).

circulation rate in the soil – plant cycle; this hypothesis has not been tested. Most basic rocks are also fine grained, and the resulting clayey textures are associated with initially higher organic matter contents. But how the organic matter is maintained at a satisfactory equilibrium level has not been explained. It is unfortunate that these exceptionally fertile soils are of very limited extent.

Two qualifications are necessary to the generalization that basic rocks produce fertile soils. First, the high clay content together with the presence of expanding-lattice clay minerals may cause physical problems of difficulty of cultivation. This is a particular problem in vertisols, and can occasionally arise with ferrisols. Secondly, ultra-basic rocks are sometimes suspected of giving problems of toxicity, through the release of relatively large quantities of nickel, cobalt, copper or lead (Fox and Hing, 1971).

The effects of limestones are distinct from those of rocks of ferromagnesian composition. Tropical rendzinas only occur on steep slopes. In the humid tropics limestones usually develop into reddish coloured soils, initially alkaline but subsequently becoming acid latosols (p. 190). They are more fertile than soils from felsic rocks but do not possess the exceptional properties conferred by basic igneous rocks. In the semi-arid zone limestones commonly give rise to vertisols. In arid conditions the calcium carbonate released in weathering may accumulate in the profile in excessive amounts, leading to the special agricultural problems of calcareous soils (p. 197).

If the percentage of total silica in the parent rock were available, it would be possible to explain, in the statistical sense, quite a large proportion of the variability in tropical soil properties by means of this index together with that of rainfall. Practical considerations dictate that whereas rainfall can be treated as a continuous variable, and reasonably accurate estimates made of its value, rock composition normally has to be treated in terms of four discrete classes. Nevertheless both are in fact continuous variables, and rock composition ranks with rainfall as one of the two main axes of soil differentiation within the tropics.

Classification

In grouping rocks on the basis of their properties as soil parent materials, the initial distinction is between crystalline rocks, sedimentary rocks and unconsolidated materials. There is no clear difference between the effects on pedogenesis of igneous and metamorphic rocks of similar composition, so these are grouped as crystalline rocks. Siliceous sedimentary rocks have already passed through one cycle of weathering, and hence frequently supply few weatherable minerals. The other bases of division are composition and

19

grain size. Composition is divided into four classes on a felsic to basic axis, with a fifth class for calcareous rocks; grain size is divided into coarse, medium and fine-grained rocks.

The three methods of division in theory yield $3 \times 5 \times 3 = 45$ classes, but many either do not exist or are infrequent. The more common rock types are given in table 3, and of these, the following eight groups account for a high proportion of soil parent materials:

Felsic crystalline rocks. These are mainly granites and granitic gneisses, with coarse to medium grain size; fine-grained felsic crystalline rocks are uncommon. Owing to the great extent of Basement Complex shields, this is the most extensive type of parent material in the tropics. The quartz grains released by weathering accumulate to give a thick regolith, and soil textures are either sandy or have a bimodal sand-and-clay particle size distribution (the latter being characteristic of soils derived from granite). In the humid tropics the combination of a poor supply of weatherable minerals with a high permeability is conducive to strong leaching, giving a relatively acid reaction, low base saturation, and a clay complex dominated by kaolinite. Few or no weatherable minerals remain within the upper two metres. Hence soils on felsic crystalline rocks are almost invariably ferrallitic in the rainforest zone and often so in the savannas. These soils have a relatively low fertility. The

TABLE 3 *Classification of common rock types on the basis of their properties as soil parent materials*

Composition	Grain size	Crystalline	Sedimentary	Unconsolidated
Felsic	Coarse	Granite	Sandstone	Alluvial, coastal
	Medium	Felsic gneiss	Conglomerate Quartzite[a]	and aeolian sands
	Fine	Rhyolite	Most shales	Most alluvial silts and clays
Intermediate	Coarse	Syenite, Diorite Intermediate gneiss		
	Medium	Many schists		
	Fine		Iron-rich shales	
Basic	Coarse	Amphibolite, Gabbro		
	Medium			
	Fine	Basalt, andesite[b]		Alluvium from basic rocks
Ultra-basic		Peridotite		
Calcareous		Marble	Limestone	Calcareous alluvium

[a] Metamorphic in origin, but as a soil parent material resembles resistant sandstone.
[b] Intermediate in composition, but as a soil parent material resembles basalt.

soils tend to be deep with free profile drainage and easy root penetration, but moderate to low moisture retention owing to sandy textures; organic matter content is rather low, and chemical properties are poor, owing to the low cation exchange capacity (and hence ability to retain added nutrients), low base saturation, relatively acid reaction and absence of weathering minerals from the profile.

If proximate parent material is distinguished from parent rock, a sub-group of *highly-weathered felsic crystalline rocks* may be distinguished. This refers to the deep regolith, greatly impoverished in primary minerals, that covers many gently undulating plateaux, particularly in the savanna zone. The properties of this as a parent material are similar to those of highly felsic rocks, giving rise to ferrallitic soils. The question of the origin of this regolith is discussed on p. 265.

Intermediate crystalline rocks. These include many schists and gneisses, e.g. hornblende – biotite gneiss. They commonly give rise to ferruginous soils in the savannas, but in the rainforest zone are not sufficiently base-rich to counteract the strong leaching and hence acquire ferrallitic properties.

Basic igneous rocks. Basalt is the main rock type in this group; andesite, although technically of intermediate composition, is fine grained and its response as a soil parent material is similar to basalt. Soils derived from basic rocks are clayey in texture, strongly structured, contain 2:1 lattice clay minerals, and retain weatherable minerals in the profile. These soils, collectively termed basisols, include the various types of ferrisols together with eutrophic brown soils and vertisols; they are of relatively limited extent. As noted above these soils are highly fertile and possess the capacity unusual among tropical soils of being able to sustain continuous dry cropping.

Coarse and medium-grained sedimentary rocks. This group includes sand-stones and conglomerates; quartzites, although metamorphosed, are also included since their properties as soil parent materials are much more similar to resistant sandstones than to gneisses. Their influence on soil formation is generally similar to that of felsic crystalline rocks except that they weather more slowly and less deeply. In the humid tropics they give rise to ferrallitic soils or to ferruginous soils of sandy texture; some quartzites weather so slowly as to give lithosols even on gentle slopes. In the dry tropics lithosols are widespread on this rock group.

Fine-grained sedimentary rocks. These comprise shales and related rocks. Most are felsic in composition, although iron-rich shales also occur; the latter weather more deeply and give more fertile soils than the former. Shales are an exception to the generalization that the regolith in the rainforest zone is many metres deep; fresh or only slightly weathered shale is often encountered at no more than 1.5 to 2m, probably because the minerals of which they are formed have already passed through one cycle of weathering. Soils derived from shales have poorer physical but better chemical properties than

21

those from sandstones. Shales often give subsoils with very firm consistency, restricting root penetration, but have higher available water capacity and somewhat higher nutrient levels. Horizons of concretionary laterite are common in shale-derived soils.

Limestones. In the humid tropics limestones commonly yield latosols of comparable or rather higher fertility to those derived from intermediate crystalline rocks. They do not confer the distinctive properties of basic igneous rocks. In drier climates, vertisols and calcareous soils are common. Coral limestone is an important parent material in the Pacific islands and parts of the West Indies.

Unconsolidated sands. There is usually a clear division between sands and fine-grained superficial deposits. The sands may be of alluvial, coastal or aeolian origin. In more humid regions alluvial sands are confined to levees and river beds, but in the dry tropics they may extend over the entire flood-plain. Sand bars, sometimes with associated dunes, are widely found on tropical coasts; they also occur on the shores of the major African lakes where beaches face the prevalent winds. Low raised beaches of unconsolidated sands occur. Besides desert dune fields, sands of aeolian origin are widespread in the present semi-arid zone, originating from drier periods in the Pleistocene. In the *dallols* (relict Pleistocene valleys) of southern Niger, the parent materials have a variety of origins but all contain over 85 percent sand (White, 1971); when working in this region one is tempted to recommend the employment of seven maids with seven mops. The soil types derived from sands are arenosols, regosols and sandy alluvial soils. They are easy to cultivate but have very low fertility and are frequently used for cassava or coconuts. These are poor soils for irrigation owing to the high permeability, low water capacity and poor retention of added nutrients.

Unconsolidated clays and silts. Fine-grained superficial deposits originate as fluvial and coastal alluvium. Fertility levels are generally moderate. Their special value lies in the flat topography, permitting irrigation and padi cultivation. Drainage is the main problem. In relatively infrequent cases where alluvium is derived from drainage basins consisting mainly of basic or calcareous rocks, alluvial soils of exceptionally high fertility occur.

RELIEF

The spatial distribution of soil types is related to relief at all levels of scale. But besides being one of the major soil-forming factors, relief possesses a wider significance in the context of soil survey and land development. Soil survey, both in the field and in air photograph interpretation, makes extensive use of relief as a means of recognizing soil patterns. Moreover, properties of landforms, particularly slope angle, are as important in land evaluation and development as are properties of soils (p. 404).

The relations between relief and soil formation are manifold. In the first place there are direct effects of relief on soil-forming processes and indirect effects arising through the influence of relief on other environmental factors. Secondly, spatial arrangements of relief exist at various levels of scale, from sub-continental to that of an individual slope. Thirdly, landforms are themselves influenced by other environmental factors, notably structure and climate.

The indirect effects of relief on soils are through climate and hydrology. Altitude bears a fairly direct relation to temperature, a rise of 1000 m corresponding to a fall of about 6 °C. This leads to altitudinal soil zonation, which is particularly important with respect to organic matter. Orographic effects frequently cause an increase in rainfall with altitude, greatly magnified where the rise in height faces rain-bearing winds, and correspondingly reduced by rain-shadow effects. These orographic influences on rainfall are sufficiently large to have a substantial effect on soil distribution, well illustrated in mountainous islands of the West and East Indies. However, since direct measurements of temperature and rainfall are available, these relations are more conveniently treated as direct effects of climate on soil formation.

The second indirect effect is the influence of relief on the position of the water table, a connection which is so close that hydrology is often not considered as an independent soil-forming factor.

The levels of scale at which relief is ordered on the surface of the earth do not form a continuous spectrum but contain a number of natural breaks. Up to seven scale levels have been distinguished, but the following five are the most important (cf. Thomas, 1969; Young, 1969*a*):

Major relief units are geomorphological divisions of up to sub-continental scale, such as may be represented on map scales of 1:1000000. Their dimensions range from the order of 100 to over 1000 km. Major relief units are partly the geomorphic expression of macro-structural elements, e.g. mountain ranges associated with fold mountain belts, and are also related to erosion surfaces of continental scale.

Relief units are regional assemblages of similar types of landforms, typically of the order of 20–200 km in extent. Examples are a plain, range of hills or scarpland zone. They cannot usually be seen as a whole from a ground viewpoint, but are readily distinguishable on air photographs. They are described in terms of the types and relative extent of the landforms of which they are composed. Relief units are the geomorphological, and principal, basis of *land systems* as defined in air photograph interpretation (p. 335).

Landforms are commonly of the order of 100 m–5 km in extent, and can often be seen in large part from one viewpoint. Examples are a valley, scarp, pediment or flood-plain. In areas of erosional relief, the valley is by far the most frequent landform. Landforms are described in terms of their horizontal and vertical dimensions and the form and angle of the slopes of which they are composed.

23

Environment

Slopes extend from the highest to the lowest points in the local landscape, most frequently from interfluve crest to valley floor. In areas of erosional relief the greater part of the land surface is made up of valley slopes. Dimensions are commonly 100 m–2 km. Slopes give rise to the differentiation of soil properties expressed in the soil catena, which dominates soil distribution patterns at scales of the order of 1:25000–50000. This category of relief is not recognizable in areas of depositional relief.

Slope units are divisions of an individual slope or other landform which possess a relatively constant angle or curvature (Young, 1972a). Typical dimensions are 50–500 m. They are the smallest units that can be distinguished in air photograph interpretation, and are the geomorphological basis of *land facets* in the land systems approach. In soil survey at other than intensive scales it is necessary to assume that the soils on any one slope unit are uniform, and quite often this assumption is reasonably well justified.

The relief units, landforms and slopes present in any region are influenced by the major elements of geological structure, by geomorphological evolution on a continental scale, and by the effects of climate and local structure. These causes of variation in relief are considered in the following sections. The relation of slopes to soil formation requires prior consideration of the hydrologic factor, and is discussed below in the context of the soil catena (chapter 15).

Macro-structures and major relief units

Major relief units form an environmental context for soil formation fully as important as that provided by the climatic zones. They result from structural evolution on a continental scale, together with the response of surface processes, expressed for example as erosion surfaces, belts of dissected relief, or areas of deposition. Whereas the detailed shapes of landforms bear some relation to climatic zones, major structural elements occur irrespective of climate, and only by coincidence are some elements more extensively represented in the tropics.

The main *world macro-structural elements* are as follows:

Basement Complex shields: exposed shields
                                           buried shields
Structural basins
Older fold mountain belts
Younger fold mountain belts
Areas of rift faulting
Lava fields
Areas of recent sedimentation: coastal plains
                                           deltas
                                         alluvial plains

The *Basement Complex shields* are areas of Precambrian rocks which have remained rigid, resisting folding, throughout post-Cambrian time. Those lying wholly or partly in the tropics are the African Shield, forming much of the African continent together with Arabia; the Brazil–Guyana Shield; the Indian Deccan Shield; and, mainly in sub-tropical latitudes, the Australian Shield, forming the western and central parts of the continent, and the South China Shield. All of these now tropical shields are currently supposed to have formed, prior to the Jurassic period, part of the Gondwana super-continent. They exhibit in common a tendency for repeated epeiro-genic uplift, as a result of which substantial parts stand at altitudes of 1000 m or more, in contrast to the generally low-lying shields of northerly latitudes. The Deccan Shield is tilted from west to east, and the Brazil–Guyana Shield dips beneath the Amazon Basin. The African Shield stands at high altitudes south-east of a line from Khartoum to Mount Cameroun and low to the north-west of this, giving divisions sometimes known as 'high Africa' and 'low Africa'. If you have grown accustomed to rainfall/soil relations in one of these, transfer to the other requires a mental 'altitude-adjustment'. It is on the shields that the major continental erosion surfaces are mainly developed.

Exposed shields are areas where Basement Complex rocks are exposed at the surface. This occurs on the more elevated shield areas, or 'swells'. The rocks are mainly metamorphic gneisses, schists and quartzites, together with granitic intrusions, and are frequently felsic in composition. In areas of metamorphic rocks a sub-parallel lineation to relief elements, especially ridges, is frequently visible on air photographs and sometimes strongly marked. Exposed shield occupies large parts of East and Central Africa; an-other extensive area covers most of the Guianas with adjacent parts of Venezuela and Brazil. It happens that large areas of exposed shield occur within the savanna climatic zone, and these regions are characterized by the extensive development of a deep and highly-weathered regolith with weathered ferrallitic soils.

Buried shields carry a cover of sedimentary rocks, either nearly horizontal or gently tilted, but not folded. The Nubian Sandstone of Egypt and Libya, the 'Continental Terminal' sandstone of north-western Africa and the sand-stones of the southern Mato Grosso of Brazil are examples. Where horizontal they give a relief of plateaux and mesas. When tilted, as for example the Voltaian Sandstone of Ghana, the result is cuesta topography, not unlike that of temperate latitudes in general form although differing in detail.

Structural basins are areas where shields have been subject to repeated downwarping. They carry a thick and approximately concentric cover of sedimentary rocks, the youngest in the centre and successively older strata outcropping toward the periphery. Basins in the humid tropics, the Amazon and Congo, have abundant rivers and swamps, and large areas of coarse sandy sediments which yield extremely infertile soils. Basins in the dry tropics have, or have had in the recent geological past, endoreic drainage. They include the Upper Niger, Chad, White Nile, Okovango and Kalahari Basins. These have shallow lakes, the extent of which varies from year to year, as well as swamps and saline areas. In the Upper Niger Basin the major river coming from humid coastal mountains discharges into a semi-arid basin, giving the 'inland delta' of Timbuctoo. Smaller structural basins, all with shallow lakes and swamps, include those of Lakes Victoria and Kioga in East Africa, the Bangweulu swamps of Zambia and Lake Chilwa in southern Malaŵi.

Older fold mountain belts are not well represented in the tropics, occurring mainly as peripheral belts to Alpine belts. Their relief varies from mountains to hill areas of moderate height.

Younger fold mountain belts comprise the Andean and Himalayan chains, with the extension of the latter into South-East Asia and the East Indies. They give a relief of mountain ranges with intervening sub-parallel valleys, often called 'intermont basins' although most are of erosional and not structural origin. Altitudinal zonation of climate, vegetation, soils and land use is a marked feature of the environment. Large areas are of little agricultural value owing to steep slopes, but the basins include good cultivable land. The soils of intermont basins have not been systematically studied. In the Great Valley of Katmandu, Nepal, the soils are entirely developed on terraced alluvium which fills the valley to a substantial depth; a local legend that the Bagmati River breached the mountain wall and drained a former lake is in accord with geomorphological evidence.

The principal *area of rift faulting*, the African Rift Valley complex, commences in the Red Sea, divides in East Africa into eastern and western branches, re-unites in Lake Nyasa and terminates in the southward continuation of Lake Nyasa, the Shire Valley. Smaller rifts are the Saõ Francisco valley of Brazil and the Narbada and Tapti Valleys of the Indian Deccan. Rift faulted areas have two main landform elements: scarps or scarp zones and valley floors. Rift margins vary from a single high scarp to a belt of dissected country up to 50 km wide; a road sign bordering the Luangwa Valley of Zambia used to read 'steep gradients and sharp bends for the next 100 miles'. Rift valley floors, where not occupied by lakes, often carry a thick cover of alluvium; since they are low-lying and often extremely hot, calcimorphic alluvial and hydromorphic soils are common.

The largest *lava fields* are those of Ethiopia–Kenya and the north-west Indian Deccan; there are smaller flows in the Andean chain, Nigeria, Lesotho and elsewhere. These have a stepped topography, with horizontal plateaux separated by bouldery scarps. The basic lavas give humic ferrisols at high altitudes (as in Ethiopia), vertisols (as in the Deccan), ferrisols and eutrophic brown soils.

Areas of recent sedimentation comprise coastal plains, deltas and alluvial plains. *Coastal plains* are more extensive in the tropics than in temperate latitudes. Most tropical coasts are low-energy environments, as a result of which cliffs are rare and the coastline is frequently prograded by a combination of marine and fluvial sedimentation. Even high coasts, such as the Western Ghats, are usually fringed by a coastal plain. Broad zones of alluvium, sand bars and mangrove swamps, with deltas at river mouths, occur along many coasts. Examples are the coastal plains of the Guianas and Malaya. The recent sediments may be backed by uplifted sedimentary formations giving broad coastal plains, as for example in Moçambique. The major *deltas* offer a distinctive environment of sandy levees and extensive swamps; they include the Niger, Zambezi, Indus, the Godavari and other deltas of the eastern Deccan, the Ganges, Irrawaddy and Mekong (Unesco, 1966). *Alluvial plains* occur as belts along the major rivers. The Indo-Gangetic plain is unique, both structurally and in respect of its extent. In these depositional areas, landforms and soils are dominated not by the alternation of valley and interfluve but by the distribution pattern of sedimentation units. The suite of soil types is also largely distinct from that on erosional topography, being dominated by alluvial and hydromorphic soils. The average world population density on alluvial areas is much above that on any other structural unit.

Erosion surfaces

The macro-structural elements described above form one basis for major relief units. The other is provided by the erosion surfaces of continental extent together with the zones of dissected country which separate them. These surfaces are developed mainly, but not exclusively, on the Basement Complex shields.

If you were placed at some random point in the African continent, the chances are that the landscape would consist of a gently undulating plain of great extent. There is a level skyline, and below it a relief of broad, gently-sloping valleys. Sometimes hill ranges or isolated inselbergs rise steeply above the plain, but often there is a level horizon in all directions as far as the eye can see.

The same type of relief is characteristic of shield areas in other continents. Although appearing as plains when you are on them, these are often extensive plateaux, separated from the coastál plain by one or more belts of steeply dissected country. This erosion surface is most extensively developed in the savanna climatic zone, where it forms a highly distinctive environmental and pedological zone. It forms the major reference datum in tropical geomorphology, and irrespective of continent it is termed the *African surface*. Its altitude varies from one part of the world to another; in 'high Africa' it typically stands at 1000–1500 m, but in 'low Africa' much is below 500 m. In the Guiana and Mato Grosso areas altitudes of 500–1000 m are common. In the Indian Deccan, overall altitude decreases from west to east.

Rising above these plains, and separated from them by steep scarps and belts of hilly country, are high-altitude plateaux of very much smaller extent. These commonly stand some 300–1000 m above the local height of the African surface, for example at 1500–2000 m altitude in Central Africa. They are sometimes developed on igneous intrusions or other more resistant rocks. Examples are the Jos Plateau of Nigeria and the Nyika and Mlanje Plateaux of Malaŵi. Relief on the plateau surfaces is gently to moderately undulating.

Below the African surface there is frequently a broad belt of dissected country or less commonly a high scarp, of which the Drakensberg Scarp of southern Africa is an extreme example. Below this belt of steep topography are other surfaces, at low altitudes, including coastal and alluvial plains and in Africa the Rift Valley floor. Early African explorers struggled across these malaria-infested hot plains, up the hilly country, subsequently to enthuse about the open, cooler country forming large parts of the continental interior. In parts of West Africa, by contrast, the African surface slopes gently down to the coast.

A chronology of these continental-scale erosion surfaces has been attempted by L. C. King (1950, 1962) (table 4). It was based initially on Africa and subsequently extended to the 'Gondwanaland' shields of other continents. The Gondwana surface is represented by high-altitude plateau remnants and is found only on major continental divides; it is considered to have been formed in Jurassic times, prior to the fragmentation of the Gondwanaland super-continent. The valleys cut a small height below these

TABLE 4   *Chronology of world cyclic land surfaces according to L. C. King (1962)*

Cycle	Age	Remarks
Gondwana	Jurassic	Of small extent. Found only on the highest continental divides. Planed prior to the fragmentation of Gondwanaland.
Post-Gondwana	Cretaceous	Of small extent. Valley floors below plateaux of Gondwana cycle.
African	Early Tertiary	Extensive. Level crests of the main shield plateaux. Cycle lasted from late Cretaceous to mid-Tertiary.
Post-African 1	Late Tertiary	Extensive. Valley floors below the level crests of the African cycle.
Post-African 2	Late Tertiary	Only represented in some areas. Planed at a relatively small height below first post-African cycle.
Congo	Pleistocene	Floors of major valleys, e.g. Niger, Zambezi. Followed by a positive sea-level movement causing coastal drowning and sedimentation.

plateaux and are ascribed to a post-Gondwana cycle of Cretaceous age. The level crests of the most extensive surfaces represent the African cycle of erosion, the greater part of which took place in early Tertiary times, whilst the broad, gentle valleys out a short way below this are ascribed to a separate post-African cycle of late Tertiary age. In some regions a second post-African cycle is distinguished. The major valley floors of low altitude belong to the Congo cycle, of Pleistocene age. The dating of these cycles, whilst in part based on the principle of 'counting upwards' from the present coast, is derived also from correlative sedimentary cycles, in marine sediments on coastal plains and off the present coastline and in terrestrial sediments deposited in the structural basins. King has produced a map showing these cyclic landsurfaces for the whole of Africa, and it may be noted that the aggradational landsurfaces, in the coastal plains, basins and parts of the buried shield areas, are as extensive as the denudational landsurfaces (King, 1962, fig. 119). Space prevents an adequate account of the evidence behind this chronology, and a study of the original account is rewarding (King, 1962, especially chapters 8 and 9).

It may be noted that both the high-altitude plateaux and the plains here termed the African surface are attributed a two-cycle origin in King's chronology, one cycle to produce what are now the level interfluve crests and one to cut the present valleys. This is unproven and seems unlikely. Furthermore, whilst there is evidence to date some of the surfaces in a few places, any confirmation of their correlation on a continental, and still more a world, scale would require very many more detailed studies. This is not to belittle King's magnificent achievement of world synthesis, but only to say that it should be regarded as an hypothesis to stimulate further critical study, and not an established fact.

What is not in doubt is the reality of the existence of these surfaces. Local geomorphological studies almost always show each of the surfaces to be composite, but these smaller differences are often of little pedogenetic significance. What is of fundamental importance to soil development, as also to land potential, is the differentiation of the landscape into relief units of two kinds: erosion surfaces, with gentle slopes and often a deeply weathered regolith; and, equally important, scarp zones, with moderate or steep slopes. These two types of major relief unit provide a regional framework for the understanding of soil development.

Landforms

The landform is a category of relief intermediate in scale between the in-dividual slope and the relief unit. Relief units are assemblages of landforms, and the nature of the latter influences the soil types present in the relief unit, their relative extent and distribution.

The classical determinants of the characteristics of landforms are geological structure, climate, and the stage of evolution attained. Some of the effects of macro-structures and of cyclic dissection have been noted above. The relations between climate and landforms, as brought about by the direct and indirect effects of climate on denudational processes, is the subject of climatic geomorphology (see e.g. Birot, 1960/68; Young, 1972a, chapter 18). The following are brief descriptions of some of the more common types of landforms and landform assemblages found in the tropics. The diversity of landforms is so great that comprehensive classification is not attempted.

The first three types described are the landforms characteristic of the main climatic zones: rainforest, savanna and dry climates. Landforms of the *rainforest zone* are characterized by a deep regolith, infrequency of rock outcrops (except as rock-dome inselbergs), a continuous range of slope angle distribution, and often a high drainage density. A dendritic pattern of quite closely-spaced first and second order streams is typical, with few points on the ground surface lying more than 400 m from a channel. Partly owing to the high density of dissection, moderately undulating country is common. Where dissection is rapid, ridge-and-ravine relief is found, made up of narrowly convex crests, steep and nearly rectilinear valley sides, and V-shaped valley floors (fig. 1(*a*)). As vertical erosion slackens, valley floors are widened by lateral erosion during storm flooding, and a characteristic type of landform is developed for which the only existing name is the French 'demi-orange' relief (fig. 1(*b*)); this consists of gently convex crest areas which steepen as the valley is approached, passing abruptly into level flood-plains. A striking

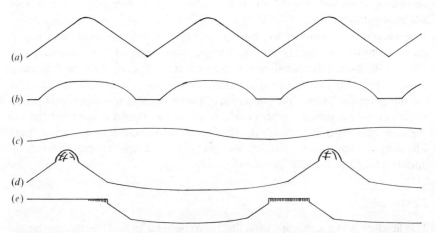

Fig. 1. Schematic cross-profiles across some common types of landforms in the tropics. (*a*) Rainforest zone, ridge-and-ravine relief. (*b*) Rainforest zone, 'demi-orange' relief. (*c*) Savanna zone, gently undulating plain. (*d*) Inselberg and pediment relief. (*e*) Plateau and mesa relief.

regularity of slope forms is encountered when traversing rainforest relief on foot.

A very extensive landform type in the *savanna zone* is that usually described as a 'gently undulating plain'. There is a low drainage density and broad, gently-sloping valleys of shallow to moderate depth; typical values are valleys 500 m–2 km wide, 50–100 m deep, and with maximum slopes of from 2° to 5°. Interfluves are broadly convex (fig. 1(*c*)). Larger rivers develop level flood plains, but the floors of smaller valleys are often continuously concave and do not have a channel. Belts of sandy soil occur along the valley-floor margins and in the lobate expansions which occur at valley heads. These broad, often streamless, valley floors, with a vegetation of hydromorphic grassland and gley soils, are known by the Ci-Cewa vernacular term *dambo* (Ackermann, 1936; Louis, 1964; Mäckel, 1974). There is a tendency towards a bimodal angle frequency distribution in this zone, slopes being either gentle or steep. Inselbergs typically have a bare rock summit above a steep debris-slope formed of large boulders set in finer soil. The depth of regolith may be considerable and is sometimes highly irregular (Thomas, 1966*a*, 1974).

In the *dry tropics*, plains are often very gently undulating, often so gentle that the pattern of valley and interfluve is not easily distinguished from the ground. As the climate becomes drier the tendency increases for slopes to be either steep or very gentle, the former being rocky or boulder-strewn, the latter with a fine regolith. Landforms relict from more humid periods in the past and widespread in deserts, whilst in the semi-arid margin there is frequently evidence of both drier and wetter periods, e.g. fixed dunes and laterite respectively.

Isolated inselbergs, invariably rising above pediments, occur in all tropical climates. It is a common misconception, however, to suppose that tropical landforms consist largely of pediments and inselbergs. Most of the land surface in the tropics is formed, as in temperate regions, of valleys – as gently undulating plains, steeply dissected country, or, less commonly, moderately sloping relief. There are, however, limited areas in which *inselberg and pediment relief* is found. These are landscapes in which frequent steep-sided inselbergs rise above pediments which lead down to valley floors (fig. 1(*d*)). In the savanna zone, pediments are typically 2°–6° and merge into the debris slope of the inselberg by a concavity; in dry climates, pediments are gentler (1°–3°) and there is a more abrupt piedmont angle. Soils on pediments sometimes show a relation to the rock type of the inselberg.

Structural and lithological influence is dominant in a number of common types of landforms. *Scarplands* of the tropics have the same basic elements of scarp, dip-slope and vale as are found in the temperate zone. The detailed form of the slopes varies according to climate; the scarps are often subject to closely-spaced gully dissection. *Plateau and mesa relief* (fig. 1(*e*)) is developed on horizontally bedded sedimentary or igneous rocks, for example on Nubian Sandstones in Egypt, as caused by dolerite sills in the Karroo Beds of semi-arid interior Natal, and by basalt flows in Lesotho. Thick and massive layers of laterite can have the same effect as a resistant rock, and there is a type of relief consisting of flat-topped hills bounded by laterite breakaways, with pediments and valleys beneath; this is well-known from its occurrence north of Kampala, Uganda. *Badlands* are areas of very closely-spaced gully dissection. Whilst they can result from severe man-induced erosion, they also occur naturally in both savanna and semi-arid climates, being caused by dissection of weakly-consolidated but impermeable sediments. They are not extensive, no doubt because they are so rapidly eroded away, and have a characteristic fine linear appearance, as of hoar frost, on air photographs. *Tropical karst* gives a variety of types of relief, typically with near-vertical bare rock walls rising from gentle slopes.

Finally among erosional landforms, mention may be made of *dissected country with steep slopes*, one of the most widespread types of landscape in all climatic zones. The landforms making up such areas are largely deep and steep-sided valleys, sometimes over 300 m deep. It is regrettably rarely possible to explore these spectacular landscapes in soil survey, since they are readily delineated from air photographs and excluded from field survey.

In *areas of depositional relief* there is a distinctive landform and soil distribution, dominated by the sedimentation pattern of active flood-plain, cover flood-plain, river terraces, levees, backswamps and other elements. This is in contrast to the interfluve and valley, or catenary, pattern which dominates detailed soil distribution on erosional landforms. Small differences in height above the river channels, and hence in water table depth and frequency of flooding, are associated with large changes in soil type, particularly in respect of texture and degree of gleying. Despite the overall flatness of alluvial areas, detailed soil distributions are no less dominated by landforms than in areas of erosional relief.

Slope angle and soil formation

The effects of slopes upon soil formation are direct and indirect. The indirect effects, including the influence of slope form upon hydrological conditions, are discussed below in the context of the soil catena (chapter 15). The direct effects concern the rate of denudation.

The steeper a slope, the faster will be the rate of natural denudation. The erosive power of surface wash increases faster than linearly with increase in slope angle. So far as is known, however, rates of weathering are unaffected by steepness of slope; it is generally assumed, although unverified, that weathering becomes slower as regolith depth increases. Hence on steep slopes, an equilibrium between rates of weathering and denudation is reached with a comparatively thin soil cover. This soil is in a state of transition, being continuously removed by denudation and replaced from below by rock weathering; the more rapidly this occurs, the less weathered will be the soil. Hence soils on steep slopes tend to be both thin and stony.

This causal connection between rate of surface denudation and degree of weathering of the regolith has consequences for the soils of moderately to gently undulating areas. Where slopes are very gentle and surface denudation slow, the insoluble materials are not removed and accumulate to considerable depths. The upper part of such a regolith becomes increasingly highly weathered, until only secondary minerals and residual quartz remain. Thus with gentle slopes, a very highly-weathered regolith can develop under only a moderate intensity of weathering.

This effect is important in the savanna zone. It is common to find highly-weathered ferrallitic soils on gently undulating plateau surfaces, bounded by less weathered soils on more steeply sloping land. If slopes are moderate the latter are likely to be ferruginous soils, with a larger content of weatherable minerals and a higher fertility than the plateau soils. On steep land the soils will be lithosols, which despite their unsuitability for cultivation on grounds of depth and erosion hazard often have a high chemical fertility. Indigenous vegetation is able to take advantage of the nutrient reserves of lithosols, and there is often a contrast between grassland or sparse tree savanna on the highly weathered soils of the plateau and biologically richer savanna woodland on the bounding scarps (Milne, 1937; Cole, 1963a). Moreover on plains, those areas which have very gently undulating relief, with maximum valley side slopes of 2° or less, tend to have more highly weathered and less fertile soils than areas where slopes reach 4°–6°.

The description of relief in soil survey

The description of relief is an integral part of soil survey. It fulfils two functions. First, landform differences are employed as a basis for soil mapping, both in air photograph interpretation and in field survey. Secondly, slope angle is in itself one of the main determinants of land use suitability, principally through its influence on the erosion hazard. Land potential is determined by slope at least as much as by soil type. Hence it is only by custom and convenience that soil surveys in the tropics are not called soil-and-landform surveys.

The two observations recorded in field soil survey are the relief at the site itself and the landforms of the surrounding area. The former can be described in terms of three parameters:

Slope angle: in degrees;

Profile form: convex, slightly convex, rectilinear, slightly concave, or concave;

Position in slope: crest, upper slope, middle slope, lower slope, or base.

Field description of the landforms of the surrounding area is given in the same terms as the description of relief units in air photograph interpretation. In both cases it is quite common to employ qualitative terms, e.g. 'gently undulating plain with broad valleys'. Complete surveys can be successfully accomplished in this way, particularly if only one surveyor is involved. For comparisons between areas, however, terms such as 'gentle slopes' and 'broad valleys' do not have a constant meaning, and it is better to supplement the descriptive terms with quantitative values, known as *landform parameters*.

In the description of erosional relief, three terrain parameters are the most important: relative relief, drainage spacing and slope angle. *Relative relief* (see fig. 2) is the difference in altitude between locally adjacent higher and lower parts of the land surface, normally interfluve crest and valley floor. Where isolated inselbergs are present the height of these is separately recorded. *Drainage spacing* is the distance between rivers, streams or lines of concentrated drainage; it may alternatively be estimated as the distance

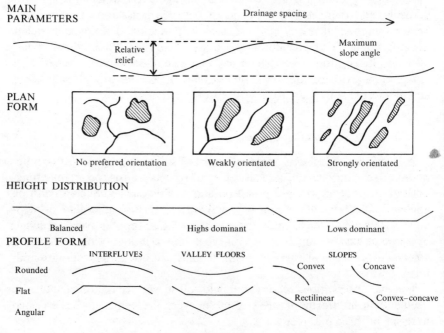

Fig. 2. Parameters for the description of landforms. Cf. table 5.

TABLE 5 *Classification of landform parameters (cf. fig. 2). Slope angle classes after Young (1972a, p. 173), profile form classes after Ollier (1967). Cf. also Van Lopik and Kolb (1959) and Speight (1967, 1968)*

Relative relief (valley depth)

<10 m	Very low	(Depositional relief)
10–30 m	Low (shallow)	(Plains of low relief)
30–100 m	Moderate	(Plains of moderate relief)
100–300 m	High (deep)	(Hills)
300 m	Very high (very deep)	(Mountains)

Drainage spacing (valley width)

<100 m	Very close (narrow)	(Badlands)
100–400 m	Close (narrow)	
400–1500 m	Moderate	
1500–3000 m	Wide	
3000 m	Very wide	

Slope angle

0°–2°	Level to very gentle
2°–5°	Gentle (gently undulating)
5°–10°	Moderate (moderately undulating)
10°–18°	Moderately steep
18°–30°	Steep
30°–45°	Very steep
>45°	Precipitous to vertical

Plan form
No preferred orientation
Weakly orientated (give direction)
Strongly orientated (give direction)

Height distribution (cf. fig. 2)

Balanced (neither highs nor lows predominant)	(Normal valley relief)
Highs predominant (>66 percent of area)	(Dissected plateau)
Lows predominant (>66 percent of area)	(Plain with hills)

Profile form

Interfluves:	rounded, flat, or angular
Valley floors:	rounded, flat, or angular
Slopes:	predominantly convex, rectilinear, or concave; combinations of these forms (from crest to base), e.g. convex–concave, convex–rectilinear, convex–rectilinear–concave.

between interfluve crests. In morphometry the standard measure of valley frequency is drainage density, the number of kilometres of drainage line per square kilometre of area; for descriptive purposes, however, drainage spacing is preferable, being capable of direct observation and interpretation. In most areas of erosional topography, relative relief and drainage spacing may be thought of as valley depth and valley width.

Slope angle is a property of a point on the ground surface, and for a relief unit it is fully represented only by an angle frequency distribution. It is useful nevertheless to describe landforms by a single slope value. The most suitable is the maximum angle excluding small areas of exceptional steepness, e.g. rock outcrops; this is the operative value in determining land capability. If a precise definition is required, the angle given can be that such that 90 percent of the ground has gentler slopes. If an angle frequency distribution is required it can be obtained by a system of slope profile sampling and survey (Young, 1972a, p. 252).

These three values, relative relief, drainage spacing and maximum slope, enable most areas of erosional relief to be briefly and accurately described. Thus the qualitative description 'gently undulating plain with broad, shallow valleys' can be recorded as 30 m — 1 km — 3°, and 'dissected low hills' as 100 m — 400 m — 30°.

Further landform descriptors are plan form, height distribution of the high and low elements in the relief, and profile form (fig. 2). *Plan form* describes the extent to which drainage lines or ridge crests show linear orientation. *Height distribution* refers to whether the greater part of the land area lies towards the upper part of the local altitude range, as in a dissected plateau; towards the lower part, as in a plain with isolated hills; or whether the relief is fairly evenly distributed between high and low parts. *Profile form* describes whether interfluves and valley floors are rounded, flat or angular, and whether valley side slopes are predominantly convex, rectilinear, concave, or combinations of these, e.g. convex–concave, convex–rectilinear, convex–rectilinear–concave.

There is no commonly accepted classification system for landform parameters, and as they are continuous variables class boundaries are necessarily arbitrary. Table 5 gives a classification, with class limits chosen to correspond to common interpretations of qualitative descriptive terms.

It should be added that the variety of landforms is such that landform parameters alone frequently do not give an adequate description, and need to be used in conjunction with descriptive terms and sketches of schematic cross-sections. For example in Relief unit B, described below, valley width is not the same as drainage spacing, a fact not apparent from the landform parameters.

Table 6 gives the data from which the following relief unit descriptions are derived:

TABLE 6   *Examples of data for relief unit description*

Landform parameter	Relief unit A	Relief unit B
Relative relief	50–100 m	0–50 m
Drainage spacing	500–1000 m	1000–2000 m
Maximum slope	3°–5°	c. 30°
Plan form	Random	Weak N–S orientation
Height distribution	Balanced	Highs predominant
Profile form:		
interfluves	Rounded	Flat
valley floors	Rounded	Angular
valley sides	Convex–rectilinear–concave	Rectilinear

Relief unit A. Gently undulating plain with wide valleys of moderate depth. Dendritic valley pattern. Most valley floors without a stream channel and continuously concave, but larger valleys have a river and flat flood plain. Slopes c. 60 percent 0°–2°, 35 percent 2°–5°. Few isolated steep-sided hills with gentle pediments.

Relief unit B. Almost level plateau dissected by narrow, steep, V-shaped valleys. Valleys commence as shallow gullies, deepening and widening progressively to a maximum of 200 m width. Valleys are separated by broad areas of undissected plateau.

HYDROLOGY

Drainage and soil formation

In soils subject to frequent or permanent waterlogging, hydrology is dominant over other soil-forming factors. That is, poorly-drained profiles have certain properties in common irrespective of climate and parent material. Gleyed soils of the tropics are broadly similar in appearance to those of temperate regions. The main processes of gleying, reduction of iron compounds to the ferrous form and their partial re-oxidation and precipitation, are not temperature-dependent, although they presumably take place more readily in hot climates. In both temperate and tropical gleys, temporarily waterlogged horizons are mottled whilst horizons permanently beneath the water table are usually very dark grey where heavy-textured and medium to pale grey where sandy. The main difference is that in the temperate zone, gleyed horizons never dry out completely, and the iron precipitated in ferric forms remains as a mottle. In the seasonally-wet tropics, gley soils dry out to substantial depths in the dry season, and there is a strong tendency for iron mottles to develop into concretions; intermediate forms, as soft concretions,

37

are common. On poorly-drained sites in tropical arid climates, salinization is liable to occur in addition to gleying.

Soils subject to frequent waterlogging, and in which features of gleying are prominent in the upper horizons, are classified as gleys irrespective of their other properties, since the moisture regime is the principal control over their land use potential. When subdividing such profiles in soil survey, the degree of drainage impedance and the texture, together with salinity in arid regions, are the most important criteria. Soils that are mottled in the lower horizons only are classified as a gleyed variant of the soil type indicated by the upper horizons, e.g. gleyed ferruginous soil.

A special case is that of soils under swamp rice cultivation, which involves flooding for a period of several months annually. This period of waterlogged conditions is sufficient to cause replacement of the natural soil profile by a man-induced soil type known as padi soils (p. 226).

The drainage status of a soil profile is affected by site drainage and profile drainage. *Site drainage*, also called external soil drainage, refers to the frequency with which the site is affected by a high groundwater table, and also to the capacity of the site for the removal of excess water across the ground surface. The latter is reduced on slopes that are concave in profile or plan form. *Profile drainage*, or internal soil drainage, is the capacity of the profile for the removal of excess water vertically downwards. It is affected by the permeability of the least permeable soil horizon or of the parent material. Profile drainage impedance causes the occurrence of a perched water table. Soils in which waterlogging is caused by poor site drainage are *groundwater gleys*; also included as groundwater gleys are valley floor soils which also have profile drainage impedance as a result of heavy textures. Soils in which waterlogging is caused by profile drainage impedance only are *surface water gleys*, sometimes called pseudogleys. Site drainage impedance caused by a high groundwater table is by far the most common cause of poor soil drainage.

In freely-drained soils, a further property conveniently classed with drainage is the rapidity at which surplus water drains through the profile. This is dependent on permeability, which in turn is a function of texture and type of clay mineral. Many soils of the humid tropics which are texturally quite heavy clays have rapid permeability and free drainage owing to the absence of expanding lattice clay minerals. Sandy textured soils are described as having excessive drainage, meaning that water drains through them rapidly and implying that they are therefore subject to strong leaching.

The description of drainage in soil survey

The US *Soil Survey Manual* (Soil Survey Staff, 1951) gives six classes of surface runoff, seven of permeability, six of internal soil drainage, and seven soil drainage classes. The *soil drainage class* describes the moisture conditions in the profile as a whole, caused by the net effect of site and profile drainage. It is inferred from observations of relief, wetness at the time of survey, and morphological features resulting from gleying in each soil horizon. It is normally sufficient to record only the soil drainage class together with an indication of the cause of impedance where present, e.g. 'poorly drained, owing to high groundwater table', 'imperfectly drained, owing to profile impedance caused by impermeable B2 horizon'.

The FAO manual gives only the soil drainage classes, which may be interpreted as follows:

Very poorly drained. The water table remains at the surface for most of the year, and there is frequently standing water. This is applied to swamps.

Poorly drained. The water table is at or near the surface for much of the year. Soils exhibit gleying in the topsoil. Applied to sites which, whilst not swamps, have their land use potential dominated by poor drainage.

Imperfectly drained. Waterlogged for significant periods. Applied to soils with a clearly mottled B horizon. Crops sensitive to drainage impedance cannot be grown.

Moderately well drained. The profile is wet for short periods, and the soil mottled in depth. This is a convenient class for soils which for most practical purposes are freely drained but which show slight indications of temporary impedance.

Well drained. Excess water is removed from the profile freely but not rapidly. No mottling is present.

Somewhat excessively drained. Water drains through the profile rapidly. Applied to sandy soils.

Excessively drained. Water is removed from the profile very rapidly. Applied to stony soils on steep slopes.

Only the first five of the above are strictly drainage classes, since drainage impedance is equally absent from the last three. The division into two classes of 'excessive' drainage is not justified, since sandy soils drain just as rapidly as stony ones. Excessive drainage is estimated from the permeability of individual horizons. The technique is to squirt a jet of water onto a lump of soil; if it disappears almost instantaneously, permeability is rapid. If the entire profile has no horizon with less than rapid permeability, then its drainage is described as excessive.

Soil moisture

The moisture properties of a soil are of considerable importance in assessing its agricultural potential. The properties involved are drainage, field capacity, wilting point, available water capacity and infiltration capacity. Drainage has been discussed above. *Field capacity* is the moisture content (as percentage by weight of dry soil) retained by soil after gravity water has been allowed to drain away; in most soils 2–3 hours is sufficient but heavy clays may require 24–48 hours. *Wilting point* is the moisture content at which plants wilt and do not recover on re-wetting; it indicates that the water still remaining in fine pores of the soil is held under such tension that it cannot be extracted by roots. *Available water capacity* is equal to field capacity minus wilting point, multiplied by the depth of soil concerned. It is convenient to express it as equivalent depth of water. To do this the field capacity and wilting point are converted to percentage moisture by volume, by multiplying them by the bulk density of dry soil (typically 1.5).

Field capacity can be determined by sampling an undisturbed block of soil, placing it on a wire tray, saturating it, allowing it to drain and determining the moisture content. It is possibly better to sample a block of soil in the field within a day of heavy rain. Wilting point is correctly determined by growing a plant, customarily sunflowers, in a pot and letting the soil dry out until it dies. Nowadays these parameters are often estimated indirectly by means of a pressure membrane apparatus; field capacity corresponds approximately to the moisture retained under a suction of $\frac{1}{3}$ atmosphere and wilting point to that retained at 15 atmospheres. Samples taken for this purpose are undisturbed cores collected in cylinders of 10 cm diameter, pressed or hammered into the side of a profile pit.

Infiltration capacity is the rate at which a soil profile can absorb or transmit water applied to the surface, measured in centimetres per hour. It is determined by the permeability of the least permeable horizon. In surveys for irrigation projects, measurements of infiltration capacity are sometimes taken *in situ*. The standard test of field permeability, that of the US Bureau of Reclamation (1953), involves emplacing two concentric cylinders 30 and 45 cm diameter into the upper 15 cm of soil. Water is poured into both of these; the outer provides a 'shield' of saturated soil so that percolation of water from the inner is mainly vertical; this latter rate of percolation is recorded. A simpler method is to drive a single cylinder deeply into the soil; this can be done by jacking up a Landrover, placing the cylinder under the bumper and lowering the jack. Field permeability measurements have a high variation, owing to the presence of structural cracks, root and termite channels.

The soil moisture regime is critical to plant growth. In humid climates the whole profile reaches field capacity during the later part of the wet season. In the dry season it dries out to wilting point from the surface downwards, initially by evaporation but below 30 cm largely through the extraction of water by roots.

In rainforest climates, the soil moisture content only falls to wilting point during dry spells, and never to any great depth. In the savannas the depth to which wilting point is reached may be as much as 2 m towards the end of a six-month dry season. The roots of indigenous grasses are more efficient than crops at abstracting moisture from deep in the profile, but maize is also quite effective. Considerably more soil moisture is retained if a fallow is kept clean-weeded.

When rain falls on a dry soil, the wilting front advances progressively downwards. In savanna climates this will eventually unite with the moist soil in depth. In dry climates the soil remains permanently dry in depth, and rain seasonally penetrates the upper horizons only.

ORGANISMS

The factor termed organisms covers vegetation and soil fauna. Macrofauna, i.e. animals, only affect soils to a minor extent through the action of burrowing animals, e.g. the ant-bear, and can be largely ignored as an independent influence on soil formation. In the case of soil micro-organisms it is difficult to separate the organisms as a factor from their effects as a process. Soil fauna are therefore discussed in connection with organic matter processes in chapter 4, and the present discussion of organisms as a soil-forming factor is confined to vegetation. It may be noted that this separation is one of convenience only. One of the most important differences between the soil-forming environments of tropical and temperate latitudes arises through the presence of termites only in the tropics.

VEGETATION

Vegetation and soil formation

In the relations between vegetation and soils there is a greater element of two-way interaction than is found with other soil-forming factors. Soils exert a substantial influence upon the type of vegetation community present in a given location. Conversely vegetation influences soil properties, both directly through the supply of organic matter and in a number of indirect ways. In general terms, vegetation communities are more influenced by climate and less by geology than are soil types.

The major influence of vegetation on soils is that caused by the rate of supply of dead plant material. This includes the subaerial supply of plant litter, comprising dead leaves and wood, together with the subterranean supply of dead roots. The rate of root exudation is impossible to measure directly, and figures quoted are approximate estimates only. In ecological studies the productivity of plant communities is usually expressed in energy terms, as kilocalories per unit area, but with respect to the supply of organic matter to the soil, oven-dry weight is a more useful unit.

The total biomass of tropical plant communities is of the order of 300–900 t/ha (metric tonnes per hectare) or more in rainforest, 60–100 t/ha in moist savanna woodland, 30 t/ha in dry savannas, and < 10 t/ha in communities of semi-arid and arid climates (table 7). Woody material accounts for the greater part of the biomass in all communities, the proportion of green parts being about 4–8 percent in rainforest and 12 percent in savannas; this is why improverishment of the tree layer by burning causes such a large reduction in biomass.

For soil organic matter supply it is not the total weight of a plant community that is significant but its rates of growth and decay. In this respect the contrasts between the main formation-types are reduced, as rainforests have lower turnover rates than savannas, whilst the rate rises to as much as 25 percent in semi-arid communities with a high proportion of annual plants. Nevertheless rainforest still has the highest net primary production and litter fall rates, of the order of 30 t/ha/year. The corresponding values for well-wooded moist savannas are approximately 10 t/ha/year and for dry savannas about 5/ha/year (table 7). Thorn-scrub and related woody communities of the semi-arid zone may show higher rates of net primary production than are found in the dry savannas. The litter production of savanna communities is very much reduced if the tree layer is reduced by clearance or burning.

For lowland tropical environments, topsoil organic matter bears a direct relation to these rates of supply. Typical values are 2–5 percent under rainforest, 1–2 percent in the savanna zone and less than 1 percent in semi-arid regions. Soil organic matter is, however, also related to temperature and soil texture (p. 103).

An important part of the interaction between vegetation and soils takes place through the medium of soil moisture. Under rainforest there is a close tree canopy, a fairly continuous, if thin, cover of leaf litter, and a stable microclimate at the soil surface. Under these conditions the soil moisture rarely falls to wilting point, a state of affairs which permits continuing dominance by evergreen species. Under savanna and steppe communities the percentage of ground cover by living vegetation decreases greatly during the dry season, there is no appreciable litter layer, and the proportion of rain lost by immediate runoff is greater. The result is that the soil dries out to substantial depths, 1–2 m according to the length of the dry season. Hence a forest cover tends to conserve soil moisture whereas grassland formations emphasize the seasonal contrast.

This interaction can be significant in regions climatically intermediate between the two formation-types. In one study of the 'forest–savanna mosaic' zone of West Africa, it was found that forest areas were located on soils of heavier texture and consequently a higher moisture retention capacity (Moss and Morgan, 1970). In an area of undisturbed vegetation in the

TABLE 7 *Estimates of the biomass, net primary production and litter fall of tropical and subtropical vegetation types. Values of weight are for oven-dry matter. Turnover rate = (net primary production/biomass) × 100*

Formation-type	Biomass (t/ha)	Green parts (%)	Net primary production (t/ha/year)	Litter fall (t/ha/year)	Turnover rate (%)	Source
Tropical rainforest	990	2	–	11	1	Fittkau and Klinge (1973)
	>500	8	33	25	<7	Rodin and Basilevič (1968)
	332–360	4–8	–	8–15	c. 3	Nye and Greenland (1960)
	320–490	–	30–50	–	12[a]	Bazilevitch et al. (1971)
	–	–	50[a]	–	–	Kira and Ogawa (1971)
Subtropical forest	410	3	25	21	6	Rodin and Basilevič (1968)
Moist savanna	67	12	12	11	18	Rodin and Basilevič (1968)
	67–100	11	10	8–15	c. 12	Nye and Greenland (1960)
Dry savanna	27	–	15–30	7	–	Bazilevitch et al. (1971)
	29	11	7	7	26	Rodin and Basilevič (1968)
	–	–	3	–	10	Nye and Greenland (1960)
Tropical xerophytic woodlands	–	–	8–10	–	–	Bazilevitch et al. (1971)
Steppes (subtropical?)	10	15	4	4	25	Rodin and Basilevič (1968)
Semi-desert vegetation	6	3	2	2	33	Rodin and Basilevič (1968)
Desert vegetation	–	–	<1	–	–	Bazilevitch et al. (1971)

[a] Maximum value for pre-climax, rapidly growing forests.

southern Mato Grosso, Brazil, the strikingly sharp boundary between semi-deciduous forest and *cerrado* (short tree and shrub savanna) corresponds to a soil texture difference, forest being found on sandy clay loams and *cerrado* on loamy sands and sands. Away from the climatically transitional zone, both in wetter and drier directions, this correspondence does not hold (Askew *et al.*, 1970, Ratter *et al.*, 1973). Forest clearance, by cultivation or repeated burning, may substantially alter the soil moisture regime, causing increased annual drying out of the upper layers; this is one of the causes of 'derived savanna'. An interesting possibility is that such changes might be irreversible, with seedlings of evergreen plants unable to become re-established even if a site is subsequently protected from burning; it has not, however, been proven that this state of affairs can obtain, and it may be doubted if the intricacies of plant succession would fail to find a way across this moisture threshold.

The products of vegetation decay affect weathering and leaching. The processes customarily termed 'chemical weathering' are in fact substantially augmented by the action of complexing organic acids (p. 70). The most distinctive effect of this type, the complexing of iron and aluminium compounds by mor humus, only occurs in the tropics at high altitudes.

Tall-grass savannas dominated by grasses of the Andropogoneae family appear to inhibit nitrogen mineralization. Nitrogen deficiency occurs in grain crops planted in the first year after clearing such tall-grass fallows (Nye and Greenland, 1960, p. 96). The carbon: nitrogen ratio in moist savannas is sometimes 15 or above, compared with 10 in adjacent forested sites, indicating some form of inhibition in the action of nitrifying bacteria. African farmers are aware of the poor cereal crops obtained by clearing such fallows, and plant yams or some other crop in the first year of cultivation.

A curious situation is reported from the Tiv region of Nigeria (Vermeer, 1970). In the traditional rotation, with a 3–6 year fallow under *Andropogon gayanus*, yams were planted in the first year and millet in the second. As a result of population pressure the fallow has been reduced in some areas to 1–2 years, with consequent soil degradation and some invasion by *Imperata cylindrica*; this fallow is apparently not long enough for the nitrogen lowering effect to develop, and although yields of all crops are lower, it makes no difference whether millet comes first or second in the rotation. These results are reported on the basis of cultivators' experience and need testing by soil and leaf analysis.

Certain leguminous trees have a beneficial effect on soil fertility. The soil beneath the trees has a higher nutrient content, and sometimes better physical conditions, than that beyond the range of their branches, and yields of sorghum, millet and groundnuts may be as much as 100 percent higher. This phenomenon has been reported mainly from the dry savanna to steppe zone, the tree in question being *Acacia albida* in the Sudan and in Senegal and an introduced Indian species, the neem tree (*Azadirachta indica*) in northern Nigeria (Radwanski and Wickens, 1967; Radwanski, 1969; Dancette and Poulain, 1969). I have noted a clear increase in the height of sorghum, comparable to that around house compounds, beneath *Parkea* sp. in the moist savanna zone of Nigeria. The nitrogen fixation associated with Leguminosae is presumably an important factor, together with increase cycling of nutrients from roots via pod and leaf fall to the soil (congregation of livestock in the shade may also contribute). In semi-arid areas it may be possible to make productive use of this by planting *Acacia* or neem as a tree fallow, giving a source of fuel and timber and at the same time improving the soil for interspersed periods of cropping (Radwanski, 1969); neem is fast-growing, but *Acacia* has the advantage that its leaf fall coincides with the start of the rains.

It is not usual for the same soil type to be recorded under different vegetation communities, and most soil surveyors would be reluctant to do so. In the Western State of Nigeria, however, there are a number of soil series recorded on both sides of the forest–savanna boundary; they have the same horizon sequences and no more internal variation than that normally permissible for a series. For each series individually, and for their mean values, the savanna representatives are lower in organic matter and nitrogen, and have a higher carbon: nitrogen ratio (table 8). Thus the savanna soils are low-humus variants of their forest counterparts. The less acid forest topsoils reflect greater accumulation of bases derived from the greater volume of litter. Assuming there is no difference in parent materials, the higher B-horizon clay contents of the forest soils could be caused either by more intense weathering under forest or by more clay eluviation (out of the profile) under savanna.

In traversing from forest 'islands' to savanna, Moss and Morgan (1970) found that the vegetation boundary coincided with a soil texture change. Murdoch (personal communication), also in Nigeria, found that surrounding the forest patches were belts with soils having properties associated with forest but now under savanna, suggesting recent retreat of the forest boundary. With delicate balances between soil, soil moisture and vegetation, together with human pressures, the forest–savanna transition zone offers exceptionally good opportunities for studying soil–vegetation relations.

The description of vegetation in soil survey

Vegetation may be described on a physiognomic or a botanical basis. Physiognomic description is in terms of the life forms of individual plants and the structure of vegetation communities, e.g. grassland with scattered deciduous trees. Botanical descriptions of communities are usually given as the genera of one or more dominant plants, e.g. *Themeda* grassland, *Brachystegia–Julbernardia* woodland.

TABLE 8 *Values of soil properties for four soil series that occur in both forest and savanna zones, Western Region, Nigeria. Values are means for the four series. I-Q = inter-quartile range. After Ojo-Atere and Murdoch (1971)*

Vegetation	No. of sites	Depth (cm)	Organic carbon Median (%)	Nitrogen Median (%)	Nitrogen I-Q (%)	Median C:N ratio	pH (water) Median	pH (water) I-Q	Clay Median (%)	Clay I-Q (%)
Forest	33	6	1.5	0.15	0.07	10	6.4	0.3	10	4
		40	0.4	0.04		10	6.1	0.4	18	9
		120	0.3	0.04		8	5.7	0.4	32	8
Savanna	22	6	0.9	0.06	0.03	15	6.1	0.3	10	3
		40	0.3	0.03		10	6.0	0.4	15	8
		120	0.2	0.03		7	5.9	0.4	24	7

Botanical communities are restricted to particular floristic provinces, continents, or smaller areas, whereas similar environments in different continents often have similar physiognomic communities. The former, botanical–physiognomic, classes are *formations*, the latter *formation-types*. Thus the lowland evergreen rainforest formation-type is divided into American, African and Indo-Malaysian formations, the last characterized by *Dipterocarpus* spp. African thorn-scrub and the Brazilian *caatinga* formation, both found in semi-arid regions, have a strong similarity in appearance and life-forms present. The *Brachystegia* tree savanna formation of Africa is matched by an Australian community of very similar appearance dominated by *Eucalyptus* spp. The environmental equivalent in South America, the *cerrado* formation, is similar to African savanna in having an open layer of woody plants above a continuous tall-grass cover, but has a higher proportion of shrubs and is botanically very much richer in species.

In soil survey, vegetation communities are described both from air photograph interpretation and in the field. In the former case description must be almost entirely physiognomic and in the latter mainly so, as there is not the time to make species identifications and take counts.

In air photograph interpretation, vegetation is the next most easily seen factor after landforms; in flatlands, i.e. depositional landforms, it is the only one. Where it is thought that natural or modified natural vegetation is present it makes an excellent basis for mapping environmental units, since vegetation is a response to all the other factors. The great disadvantage compared with landforms is that vegetation is very easily altered by man's activities.

Photo-interpretation of vegetation is basically in terms of the percentage cover of trees (with shrubs), grasses (or other herbaceous vegetation) and bare ground. Both tone (shade of grey) and photographic texture (fine pattern) are used to distinguish communities. Tone results from the diffused reflectance of the leaf or ground surface. Coniferous trees have a low reflectance and a darker tone, but these are infrequent in natural tropical communities. There is no clear difference between evergreens and deciduous trees in leaf, but they can, of course, readily be distinguished by dry-season photography. Grassland is normally paler than trees, whilst bare ground has a high reflectance and appears pale. Photographic texture results from discontinuities in the cover of one or more vegetation strata, e.g. an open tree cover with (paler) grass visible beneath, or open shrubs with bare ground between. It is sometimes beneficial to study the spectral reflectance curves of different types of leaf (see, e.g. Howard, 1971, chapter 6). Much more useful, however, is a rapid field reconnaissance, photographs in hand, prior to interpretation. Individual species can only be identified on air photographs in special cases, e.g. baobab trees. Attempts at forest inventory of rain-forest, using large-scale colour photography, have not been notably successful.

47

In field soil survey, vegetation description is in terms of the strata, their percentage cover and the life-forms present, supplemented by notes of the most common or distinctive species. The Raunkaier system of life forms, based on the height of the perennating organs, is not as convenient for rapid description as one based on the form of the plant as a whole together with its size and leaf characteristics.

The best physiognomic classification is that of Küchler (1949; 1967, chapter 15), given in modified form in table 9. The basic distinction is between woody vegetation, herbaceous vegetation and special life forms. Within the woody plants, Küchler's distinction between evergreen and deciduous needleleaf trees is unnecessary in the tropics. It is, however, frequently desirable to distinguish between trees and shrubs, the former having a single stem to a height of at least a metre and the latter either

TABLE 9   *System for the description of plant communities on a physiognomic basis. Modified from Küchler (1949; 1967, p. 191)*

COMMUNITY STRUCTURE
For each stratum present, record:

Height Record average, in metres; classify as:

Trees:	tall	> 25 m
	medium	10–25 m
	short	< 10 m
Shrubs:	shrubs	> 1 m
	dwarf shrubs	< 1 m
Herbs:	tall	> 2 m
	medium	0.5–2 m
	short	< 0.5 m

Cover Record in percent; classify as:

Continuous	> 75 %
Interrupted	50–75 %
Moderately open	25–50 %
Open	6–25 %
Sporadic	1–5 %
Occasional	< 1 %

Interrupted, moderately open or open may be qualified by
Parklike: grows in dense patches separated by areas in which stratum is absent   p

LIFE FORMS
Woody plants

Trees (single stem to > 1 m):	
Broadleaf evergreen	B
Broadleaf deciduous	D
Needleleaf	N
Leafless	O
Shrubs (multiple stem or branching below 1 m):	
Shrubs (height > 1 m)	S
Dwarf shrubs (height < 1 m)	D

Herbaceous vegetation

Grasses	G
Other grass-like plants	R
Forbs (broadleaf herbs)	H
Ferns	F
Lichens and mosses	L

Special life forms

Palms	P
Lianes (woody climbers)	C
Succulents	K
Cushion plants	T
Bamboos	V
Epiphytes	X
Tree ferns	Y

Leaf characteristics

Normal (soft)	n
Hard (sclerophyll)	h
Succulent	k
Large (> 400 cm²)	l
Medium (4–400 cm²)	m
Small (< 4 cm²)	s

Record mixed strata by combined symbols, e.g. BD.

having multiple stems or branching in the lowest metre. Küchler's class of graminoids is here divided into true grasses (Gramineae) and other grass-like plants (sedges, reeds, etc.). A herbaceous class of ferns is added. The class of dwarf shrubs (< 1 m high), originally intended mainly for heathers, applies also to various tropical desert plants. In the classes of percentage cover, the 20–50 percent class is here renamed 'moderately open', since it is desirable to use the term *parklike* to qualify cases in which the cover consists of relatively dense clumps separated by areas in which the stratum is absent, e.g. clumps of trees amid grassland.

To use the system, first identify the strata present, then record their average height, in metres, and cover, as a percentage (qualified by 'p' if parklike). Next record the life forms present in each stratum, and finally the leaf characteristics, using letter-symbols if wished. Further details, and proformas, are given by Küchler (1967, chapter 15).

As a result of the common practice of using the natural plant succession as a method of fallowing, the distinction between vegetation and land use is not always clear in the tropics. It is often convenient, both in photo-interpretation and in the field, to describe them as a joint unit, 'vegetation and land use', e.g. open short-tree savanna with 20 percent cultivation patches. In the field, always record carefully the vegetation or land use at the site of the soil pit itself, separately from that of the surrounding area. This is important in the interpretation of analytical results, particularly for soil organic matter and nitrogen.

It is frequently useful to identify *indicator species*, plants which are associated with particular soil or soil moisture conditions. The most common uses are to indicate shallow soils (caused by rock or laterite), imperfect drainage or salinity. In some countries there are already published or manuscript lists of such species, but in others they have never been put to paper. Moreover, indicator species are often of very local significance; for example a plant found in one area only on sites with imperfect drainage may grow on freely-drained sites in an area with a higher rainfall. In a soil survey of six months or more duration it is worth attempting to learn such species. This may involve making proper collections, in a plant press, for subsequent identification by a herbarium. A more rapid if less reliable way is to find some local inhabitant or labourer with an encyclopaedic knowledge of verna-cular names, which can then be compared with a check list. Pending identifica-tion, field description proceeds either with collection numbers or vernacular names. One's first inclination is to use trees. Grasses, however, can be equally good indicators and are not so readily altered by cultivation; in particular, there may be grass species characteristic of patches of shallow soil, and of soil exhausted by over-cropping. Having identified indicator species, they can be used in traversing between auger observations.

Classification

Unlike the situation with respect to plant species, but like that in soils, there is no universally accepted classification of vegetation communities. In 1956 a meeting was held in Yangambi in the Belgian Congo, under the auspices of the Commission for Technical Co-operation in Africa South of the Sahara, at which a special attempt was made to co-ordinate French and English terms. The results are embodied in part in the Vegetation map of Africa (Keay, 1959; cf. also Boughey, 1957a). Whilst the botanical composition of the communities described in the African map are peculiar to the continent, the physiognomic classification is applicable throughout the tropics. Among world vegetation classification systems are those of Fosberg (1967), used in the International Biological Programme, and the Unesco (1973) system. The former has 31 formation classes and 193 formation groups; the latter is an hierarchical system with five categories and 225 classes in all.

A substantially modified version of the 'Yangambi' classification is given as table 10. This can be used to classify vegetation communities observed both in the field and on air photographs. There are 6 main groups and 25 classes, 6 of which are subdivided. The main groups are partly physiognomic and partly environmental. Groups I, II and III, respectively forest, savanna and dry-climate formations, are physiognomically defined but, with the exception of swamp and riparian forest types, correspond to decreasing rainfall. The grassland formations are collected in group IV, although each also appears in one of the other groups. Groups V and VI, hydromorphic and montane formations, represent specialized environments, again with an overlap with the previous physiognomic classes.

The distinction between forest formations and savanna formations, the latter including woodland, is a clear one. Forest formations, whether largely evergreen or partly deciduous, have a multi-layered structure with climbers present; the canopy and lower woody strata are sufficiently dense to inhibit grass, and the canopy trees have a tall unbranched trunk equal to or greater in height than that of the crown. The various savanna formations are characterized by a tall grass cover, with tree or shrub strata of varying degrees of openness. Trees typically have a Y-shaped form, branching at about one third of their height. The woodland formation, consisting of a fairly continuous canopy of deciduous trees but with some grass beneath, belongs to the savanna group, being a savanna woodland with a denser canopy and not an open type of forest.

Lowland rainforest is divided into an evergreen type and a semi-deciduous ('monsoon') type. In the latter the canopy is partly or entirely deciduous whilst the short tree and shrub strata are largely evergreen; conifers are occasionally present. Montane forest types are lower in height and have an abundance of mosses. Of the two hydromorphic forest formations, swamp forest occupies broad wet areas whilst riparian forest (gallery forest) occurs

TABLE 10   *A classification of tropical vegetation types. Based in part on the 'Yangambi' classification (Keay, 1959; Boughey, 1957a) and in part on the system used for air photograph interpretation by the Land Resources Division, UK (unpublished). Subdivision of savanna types modified from Cole (1963b)*

I. *Forest formations.* Continuous tree canopy, wholly or partly evergreen, several strata, climbers present, grasses absent or rare:
    Evergreen rainforest (lowland)
    Semi-deciduous rainforest (lowland)
    Montane forest (subdivisions: evergreen, semi-deciduous)
    Swamp forest
    Riparian forest

II. *Woodland and savanna formations.* Trees wholly or mainly deciduous; grass stratum present, in which perennial grasses >80 cm high with flat leaves are predominant or important:

Woodland	Tree canopy continuous but light (75–95 %), climbers absent or rare, grass cover open or sporadic
Savanna woodland	Tree canopy interrupted (50–75 %), crowns frequently touching, grass cover well developed
Tree savanna ⎫ Tree and shrub ⎪ savanna ⎬ Shrub savanna ⎭	Continuous grass cover with 6–50 % cover of trees and/or shrubs; can be subdivided on the basis of height of trees (medium, short) and density of woody plant cover (moderately open, open)
Savanna parkland	Trees and/or shrubs in dense patches with grass areas between
Grass savanna	Tree and shrub cover <5 %

III. *Steppe, semi-desert and desert formations.* Annual plants abundant; perennial grasses where present are <80 cm high and have narrow, rolled or folded leaves; woody plants mainly with small leaves and/or thorny:

Thorn thicket	Thicket with >50 % cover of short trees and/or shrubs
Tree and/or shrub steppe	Woody plants cover 6–50%; 'thorn scrub'
Dwarf shrub steppe	
Succulent steppe	
Grass steppe	Tree and shrub cover <5 %
Semi-desert formations	Bare ground usually >50 %; various subdivisions, including halomorphic vegetation
Desert formations	Various subdivisions

IV. *Grassland formations.* Woody plants <5 %:

Grass savanna	Cf. above
Grass steppe	Cf. above
Hydromorphic grassland	
Montane grassland	

V. *Hydromorphic formations.*

Swamp forest	Cf. above
Riparian forest	Cf. above
Hydromorphic grassland	Cf. above
Mangrove	
Herbaceous swamp	

VI. *Montane formations.*

Montane forest	Cf. above
Montane grassland	Cf. above
Bamboo thicket	
High montane formations	

in strips along river banks, either amid forest or savanna. The basic division of the savanna formations is on the basis of the density of tree and shrub cover, with a continuous spectrum woodland – savanna woodland – tree savanna – grass savanna, beside which are the variant forms representing presence of shrubs and clumping of woody plants in patches. No implications about the ecological status of the formations are contained in the classification.

The distinction between savanna and steppe types of grassland follows that of the Yangambi system. Savanna is dominated by tall perennial grasses with flat leaves, steppe by annuals or by shorter perennials with narrow leaves. The meaning of 'perennial' in both cases is that the underground parts of the grass remain living whilst the leaves and stems die and dry out during each dry season and re-grow annually. In Africa the most widespread savanna types of grassland are those dominated by *Hyparrhenia* and *Andropogon* spp., whilst the most widespread steppe grassland dominants are *Eragrostis* and *Aristida* spp. (Rattray, 1960). This difference in grass form and habit is associated with a contrast in the woody vegetation where present. On trees and shrubs of the savannas the leaves are mainly broad and of medium size. In the steppe formations the woody plants exhibit xeromorphism, most commonly as small leaves (or small leaflets to bipinnate leaves), and are often thorny. The most widespread African savanna types are *Brachystegia–Julbernardia* savanna or savanna woodland (Swahili *miombo*), and the most widespread types of steppe are *Acacia* thorn thicket or tree steppe. The corresponding South American forms are known as *cerrado* and *caatinga* respectively, and the fact that the woodland formation is not forest (*mato*) but a taller variant of *cerrado* is recognized by its being called the augmentative form of the latter word, *cerradão*.

The treeless end-members of the savanna and steppe groups, grass savanna and grass steppe, may alternatively be placed in a group of grassland formations. Hydromorphic grassland occupies valley floors throughout the savanna zone, whether extending right across the valley floor as is normal in African *dambos*, or as belts on either side of a strip of gallery forest such as is found in the wetter parts of South American *cerrado*. The hydromorphic formations include two forest types and one grassland, mangrove swamp which is structurally closer to woodland, and herbaceous or reed swamp, e.g. papyrus swamp. The montane formations contain forest, grassland (often fire-induced), bamboo brakes, which occasionally form an altitudinal belt, and a variety of high montane types with peculiar life forms.

TIME

Time is of a different nature to the other soil-forming factors, influencing soil properties not through effects on the type of processes but on the stage of evolution attained. The main significance of the time factor arises with respect to immature soils. A related topic is that of palaeosols.

Little is known about absolute rates of soil formation. Two sets of processes are involved: rock weathering, the conversion of rock into regolith, and the development of a soil profile from either weathered regolith or unconsolidated sediments. These two rates differ by several orders of magnitude. An indirect indication of rock weathering rates is provided by rates of total surface denudation and slope retreat. World average rates are of the order of 50 mm per thousand years on gently sloping relief and 500 mm per thousand years on mountainous areas or steep slopes (Young, 1969b, 1974c). The former value is sufficient to produce one metre of regolith in 20 000 years, and rates in the humid tropics may well be higher. Basalt flows under a 3000 mm rainfall in Samoa have developed only a thin cover of regolith in 60 years (Schroth, 1971). From analysis of seepage water on the Ivory Coast, Leneuf and Aubert (1960) estimated that complete ferrallitization of one metre of granite could take place in 22 000–77 000 years. Profile differentiation takes place very much more rapidly, and hence on consolidated rocks it will keep pace with rock weathering. On unconsolidated materials, appreciable profile development can take place in periods of the order of 100 years or less. Under a rainforest climate in Ecuador, volcanic ash can be transformed into productive soils, with a deep, mature profile, in as little as 10 years (Colmet-Daage, 1967).

Even for rock weathering, these orders of time are short compared with the duration of the Quaternary period. This explains a major contrast between tropical and temperate soils. In many parts of the temperate zone, weathering and soil formation recommenced at the conclusion of the last glacial stage, 10 000–20 000 years ago, on slopes glaciated or modified by periglacial processes. Notwithstanding oscillations of climatic belts, Pleistocene renovation of parent materials did not take place widely in the tropics. Hence there has been ample time for mature profiles to develop, and immature soils are confined to areas of active or very recent deposition.

Immature soils

An *immature soil* is a soil in which horizon differentiation which could develop in the environment concerned has not taken place owing to the short time over which the parent material has been exposed to soil-forming processes. The main types of immature soil are alluvial soils and soils derived from recent volcanic materials. Lithosols on steep slopes, although they have little horizon differentiation, are not immature since time will not lead to further profile development. In alluvial soils, insufficient time has elapsed since the most recent occasion of sedimentation for horizon differentiation produced by the sequence of deposition to be transformed by

53

translocation processes. The distinction between alluvial soils which are and are not subject to active, recurrent deposition is one of agricultural importance, although in practical soil survey it may be difficult to distinguish between a high flood-plain and a low terrace. It may be possible to map a river terrace sequence of increasing soil maturity with height above the flood-plain; this is one of the methods of differentiating areas of Indo-Gangetic alluvium.

Recently deposited volcanic materials are of two kinds, volcanic ash and lava flows. In the case of the former the material itself has special properties, principally a high proportion of amorphous silica, and the resulting soil type, andosols, owes its distinctive properties as much to parent material as to immaturity. Volcanic ash showers cover substantial areas and recur intermittently. Those few areas covered by recent lava flows carry lithosols, and represent a situation rare in the tropics, of soils on consolidated rocks that are immature in the true, temporal, sense of the term.

Chronosequences

A *chronosequence* is a sequence of soils of increasing maturity caused by difference in the periods over which they have been exposed to soil-forming processes. The most clearly established case is the development of volcanic ash into latosols. E. C. J. Mohr put forward a number of development sequences for ash of basic and felsic composition, on freely and poorly-drained sites, under various climatic conditions in Java. The sequence on basic volcanic ash with free drainage under a permanently humid lowland rainforest climate is as follows.

Following the deposition of fresh ash, bases and silica are leached, iron oxidizes to hydrated ferric forms, and liberated aluminium combines with silica and water to form kaolinite. The pH falls initially to 6.0–7.0. Leached silica is precipitated in depth, where conditions are more alkaline, to give a siliceous hardpan. These processes lead from the stage of a juvenile soil, with weak horizon development, to a brown earth. With continued leaching the pH is reduced to 5.0–6.0. The residual sesquioxides become first yellowish brown and then, through dehydration, red. Above the impervious silica pan drainage is impeded, and two layers form in this zone: a whitish kaolinite horizon above which, in the zone of fluctuating water table, is an horizon of iron concretions. This is the red earth stage. The kaolinite layer thickens, the iron concretion layer rises, and the red soil above it is lost by erosion. This brings the final stage of a red earth with laterite. It is accompanied by a degeneration in the vegetation, causing loss of the humic A horizon.

The above chronosequence, together with others for different types of material, drainage conditions, temperature and rainfall, is described in the 1954 edition of Mohr and van Baren's *Tropical soils* (regrettably, it is omitted from the revised 1972 edition). These sequences have neither been reported from elsewhere than in Java nor independently tested within that country. They repay study, containing many ideas of interest on soil genesis in the humid tropics.

There is an interaction in Java between altitude, time and soil development. Andosols form on fresh ash on all sites. At high altitudes, when temperatures are lower, andosols remain as a stable soil type, whilst at lower altitudes they are transformed successively into brown, dark reddish brown and dark red latosols (Dudal and Soepraptohardjo, 1960).

On Samoa, under a rainforest climate, there is a chronosequence on lavas of different ages. Flows of 1911 still carry lithosols. Flows of 'last glacial' age have soil profiles less than 60 cm deep, bouldery, relatively high in exchangeable bases, and with a clay fraction composed mainly of X-ray amorphous material. Two earlier flows (late Tertiary) have deep profiles, are low in cations, and have a clay fraction consisting of gibbsite, kaolin and hydrated iron oxides (Schroth, 1971).

Micropedology is a powerful technique for studying chronosequences, since it enables the result of both mineral alteration and clay illuviation to be seen. The stages of pedogenesis on basalts under rainforest in Nicaragua have been described by this means. In the early stages the soil possesses a marked plasmic fabric, produced by weathering *in situ*, with only weak clay illuviation. At the second stage ferri-argillans (clay skins rich in iron compounds) are prominent, indicating much clay illuviation. In the final stage, that of a ferrallitic soil, this clay has been lost, and illuviation ferri-argillans are found only in the B3 and C horizons (Eswaran, 1970).

Andosols and basalt soils are special cases in which parent materials are depositional. Of more general interest is the possibility that some of the main types of latosols are successive stages of development. For normal rocks, e.g. Basement Complex, Sys (1967) gave the following chronosequence for areas with more than 1000–1200 mm rainfall:

Stage 1. Organic surface horizon over partly weathered rock.

Stage 2. Organic surface horizon over cambic horizon (weathered *in situ*), with a reserve of weatherable minerals and medium or high activity of the clay (cation exchange capacity 25 m.e./100 g of clay).

Stage 3. Organic surface horizon over argillic horizon, with medium or high activity of the clay.

Stage 4. Organic surface horizon overlies argillic horizon, with no weatherable minerals and low activity of the clay (cation exchange capacity < 25 m.e./100 g of clay).

Stage 5. Organic surface horizon over oxic (ferrallitic) horizon.

55

With rapid drainage stages 2 and 3 are elided into a cambic horizon with low activity of the clay. With a rainfall of less the 1000–1200 mm, the sequence ends at stage 3. This suggested sequence of cambic → argillic → oxic horizons has subsequently received some support from micromorphological evidence (Bennema *et al.*, 1970).

This theory is the basis of the INEAC system of soil classification (p. 252). It contains in fact several related hypotheses: (i) that existing ferrallitic soils have been through a stage of having a textural B horizon; and *either* (ii) that existing soils with such an horizon (ferruginous soils and ferrisols) will in course of time, under a rainfall of 1200 mm, attain the oxic (ferrallitic) stage; *or* (iii) that they would do so were it not for removal of material from the soil surface by natural erosion. The third of these suggestions refers to the fact that denudation is faster on steeper slopes, and hence to the action of landforms in 'retarding' soil maturity. These are unproven hypotheses of much pedogenetic interest.

MAN

Which soil-forming factor has the greatest effect on a soil's productivity? In his article 'The influence of man on soil fertility', G. V. Jacks (1956) put forward an unorthodox argument. Fertility is the capacity of soils to produce crops. Of the various ways of determining this, one is the actual production obtained, which although open to objection at least has the merits of objectivity and ease of determination. On this definition the most fertile soils are on which production per unit area is highest. If these premises are accepted, Jacks argued that the strongest influence on soil fertility is towns, since it is in belts surrounding towns that the highest crop production, whether measured in calorific or cash terms, is found. This generalization is as true of India or Africa as it is of Western countries. 'Towns increase a country's soil fertility by enabling farmers to afford to put more into the soil than they take out of it . . . As every farmer knows, it pays to fertilize when the market is good.' If this argument, or rather the definition of soil fertility on which it is based, contains special pleading, it nevertheless serves to emphasize that the soil is not a fixed resource, to be drawn upon, but one which can be changed for the better or worse by the treatment it receives.

One possible framework for classifying man's influence is in terms of the effects, beneficial and detrimental, on each of the other soil-forming factors (Bidwell and Hole, 1965). Alternatively, soil formation can be regarded as having two stages, pedogenesis and metapedogenesis (Yaalon and Yaron, 1966). In this latter model, pedogenesis is regarded as the operation of soil-forming processes caused by the natural environment, operating over a long period of time to produce a 'natural' soil profile; the natural profile is then acted upon by 'added metapedogenetic process factors', i.e. man's influence, over a much shorter duration, to produce the soil profile as observed at the present. There are feedback loops in the model, representing the influence of man-induced changes on natural processes. It may sometimes be difficult in practice to determine what would have been the 'natural' soil profile.

In some circumstances, the effects of man's activities may be at least as important in determining soil properties and productivity as the processes of the natural environment. Examples are irrigated desert soils, soils under padi cultivation and land laid bare by accelerated erosion.

The effects of man on soil formation can be listed either in terms of types of activity or of soil properties affected. Taking the former as the basis, the main activities and their effects are as follows:

Human activity	*Main actual or possible effects on soils*
Vegetation clearance	Loss of organic matter and nutrients
Savanna burning	Loss of organic matter and nutrients
Addition of farmyard manure	Gain in organic matter and nutrients
Green manuring, grass leys	Gain in organic matter
Addition of fertilizers	Gain in nutrients
	Possible change in pH (liming or acidification)
Irrigation	Change in moisture regime
	Siltation
	Possible gain in organic matter (arid soils)
	Possible gain in nutrients
	Possible salinization
Drainage, land reclamation	Change in moisture regime
Flooding of padi fields	Change in moisture regime; consequential change in redox potential and soil profile
Cultivation (ploughing, hoeing)	Replacement of O/Ah horizons by the Ap horizon
Terracing	Effects of cut-and-fill on profile
Soil conservation works	Modifications to relief and runoff
'Overcropping'	Accelerated erosion
'Overgrazing'	Accelerated erosion, compaction
Addition of toxic chemicals	Soil pollution

In terms of properties affected, the main changes concern soil organic matter, chemical composition including nutrient content, moisture regime, and changes in texture or the mineral fabric. Of these changes, that of greatest significance in the tropics is the decrease in organic matter content on cultivation.

Some of these effects are discussed below. They are listed among the soil-forming factors since it is difficult, and often unrealistic, to separate natural from man-induced processes of soil formation.

COMPARISON BETWEEN THE SOIL-FORMING FACTORS

The meaning of statements that one soil-forming factor is 'more important than' or 'dominant over' another requires clarification. One special case is the statement that if factor *A* has value *v*, certain properties of the soil are similar irrespective of the values of factors, *B, C, D* . . . Examples are that soils on poorly-drained sites will exhibit properties of gleying irrespective of climate and parent material, and that very steep slopes usually carry litho-sols.

A second special case is the statement that if factors *B, C, D* . . . have 'normal' values, then soil properties are largely determined by factor *A*. The major example of this type of proposition is the concept of zonal soils, which requires that all factors other than climate should be normal. The meaning of 'normal' in this context is more nearly expressed by 'not abnormal', and is tacitly assumed to be as follows:

Climate: not applicable
Parent material: *not* heavy clays, sands, basic igneous rocks, lime-
                    stones, volcanic ash, or rocks very resistant to weather-
                    ing (e.g. quartzite)
Relief: gently sloping, *c.* 0°–5°
Hydrology: free drainage
Vegetation: not applicable; assumed to be dependent on climate
Time: sufficient for the development of a mature profile
Man: no influence

This interpretation of normal values is acceptable with respect to relief, hydrology, time and man, but becomes strained in the case of parent material. Rock types that are definitely abnormal are readily identified, but it is harder to specify the range of grain size and composition falling within the normal range; granite, felsic to intermediate gneisses and schists, shales, and fine and medium-grained sandstones perhaps qualify. A zonal soil is thus the mature soil which develops on any of these normal parent materials, on gentle slopes with free drainage, under a specified climate with its associated vegetation.

The general form of propositions about the dominance of one soil-forming factor is as follows:

If the values of factors *B, C, D* . . . do not fall outside specified limits, then soil properties are largely determined by factor *A*.

The statements given in the first two paragraphs above are special cases of this general proposition. The limits specified for relief, hydrology, time and man are usually those given above as 'normal'. An example of this type of proposition is the statement that within the moist savanna climatic zone, within a range of 900–1200 mm annual rainfall, soil properties are largely determined by parent material; that is, differences of parent material have a greater effect on soil properties and soil types than differences of rainfall *within this range*. The zonal concept states that within the range specified (tacitly or otherwise) for normal parent material, soil type is largely determined by climate.

In comparing the relative importance of the different factors with respect to their influence on tropical soils, vegetation, hydrology, time and man may be briefly dismissed. Vegetation will be taken as the climax vegetation corresponding to climate, and hence not an independent variable. In all except arid climates, poor drainage leads to the development of a gley, albeit that whilst characterized by reduced iron compounds, mottling or grey colours, gleys have a wide range of texture, reaction and other properties. Similarly the importance of the time factor is clear in the special circumstances of immature soils. Man becomes an important influence in the special case of soils affected by accelerated erosion. In these instances it is clear that the one factor is dominant but that these are abnormal values for the factor.

This leaves three factors, climate, parent material and relief, as the main controls on soil formation in other than special circumstances. Their influence will be examined with respect first to individual soil properties affected and secondly, to soil types and their distribution.

Factors and soil properties

The relative importance of climate, parent material and relief in the determination of soil properties in the tropics is shown in table 11. Climate, either as temperature or rainfall, has major effects on organic matter (and therefore indirectly nitrogen), reaction, base saturation, and contents of carbonates and soluble salts. Among the few cases where good correlations between environmental parameters and soil properties can be obtained are those of organic matter with both temperature and rainfall (see p. 104), and of both pH and cation saturation with rainfall. Parent material is the dominant factor over texture. Moreover, of all soil properties, texture is that which has the strongest influence on other properties, affecting structure, consistence, organic matter and properties associated with it, and cation exchange capacity; hence these properties are indirectly influenced by parent material. Type of clay mineral, which also has important secondary effects on other properties, is influenced both by climate and parent material.

The effects of relief indicated in table 11 are those of slope angle. Indirect effects, such as that of altitude upon temperature or valley form upon drainage, are excluded. Slope angle, through its influence on the rate of natural erosion, affects two sets of properties; those associated with the weathering of rock to fine material, namely profile depth, stoniness and texture, and those related to stage of weathering, namely content of weatherable minerals and to a minor extent clay mineral type.

TABLE 11 *Effects of climate, parent material and relief on soil properties in the tropics. The table refers to freely-drained sites.* ++ = *major effect,* + = *substantial effect. Minor and indirect effects not shown*

Soil property	Factor			Dependence on other soil properties		
	Climate	Parent material	Relief	Texture	Clay mineral type	Oth
Colour	+	+				
Texture	+	++	+			
Stoniness	+	+	++			
Profile depth	+	+	++			
Structure				+	+	
Consistence				+	+	
Organic matter	++			+		
Nitrogen						a
Reaction	++					
Cation exchange capacity				+	+	
Base saturation	++					b
Total exchangeable bases						
Carbonates	++	+[c]				
Soluble salts	++[d]					
Clay mineral type	+	+				
Weatherable minerals	+	+	+			
Silica: sesquioxide ratio	+	+				

[a] Organic matter.
[b] Cation exchange capacity and base saturation.
[c] In the special case of limestones.
[d] In conjunction with drainage.

Soil types and distribution in relation to factors

It is clear from the above discussion of the controls upon individual soil properties that with respect to soil types, climate and parent material are likely to be of approximately equal importance. If a soil classification were primarily dependent on, for example, organic matter and reaction, then climate would exert the greater influence on distribution of soil classes. Conversely in a classification greatly dependent on texture, parent material would be the more important (the 'classifications' used by farmers to refer to soils on their own land are of this latter type, e.g. 'sands', 'loams'). But it is significant that the natural soil types of the tropics are particularly related to the types of clay mineral present, and that this is a property influenced by climate, parent material and to a lesser extent relief.

The well-established zonal concept remains valid, within the range of normal parent materials, within the tropics. The succession of zonal soils in the lowland tropics, in order of decreasing rainfall from rainforest to desert, is leached ferrallitic soil, ferruginous soil, brown calcimorphic soil, sierozem, grey and red desert soils, and desert detritus. Surprisingly, in view of the fact that the concept is a century old, there have been few attempts to compare properties of soils under different climates on parent materials matched as closely as possible; granites provide opportunities for such studies.

In contrast there is abundant evidence of the effect of parent material, since many detailed soil surveys fall within one climatic zone. The importance of basic composition of rocks is particularly great in the tropics, by reason of the fact that base-rich parent materials counteract the leaching tendency present in hot climates. There is a climatic sequence of soils on basic parent materials, all non-zonal, consisting in order of decreasing rainfall of strongly leached ferrisols, (normal) ferrisols, eutrophic brown soils, vertisols, followed by the same semi-desert and desert soils as in the zonal sequence.

Parent material influence is well illustrated in the soils of Malaya. The whole country has a modal rainforest climate and the internal variation in annual rainfall, from 1770 to over 2500 mm, has no detectable influence on soil properties. Granite, sandstone, shale and andesite are each associated with readily distinguishable soil series. Granite, sandstone and shale all develop into leached ferrallitic soils. The profile on granite is deep yellowish red, has a sandy clay texture and a well-developed and deep textural B horizon. That on sandstone is a less deep brownish yellow sandy clay loam. The soils developed on non-ferruginous shales are relatively shallow, with weathering rock at about 150 cm, and have a compact, plastic B horizon; in one soil series there is a 20 cm thick horizon of nodular laterite at a depth of about 50 cm. Andesite develops into a strongly leached ferrisol, a deep, uniform, very red clay with a moderately to strongly developed blocky structure. Morphologically and in respect of physical properties these four soils differ widely. In respect of most analytical properties they are almost indistinguishable; in particular, all are strongly acid and have an apparent base saturation of 10 percent or less. The only chemical property which differentiates them is free iron oxide content.

Examples showing no less a measure of variation in properties could be cited from the savanna climatic zone. Within this zone certain chemical properties, notably reaction, base saturation and organic matter content, show only a limited range of variation, whilst colour, texture, clay mineral type, grade of structure and cation exchange capacity vary widely with parent material.

Generalizations on the relative effects of the different factors can be made in terms of the scale of mapping. Soil maps on a world and continental scale show climatic influence most strongly. Climatic zonation is apparent on the *Soil map of Africa* (D'Hoore, 1964) and the sheets of the FAO *Soil map of the world*, although one cannot take this as firm confirmation of zonality since substantial parts of these maps still rest on inference from climatic data. Soil maps of a single country show the influence of parent material prominently. In countries with a restricted range of climate, for example Malaya and Rhodesia, the geology dominates the soil distribution pattern. In other countries, for example India, Kenya and Ghana, geologically-controlled soil boundaries are superimposed upon the progressive of climatic zonality. On maps at scales of 1:50000 and smaller, the catenary effects of relief frequently dominate the soil pattern; a few geological boundaries may cut across to give major regions, whilst climatic variation is insignificant. In depositional landscapes, there is a non-catenary but again relief-dominated detailed distribution pattern of soil types. For a case study of the relative importance of factors at different scales and in differing types of landscape, reference may be made to the discussion of soil distribution patterns in Rhodesia, Zambia and Malaŵi by Ellis (1958), Webster (1960) and Paton (1961).

The relations of relief to soil distribution are complicated by the different scales on which this factor is ordered. Major relief units are not only differentiated from each other with respect to predominant landforms but are also frequently associated with differences in parent material and, through altitude, climate. This scale of differentiation is illustrated in Malaŵi. Three of the major relief units of Malaŵi are erosion surfaces: the plains of medium altitude, 900–1500 m (the African surface), the small, high-altitude plateaux remnants at 1700–2500 m, and the Rift Valley floor at 60–500 m. The plains of medium altitude are developed on Basement Complex rocks, mainly gneisses; they have gentle slope, a moist to dry savanna climate, and a pattern of weathered ferrallitic soils where slopes are very gentle and rocks felsic, and ferruginous soils on areas with less gentle slopes and less felsic rocks. The high-altitude plateaux are partly on igneous intrusions; they have moderate slopes and a climate sufficiently cool to give humic latosols. The Rift Valley floor consists largely of alluvium, and the high temperature and impeded to poor drainage combine to give a pattern of calcimorphic and hydromorphic soils. Of the two relief units separating these surfaces, the hills and the Rift Valley scarp zone, the former has moderate or steep slopes, with some areas of ferruginous soils and some lithosols, whilst the latter is formed largely by steep slopes lithosols (Young, 1960/61, 1969a; Young and Brown, 1962, 1965).

The importance of relief in detailed soil distribution has already been noted. Intermediate in scale between the landform and the major relief unit are the relief units which are the basis of the land systems method of air photo interpretation. At this scale it is such features as the relative frequency of slopes at different angles and the extent of valley floors that affect soil types. In land systems interpretation, however, landforms are to some extent being used as a means of identifying, on air photographs, differences in other factors.

It is clear that climate, parent material and relief, together with hydrology in the special case of poorly-drained sites, may each exert a strong influence on tropical soil formation. What is in need of emphasis is that parent material is as important as climate. There has been a tendency to treat the zonal influence as the primary determinant of soil type, perhaps because it operates at the continental scale with which introductory accounts are concerned. In field soil survey the climatic zone provides a necessary frame of reference, but a real differentiation of soil types is very much more dependent on the pattern of geological outcrops. Thus, freely-drained soils of the lowland tropics are better regarded as arranged not on a single, zonal, spectrum but on two perpendicular axes, those of rainfall and parent material composition.

CHAPTER 2
PROCESSES

The properties of a soil at any one time constitute its *form*. *Soil-forming processes* cause change in soil form. Form is static, whilst processes involve change. The processes are themselves influenced by the environment, normally described as the *factors of soil formation*. The change in form over time, as brought about by processes, is the *evolution* of the soil. With respect to the environment as a whole, the factors are themselves properties of form. The distinction between factors and the properties of soils is the result of treating soil as the dependent variable.

The term process is commonly applied in two senses: *composite processes*, such as ferrugination, podzolization and gleying; and *specific processes*, such as hydrolysis, oxidation and clay translocation. The former type are made up of groups of the latter; thus gleying involves oxidation, reduction and the solution, movement and precipitation of iron compounds. Intermediate terms exist, describing groups of specific processes, e.g. weathering. The specific processes themselves result from the operation of relatively few active *agents*, principally water and biological agents.

WATER AS AN AGENT OF PROCESSES

The main agent in bringing about changes in soil properties is water, together with substances dissolved within it. The attributes of water (or, more strictly, the soil solution) that affect the resulting processes are its temperature, acidity, content of dissolved materials, and movement.

Water temperature affects rates of chemical weathering processes, and in particular hydrolysis. The dissociation constant, and hence the concentration of hydrogen ions, increases with rise of temperature; for pure water, the hydrogen ion concentration at 25 °C is 1.85 times that at 10 °C, although for carbonic acid the difference is smaller. The main cause, however, is the effect of the increased heat supply on chemical reactions. The generalization known as van't Hoff's temperature rule is commonly quoted; that for every 10 °C rise in temperature, the velocity of a chemical reaction increases by a factor of from to 2 to 3 times.

The temperature of rainfall, on striking the ground, is on average considerably higher in the tropics than in temperate latitudes. This is not the main cause of difference, however, for soil water largely takes on the temperature of the soil through which it passes, the higher specific heat of water being outweighed by the greater volume of soil. Thus, records of soil temperature provide a guide to that of the water. Soil temperatures at 1 m depth and below approximate to mean air temperatures. World maps of soil temperatures (Chang, 1957) show that between latitudes 30° N. and S., at low altitudes, mean annual soil temperatures are 25–30 °C. This compares with about 10 °C in the temperate belt. Thus if van't Hoff's rule holds, chemical reactions are some (1.5 to 2.0) × (2 to 3) = 3 to 6 times faster in the tropics than in the temperate zone.

Rainfall becomes a weak solution of carbonic acid through dissolving carbon dioxide from the atmosphere, and subsequently from soil air. It acquires additional acidity by taking into solution organic substances, loosely termed humic acids. The soil solution takes on the acidity of the soil itself; there is thus a positive feedback mechanism, the water draining through an acid soil becoming itself more acid, and thereby a more powerful leaching agent.

Rainwater is a dilute solution, having acquired from the atmosphere appreciable quantities of dissolved matter. In passing through the soil it dissloves additional substances, including salts, the exchangeable bases, silica and iron compounds. The higher temperature of soil water in the tropics increases the quantities of dissolved substances that can be held in solution at saturation; and for a given absolute amount of a substance in solution, the relative concentration is lower at high temperatures. In accordance with mass action principles, a decrease in concentration of the solute increases the tendency for matter to pass between solid and solute; this remains valid even though it is unlikely that a condition of chemical equilibrium is attained because of the slowness of most reactions in soils (at what are in chemical terms low temperatures). The presence of warm water therefore increases the effectiveness of removal in solution.

Thus for given moisture conditions, higher temperatures increase the rates of both weathering and leaching in the tropics as compared with temperate latitudes by a factor of about 3. This is the major cause of differences between the soils of temperate and tropical areas.

Processes are affected by the amount, changes in amount and movement of water in the soil. The following are the possible conditions that may be experienced by a given soil horizon over any period:

WATER CONTENT

Wet (waterlogged). Above field capacity; below the water-table.
Wet–moist. Alternately above and below field capacity; fluctuating water-table.
Moist. Between field capacity and wilting point; moisture content static or increasing; may rise above field capacity for short periods following rain.
Moist–drying. Between field capacity and wilting point, with moisture content decreasing.
Dry. At wilting point.

WATER MOVEMENT

Rapid downward flow. More than 50 mm/month flowing downwards, approximately vertically, through the soil.

Slow downward flow. Less than 50 mm/month flowing downwards.

Static. Little or no water movement.

Upward flow. Water moving upwards.

Throughflow. Lateral flow beneath, and approximately parallel to, the ground surface.

Inward flow. A form of throughflow with large amounts of water coming into the site, incoming flow intermittently exceeding outgoing flow.

Surface flow. Lateral flow across the ground surface.

Inward surface flow. Surface flooding.

The distinction between *moist* and *moist–drying* is made because in the latter case there will be a tendency for precipitation of dissolved substances. The value of 50 mm/month separating *rapid* from *slow downward flow* is suggested as an arbitrary limit, separating conditions liable to cause intense and moderate leaching respectively. It has not been measured, but must be estimated from the surplus of rainfall over potential evapotranspiration, with allowance for surface runoff. *Upward flow* can occur only very slowly by capillary forces, and is more likely to be caused by inflow to a depression. *Throughflow* is the agent activating the process of lateral eluviation. *Surface flow* causes surface denudation or natural erosion. *Inward surface flow*, or flooding, is the agent causing sedimentation, as in alluvial soils.

By estimating the seasonal or monthly conditions of water content and movement, an indication of the probable effects upon processes can be made. Some representative environments are as follows.

Lowland rainforest climate, free drainage. Moist with rapid downward flow throughout the year; throughflow and surface flow on slopes. High rates of both weathering and leaching.

Lowland rainforest climate, poor drainage. Wet throughout the year. In valley-floor sites, inward flow from valley sides, down-valley throughflow, and intermittent down-valley surface flow. Permanent reducing conditions, except for intermittent oxidation close to the ground surface.

Savanna climate, free drainage. Moist with downward flow during the wet season, and rapid downward flow in some months; moist, drying and static in the early dry season; in the late dry season becoming dry to 1–2 m depth, remaining moist below. Throughflow and surface flow on slopes. Weathering rates of bedrock remain high, but within the profile the weathering intensity falls through the dry season; leaching in the wet season alternates with some precipitation of dissolved substances in the dry season.

Savanna climate with high water-table. Wet with limited throughflow in all horizons in the wet season, in depth remaining wet throughout the year. The upper horizons experience wet–moist, moist or moist–drying conditions with inward flow in the dry season, possibly becoming dry near the surface.

Reduction in the wet season alternates with precipitation of dissolved substances during moist-drying periods.

Semi-arid climate, free drainage. Moist, with slow downward flow, for 1–3 months, followed by moist–drying and a prolonged period of dry conditions. Substances taken into solution in the wet season are subsequently precipitated lower in the profile.

Arid climate, depression site. Dry, with occasional periods of inward and/or upward flow. Precipitation of substances brought to the site in solution.

SPECIFIC PROCESSES

There are few specific processes found only in the tropics, and absent in temperate and polar latitudes. Where differences exist, it is largely in respect of absolute and relative intensities or rates of operation.

This similarity in kind originates from the fact that the same agents activate the processes in both latitudinal zones. Water and its dissolved substances do not differ qualitatively as between hot and cool climates, there are broad similarities between plant litter and humus forms, and many soil microorganisms are similar. Only in the case of termites is there an agent, and therefore a soil-forming process, exclusive to the tropics.

In treating specific processes, discussion is here limited to aspects that are of particular significance in the tropics. For general accounts of their manner of operation, reference may be made to textbooks. Table 12 gives a list of specific processes, for reference and to indicate the terminology used here.

Chemical weathering in its broad sense includes the breakdown of primary minerals and the synthesis of secondary minerals. *Translocation* refers to more or less vertical movements within the soil, in solution and in colloidal suspension, as distinct from *lateral movement* approximately parallel to the ground surface. Translocation in solution includes *leaching*, the taking into solution of substances and their transportation in the soil solution, and *precipitation*. Clay *eluviation* is the removal of clay particles in colloidal suspension from a horizon, clay *illuviation* (also called argillation) is their deposition in a lower horizon, the two processes together being termed *clay translocation*. The processes of lateral, or downslope, movement, include those which have received particular attention in geomorphology, the *surface processes*: soil creep, surface wash and rapid mass-movements (landslides). Included for completeness in the listing are structure formation and termite activity, although the former term begs the question of what are the processes involved whilst the latter is more an agent than a process.

Weathering, leaching and clay mineral synthesis are considered in this chapter. Clay translocation and structure formation are discussed in connection with texture and structure as properties in chapter 3, and biological processes in chapter 4.

Processes

WEATHERING AND LEACHING

A central question in tropical soil formation is that of the changes that take place in the mineralogical and chemical composition of the profile. The two main aspects of this question are the relative removal from the profile of silica and sesquioxides, and the types of secondary minerals that are formed. These two aspects involve leaching and weathering respectively, but these processes cannot be wholly separated. It is in part the release

TABLE 12 *Specific processes of soil formation*

WEATHERING
Physical
Chemical: Hydration
Carbonation
Hydrolysis
Oxidation
Reduction
TRANSLOCATION
In solution: Leaching } of: soluble salts
Precipitation
carbonates and sulphates
exchangeable bases
silica
iron
aluminium
other mineral substances
organic matter
Chelation, of sesquioxides complexed by organic acids
In suspension: Clay eluviation } clay translocation
Clay illuviation
CLAY MINERAL SYNTHESIS
LATERAL MOVEMENT
On the soil surface: Surface wash
Sedimentation
Of the soil material: Soil creep
Rapid mass-movements
Within the soil: In solution: lateral eluviation
In suspension
STRUCTURE FORMATION
BIOLOGICAL PROCESSES
(Plant growth)
Accumulation of dead plant material (litter, roots)
Humification of dead plant material
Loss of organic matter by: oxidation of humus
leaching of humus
denudation of litter and humus
(Chelation, see above)
Nitrification
Nitrogen fixation
Denitrification
Termite activity

of elements by weathering that determines the extent to which they are removed by leaching; conversely, the intensity of leaching, through its removal of substances from the weathering matrix, affects the types of secondary minerals that are formed. Thus, the leaching process can in some respects be regarded as interposed between the two components of weathering, the breakdown of primary minerals and the formation of secondary minerals.

Weathering

Physical weathering is unimportant in the tropics except in special circumstances, e.g. sheeting on inselbergs. Mineral breakdown, including in its early stages the transformation of rock to regolith, is accomplished by the processes of chemical weathering, particularly hydration and hydrolysis. The extent and rate of rock weathering depends on a passive and an active factor: the susceptibility of the minerals, and the intensity of chemical attack.

The relative susceptibility of minerals to chemical weathering depends on the elements of which they are composed and the crystal structure. All common rock-forming minerals include silica tetrahedra, and the greater the degree to which these are directly linked, the greater the resistance to weathering. On this basis, the main groups of primary minerals are:

1. Isolated tetrahedra   Olivine
2. Chain structures:
    Single chain   Augite
    Double chain   Hornblende
3. Sheet structures   Biotite
                                 Muscovite
4. Framework structures:
    With substitution of Al^{3+} for Si^{4+} Feldspars
    Without substitution of Al^{3+} for Si^{4+} Quartz

Various indices of mineral susceptibility to chemical weathering have been constructed (see Loughnan, 1969, pp. 53–60; Ollier, 1969, pp. 59–63). There is agreement that quartz and muscovite are the most resistant, ferromagnesian minerals the most susceptible to weathering, and that the feldspars together with biotite occupy an intermediate position, probably with orthoclase as the most resistant. The high resistance of muscovite is unexplained; in terms of both composition and crystal structure it should fall in the second group, yet it persists in soils at an advanced stage of weathering.

Organic acids are always present in the soil solution, and this may affect

69

the relative susceptibility of different elements and minerals to weathering, in extreme cases to the extent of changing the sequence of mineral stability. Thus, strongly-complexing organic acids can dissolve alumina preferentially over silica (Huang and Keller, 1972).

There is no evidence that the order of mineral susceptibility differs between the tropics and other climatic zones. The greater intensity of weathering, however, results in a qualitative difference in the extent of mineral breakdown. 'The higher the temperature, and the stronger the leaching, the more complete is the breakdown of the original crystal lattice of the minerals being weathered, so the more different crystallographically are the residual products of weathering likely to be' (Russell, 1968). Duchaufour (1960) presents this as a difference in the extent to which oxides are released from their parent minerals and pass into the weathering complex. In the (composite) process of ferrugination there is a 'moderate loss of silica, liberation of ferric oxides but not alumina'; whilst in ferrallitization there is a 'maximal loss of silica, completely liberated in the course of weathering, liberation of sesquioxides Fe_2O_3 *and* Al_2O_3'.

Leaching

Components released from the crystal lattice by weathering may be taken in solution and removed by leaching. The solubilities are influenced by both reaction (pH) and redox potential (Eh). The main groups are as follows (fig. 3):

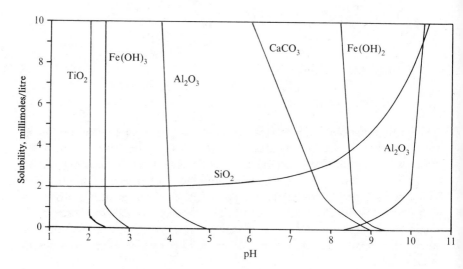

Fig. 3. Solubilities of soil components in relation to pH. After Loughnan (1969).

Soluble salts (chlorides, sulphates). Very highly soluble; removed wherever there is even slight and infrequent leaching.

Exchangeable bases (Ca^{2+}, Mg^{2+}, Na^+, K^+). Highly soluble; readily lost under moderate leaching intensities.

Silica. It was formerly thought that silica became more soluble with increasing acidity, and some earlier theories of tropical soil evolution were based on this assumption. It is now generally held that silica has a low but constant solubility in the pH range 3.5–8.0, rising rapidly in strongly alkaline soils (McKeague and Cline, 1963). The solubility of quartz is very low, about one tenth that of amorphous silica, although it is incorrect to regard it as completely insoluble.

Iron. Solubility is dependent on degree of reduction. In an Eh/pH diagram, the line of equilibrium between ferrous oxides (Fe^{2+}, as FeO) and ferric oxides (Fe^{3+}, as Fe_2O_3 or FeOOH) passes through the range of natural soil environments. In the reduced (ferrous) form, iron fairly readily passes into solution. Under free drainage, most iron remains in the oxidized (ferric) state, in which it is almost insoluble. Thus iron movement takes place more readily from the groundwater zone. In the special case of tropical podzols, iron mobility is thought to be increased by chelation, organic complexing associated with strongly-acid mor-type humus.

Alumina. The Eh range of soils is such that aluminium invariably remains in the form Al_2O_3. As such, it is immobile at pH values about 4.5, becoming soluble in extremely acid soils. It may also be affected by chelation.

Titanium. This element is sometimes taken as constant (i.e. not subject to any leaching loss) when deriving absolute losses from data on relative element concentrations. This is justifiable if it is in the form of TiO_2; if released from the parent mineral as $Ti(OH)_4$, however, it may show limited mobility (Loughnan, 1969, p. 52).

In terms of order differences, the soluble salts are about 30–100 times more mobile than the exchangeable bases, the bases 5–10 times more mobile than silica (in forms other than quartz), and non-quartz silica 5–10 times more mobile than quartz and the susquioxides.

There is thus an approximate leaching sequence, with increases in the frequency and amount of water passing through the profile, from the leaching of soluble salts, through bases and silica, to the sesquioxides; where any group is leached, removal of all the preceding groups is greater. The division between conditions in which calcium is removed or accumulates (as calcium carbonate) marks a major pedogenetic boundary, between pedalfers and pedocals; in the lowland tropics this transition occurs on freely-drained sites at a mean annual rainfall of about 600 mm. A further qualitative distinction is sometimes made between two types of leaching within the humid tropics, associated respectively with the composite processes of ferrugination and ferrallitization, and with savanna and rain-

71

forest climates respectively. There is no sound basis for such a distinction The leaching processes in savannas (during the wet season) and in rainforest both involve removal of bases and silica; they differ only quantatively and by a gradual transition of intensity.

Precipitation

The order of precipitation of components from the soil solution is approximately the reverse of their susceptibility to leaching. Chlorides and sulphates are precipitated only where there is net evaporation of upward or inward moving groundwater, as in depression sites in arid climates. Under semiarid conditions the occasional leaching is sufficient to remove salts, but calcium is precipitated as carbonate. The drier the climate, the nearer to the surface the calcium carbonate horizon occurs. The mechanism is presumably that in a soil which at depth is normally below field capacity, leaching of carbonates occurs in the upper horizons only; the amount and frequency of rains controls the depth at which this leaching ceases and carbonate precipitation occurs. There is a need for quantitative studies of the relations between field capacity, soil moisture regime and depth of carbonate precipitation.

Besides involvement in clay mineral synthesis, discussed below, the exchangeable bases are adsorbed onto the negatively-charged surfaces of clay minerals and clay–humus molecules. Apparent base saturation is thus a good indicator of how intensely a soil is being leached (but see p. 95). Most freely-drained soils in rainforest environments, with a rainfall exceeding 2000 mm, have subsoil saturations below 10 percent, and in savanna climates there is a general increase in saturation with decrease in rainfall until values of 90–100 percent are found where 600–800 mm rainfall occurs in a wet season of 3–4 months. Hence subsoil base saturation provides a good differentiating property indicative of genetic soil conditions.

Silica is held in the soil solution as monosilicic acid (H_4SiO_4), typically in concentrations of 5–50 p.p.m. The equilibrium concentration decreases considerably with increase in alkalinity (McKeague and Cline, 1963). Hence the passage of soil solution from a more acid to a less acid environment may cause precipitation of silica. This mechanism was used by Mohr and Van Baren to explain the formation of a B horizon of silica deposition during the weathering of basic tuff in Indonesia (p. 54). On being precipitated, silica is adsorbed by ferric hydroxide and by mixed silicon–iron ('ferrosic') hydroxides (Herbillon and Tran Vinh An, 1969).

The main cause of precipitation of iron oxides is oxidation into the ferric form, caused when soil solution containing dissolved ferrous iron passes into an environment of higher (more oxidizing) redox potential. This occurs in savanna soils in the zone separating the levels of the permanent water table in the dry and wet seasons. A second mechanism is that for a constant redox potential, a rise in pH can increase the proportion of iron in the oxidized form, causing precipitation of iron oxides from solution. This could account for the iron deposition that frequently occurs in acid latosols within the rock/soil transitional horizon. Alumina, since it is only soluble in extremely acid conditions, is presumably also precipitated where acidity falls.

CLAY MINERALS

Clay mineral synthesis

The types of secondary minerals exercise a powerful effect upon the properties of tropical soils, comparable only to the effects of texture and organic matter. The predominance of 1:1 lattice and absence or rarity of 2:1 lattice clay minerals is one of the main features that distinguishes soils of the humid tropics from those of temperate regions.

The most common types of secondary minerals and other secondary products of weathering found in soils are as follows:

Clay minerals (silicate minerals)

Kaolinite group (kandites)	1:1 lattice (one sheet of silica tetrahedra and one of alumina octahedra), non-expanding
Kaolinite	$Al_2Si_2O_5(OH)_4$
Halloysite	$Al_2Si_2O_5(OH)_4.2H_2O$
Metahalloysite	$Al_2Si_2O_5(OH)_4$
Illite group ('mica minerals')	2:1 lattice (two tetrahedral and one octahedral sheet), non-expanding
Illite	$KAl_2(AlSi_3)O_{10}(OH)_2$ with partial substitution of Mg^{2+}, Fe^{2+} or Fe^{3+} for Al^{3+}
Montmorillonite group (smectites)	2:1 lattice, expanding
Montmorillonite	$(Al_{3.33}Mg_{0.67})Si_8O_{20}(OH)_4 \ldots M_{0.67}^+$
Chlorite–vermiculite group:	Two tetrahedral and one octahedral sheet separated by an octahedral Mg, Fe or Al sheet, non-expanding

Amorphous silica ('allophane')

Iron Oxides

Non-hydrated	
Goethite	$FeO.OH = Fe_2O_3.H_2O$
Haematite	Fe_2O_3
Magnetite	$Fe_3O_4 = FeO.Fe_2O_3$
Hydrated	
Various minerals of the form $mFe_2O_3.nH_2O$	

Hydrated aluminium oxide

Gibbsite	$Al(OH)_3 = Al_2O_3.3H_2O$

73

Processes

The secondary minerals synthesized within the soil are affected by the nature of the weathering complex and the types of primary (parent) minerals. As regards the former influence, there are two overlapping theories, those of the weathering/leaching ratio and of the silica potential. These two theories have many features in common, but differ in the emphasis placed on the amounts of exchangeable bases and of silica within the weathering complex.

The concept of the weathering/leaching ratio was enunciated by Crompton (1960, 1962) as a general theory of soil formation. The *richness of weathering* 'is the product of the quantity and variety of elements in the parent material capable of being brought into the soil solution and those effects of temperature, site, vegetation, etc., which tend to bring them into solution'. Thus basic rocks, high temperatures and a permanently moist soil tend to give rich weathering, whilst with felsic rock, lower temperatures and soils frequently dry the supply of bases and silica is less. The richness of weathering 'is offset by *intensity of leaching*, an outcome of the rainfall/ evapotranspiration balance and soil permeability' (Crompton, 1962, p. 211). Thus a large excess of rainfall over evapotranspiration combined with

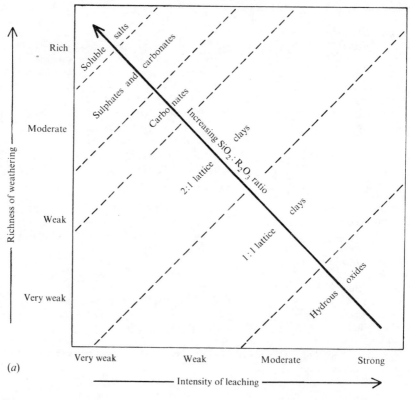

(*a*)

sandy, permeable soils results in a large volume of water passing downwards through the profile and hence intense leaching. The bases and silica are removed by leaching as soon as they are released by weathering, and do not remain in the weathering complex long enough to participate in clay mineral synthesis. The effect of these combined influences on secondary mineral formation is shown in fig. 4(*a*). Where weathering is least rich and leaching intense, only iron and aluminium oxides and hydroxides remain. With a slightly less depleted weathering complex, only kaolinitic clay minerals are formed; as the availability of bases and silica becomes greater, 2:1 lattice minerals are formed in addition to 1:1 lattice. The lines separating pedalfers from pedocals and, finally, halomorphic soils are then crossed.

Fig. 4. (*a*) The weathering/leaching ratio. After Crompton (1962). (*b*) Semi-quantitative interpretation of the weathering/leaching ratio applied to tropical soil types. The points marked indicate combinations of climate and rock type. Climates: R = rainforest, T = rainforest–savanna transitional, S = savanna, D = semi-arid. Rock types: B = basic, I = intermediate composition, F = felsic, V = very felsic or highly weathered. W/L = weathering/leaching ratio.

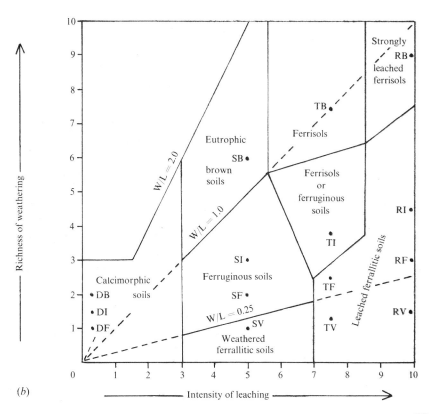

(*b*)

Fig. 4(*b*) is an attempt at a partial quantification of this concept. Richness of weathering and intensity of leaching are both allotted scales of 0–10. Based on the need to accommodate to these scales, the following assumptions are made:

Richness of weathering = rock composition factor × climatic weathering factor

Rock composition factors		*Climatic weathering factors*	
Very felsic or highly-weathered	0.5	Arid	0
Felsic (felsic igneous or siliceous sedimentary)	1.0	Semi-arid	1.0
Intermediate composition	1.5	Savanna	2.0
Basic (basic igneous or calcareous sedimentary)	3.0	Transitional	2.5
Ultra-basic	3.3	Rainforest	3.0

Intensity of leaching = climatic leaching factor × permeability factor

Climatic leaching factors		*Permeability factors*	
Arid	0	Low permeability	0.75
Semi-arid	0.5	Normal permeability	1.0
Savanna	5.0		
Transitional	7.5		
Rainforest	10.0		

Thus, for example, for a felsic igneous rock in a savanna climate:

$$\text{Richness of weathering} = 1.0 \times 2.0 = 2.0$$
$$\text{Intensity of leaching} = 5.0 \times 1.0 = 5.0$$
$$\text{Weathering/leaching ratio} = 2.0/5.0 = 0.4$$

Points for the main climates and rock types are plotted in fig. 4(*b*). Lines of equal weathering/leaching ratio do not run diagonally and parallel, as they do on Crompton's diagram, but radiate from the origin. Zones occupied by some major soil groups are indicated. The diagram is a gross over-simplification insofar as numerous variables are represented on only two axes. It nevertheless gives some indication of the relative role of climate and parent material in affecting clay mineral synthesis. Predominantly kaolinitic clays are found both with high intensity of leaching, in rainforest climates, and also in savannas where a deeply or highly-weathered parent produces a very low richness of weathering. Rock has its greatest effect in the savanna and transitional savanna–rainforest zone, causing an increase in the proportion of 2:1 lattice minerals, through ferruginous soils to basisols, as the composition becomes more basic.

The theory of control of clay mineral synthesis by the silica potential differs from the weathering/leaching ratio concept in that emphasis is placed upon the supply of silica rather than of exchangeable bases (Jackson, 1968; Jackson and Sherman, 1953). Weathering reactions follow mass action principles, and are thus dependent on the solute concentration of the soil solution. 'The silica potential supplied to the matrix solution by the primary mineral is highly important in determining the mineral produced by weathering.' Basic, ferromagnesian, minerals supply enough silica to sustain the formation of montmorillonite; felspars lead to kaolinite formation but have insufficient silica potential to form montmorillonite unless silica concentration occurs, e.g. through evaporation or poor drainage. Jackson gives the following weathering index, or sequence of clay-size minerals, comprising both primary and secondary minerals:

1. Gypsum
2. Calcite
3. Hornblende, olivine, pyroxenes
4. Biotite
5. Albite plagioclase
6. Quartzite
7. Dioctahedral micas, muscovite
8. Vermiculite
9. Montmorillonite, pedogenic dioctahedral chlorite
10. Allophane, kaolinite, halloysite
11. Gibbsite
12. Haematite, geothite, magnetite
13. Anatase, rutile, zircon

Desilication can lead to transformations of one clay mineral into another: montmorillonite to kaolinite or halloysite, muscovite to illite, illite to kaolinite, kaolinite to gibbsite. Moreover, in accordance with mass action principles, such reactions are reversible. Hence by *resilication*, gibbsite may be transformed into kaolinite.

The effects of both silica and exchangeable bases are combined if clay mineral synthesis is considered in terms of thermodynamic stability fields (Loughnan, 1969, pp. 62–3; van Schuylenborgh, 1971). The products of weathering depend both on the molecular concentration of silicic acid $[H_4SiO_4]$, and on the ratio of the concentrations of each individual cation to hydrogen ions, e.g. $[K^+]/[H^+]$. Fig. 5 gives some examples. Thus with a low concentration of silicic acid, gibbsite will form; with a soil solution richer in silicic acid, whether kaolinite or montmorillonite will form will depend on the concentration of bases. A high hydrogen-ion concentration, i.e. in a strongly acid soil, has the effect of lowering the relative concentration of bases. Thus for the same concentrations of both silica and bases, kaolinite will tend to form in an acid soil, and montmorillonite in neutral to alkaline conditions.

Reactions in soils do not proceed to conditions of thermodynamic equilibrium, owing to the low temperatures, spatial inhomogeneity and temporal variability of conditions. Nevertheless, a consideration in these terms is a useful indication of trends in clay mineral synthesis and transformation,

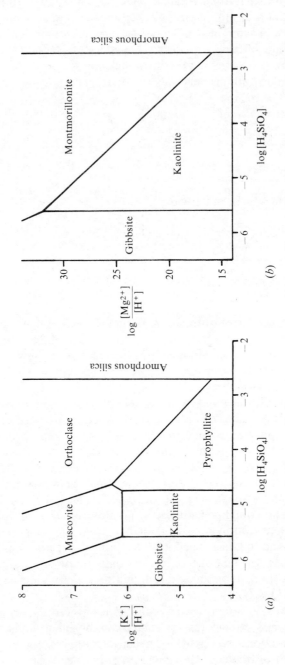

Fig. 5. Stability fields, at 25°C and 1 atmosphere pressure, of the systems: (*a*) $K_2O-Al_2O_3-SiO_2-H_2O$; (*b*) $SiO_2-Al_2O_3-MgO-CaO-H_2O$. After van Schuylenborgh (1971).

corresponding well with observed form. Low concentrations of silica and/or exchangeable bases in the soil solution, and strong acidity favour the formation of kaolinitic clay minerals or, if these conditions exist in extreme form, gibbsite and free iron oxides. Such conditions may be brought about either by a low rate of supply, caused by slow weathering or base-poor rocks, or rapid removal by leaching. Low permeability, imperfect drainage, or concentration of the soil solution by inward drainage or evaporation, lead to an increase in the weathering/leaching ratio, and consequent tendency toward the formation of montmorillonite.

In addition to the composition of the soil solution, the second main influence on secondary minerals is the nature of the parent material. On basic crystalline rocks, montmorillonite is the first weathering product; if leaching conditions are sufficiently intense it may subsequently be transformed to kaolinite (or halloysite) and ultimately to gibbsite (Kantor and Schwertmann, 1974). On felsic crystalline rocks, kaolinite and halloysite are the main first weathering products, together with inherited quartz. Gibbsite may also be found as a primary weathering product of felsic rocks, subsequently to be transformed into kaolinite (Watson, 1962b). Argillaceous sedimentaries usually contain quartz and illite; the former is inherited by the weathering complex, the latter partly inherited and partly transformed to kaolinite.

Thus, the clay-size minerals present in a soil may be:

(i) inherited,
(ii) transformed,
(iii) newly synthesized.

Inherited minerals are those present in the parent rock which are released by the breakdown of cementing materials; quartz and magnetite are common inherited minerals, and feldspars sometimes occur as such. Transformed minerals are those which suffer alteration in the course of their being released, e.g. the transformation of illite to kaolinite. The third group, that of minerals newly synthesized from the soil solution, includes both the silicate clay minerals and iron oxides.

Weathering in the humid tropics may be illustrated by data of West and Dumbleton (1970) for soils over three rock types in Malaya: basalt, granite and shales (table 13). The soils are freely-drained, under rainforest, and the climatic 'variable' is effectively constant, with lowland, permanently-humid conditions (rainfall 1800 mm) at all sites. Pairs of profiles on both granite and basalt show substantially different mineralogical compositions; these are assumed to represent moderate and advanced stages of weathering respectively, although it is clear that, in addition, the lithologies were not identical. Over basalt, the moderately-weathered soil contains mainly kaolinite and goethite, with very little gibbsite; haematite occurs in

TABLE 13 (a) Minerals in soils on freely-drained sites under a rainforest climate, Malaya. Relative abundance is estimated by X-ray diffraction. +++ = much, ++ = considerable, + = some, t = trace. cn = material in concretions. After West and Dumbleton (1970)

Parent material	Inferred stage of weathering	Depth (cm)	Quartz	Feldspar	Kaolinite	Illite	Goethite	Haematite	Magnetite	Gibbsite
Granite	Moderate	60	+++	+	++	++				
		180	+++	+	++	++				
		180 cn	+++cn	t	+cn	+cn	++cn			
		230	+++	+	++	++				
Granite	Advanced	80	+++		+++					
		210	++		+++					
		370	++		+++			t		
		370 cn	+++		+			+		
Basalt	Moderate	150	t		+++		+++			t
		240	t		+++		++			t
		240 cn			+		+	+++		
Basalt	Advanced	60	t		+		+++		t	+++
		60 cn					t	+		+++
		180	t		++		+++		++	+++
		760	t		++		++		++	+++
Shale	—	60	++		++	++				
		60 cn	+cn		++cn	++cn	+cn			
		150	++		+++	+++	t	++cn		
		610	++		++	+++				
Shale and sandstone	—	60	+++		++		+			
		340	+++		++		+			

(b) *Percentage composition of parent rocks*

Rock type	Quartz	Feldspar	Muscovite and biotite	Illite and chlorite	Augite and olivine	Iron oxides	Calcite and dolomite	Other minerals
Granite	31	52	12			2		3
Basalt		46			45	7		2
Shale	32	18	18	16		5	8	3
Sandstone	70	8	1	8		2	11	

concretions. At the advanced stage of weathering the kaolinite has largely been replaced, through desilication, by goethite; the magnetite on this site is inherited. On granite, the soil at a moderate stage of weathering contains inherited quartz and feldspar, and secondary illite and kaolinite, but almost no iron oxides; at the more advanced stage, the feldspar has gone, quartz remains, kaolinite is the only silicate mineral, and haematite nodules are accumulating in depth. The two sites on shales contain illite, probably inherited; kaolinite, probably transformed from illite; and some accumulation of haematite and goethite, mainly as nodules. Montmorillonite is absent from freely-drained soils in such a strongly-leaching climate, even over basic rocks.

Clay minerals and soil properties

For a given texture, or clay content, the type of clay mineral present exercises a profound effect upon soil properties, including structure, consistence, cation exchange capacity, response to moisture, and engineering properties. These effects are of particular importance in the tropics where the percentage of clay tends to be high.

Most soils of the rainforest zone, and some highly-weathered profiles in the savannas, have clay fractions dominated by kaolinite, with gibbsite and iron oxides also present, but 2:1 lattice minerals entirely absent. By comparison with clays of the temperate zone, the physical properties of such soils are remarkable. They remain friable, or at least labile, despite clay contents of up to 85 percent. There is an aggregation into fine structural units, a feature possibly associated with free iron oxides, imparting a 'floury' feel to the soil. Cation exchange capacities are low, often less than 15 m.e./ 100 g (milliequivalents per hundred grams) of clay. Hence such clays provide physically good conditions for root penetration, and are freely permeable, but have a low capacity for retaining nutrients. The unfavourable engineering properties associated with high clay contents are mitigated in part by the lower values of activity (= plasticity index/clay content) for this type of clay. The presence of inherited illite, as on shale-derived soils of the rainforest zone, results in a more plastic clay.

Where even small amounts of 2:1 lattice minerals are present the properties of clays change substantially. This is the case for ferruginous soils of the savanna zone, the amount becoming greater as rock composition becomes less felsic. The clay then becomes plastic, and forms larger structural aggregates. The cation exchange capacity of the clay fraction rises to moderate values.

Montmorillonite is commonly present in gleyed soils, producing a more plastic, sticky consistency and a coarse prismatic or blocky structure. Its property of swelling and shrinking on wetting and drying may produce grooves on ped surfaces, to which the borrowed term 'slickensides' is some-times applied. A combination of basic rocks with weak leaching condi-tions and/or impeded drainage may produce a clay dominated by mont-morillonite, which has a sufficiently dominant effect on other properties for a soil type, vertisols, to be identified on this basis. Here is found the reverse of conditions in the kaolinite-dominated clays: a high cation ex-change capacity, but difficulties for agriculture (and still more so for engineering) caused by excessive swell-shrink and stickiness when wet.

COMPOSITE PROCESSES

Composite processes are groups of specific processes that, acting jointly, lead to the formation of particular soil types. The groups are not just col-lections of separate processes acting simultaneously but have a con-siderable degree of causal interdependence; for example, in the composite process of ferrallitization, it is the strong leaching of cations and silica that restricts clay mineral synthesis to the formation of 1:1 lattice minerals.

The use of terms for composite processes involves a strong element of circular argument. Thus the process that forms a ferrallitic soils is some-times given as ferrallitization. But what is ferrallitization? Its definition is the process that forms a ferrallitic soil. If these terms have any value, it is as a means of referring to groups of specific processes.

In the treatment here of composite processes, podzolization and solodiza-tion are omitted as of minor importance (cf. pp. 171 and 208), and it has never been suggested that there is a process of 'vertisolization', although it would be no less tautologous to do so. The 'process' of laterite forma-tion is omitted since the specific processes involved are not established. Ferrallitization is used in preference to 'laterization', to avoid confusion with laterite formation. These composite processes are further discussed in the context of the relevant soil types in Part Two.

The following are the more important composite processes in tropical soils:

Ferrugination. Chemical weathering of moderate to high intensity. Com-plete leaching of soluble salts and carbonates, moderate leaching of ex-changeable bases, partial leaching of silica. Release and dehydration of iron oxides and their deposition as coatings to other soil constituents (rube-fication). Sometimes deposition and hardening of iron oxides as con-cretions. Synthesis of kaolinite with some goethite and haemitite, small amounts of illite or montmorillonite, but not gibbsite. Moderate clay trans-location. Formation of blocky structure with illuvial clay skins, particularly of iron oxides.

Ferrallitization. Intense chemical weathering with complete breakdown of all minerals except quartz. Complete leaching of soluble salts and carbonates, strong leaching of exchangeable bases, partial leaching of silica and to a lesser extent iron and aluminium oxides. Sometimes deposition of hardening of iron oxides as concretions. Synthesis of kaolinite with some goethite and gibbsite, sometimes haematite in concretions, but not montmorillonite. Clay translocation absent in iron-rich soils, present in others. Formation of weak blocky structure, superimposed on fine crumb structure in iron-rich soils.

Calcification. Chemical weathering of moderate intensity. Leaching of soluble salts. Weak leaching of carbonates from upper horizons and precipitation in lower horizons. Weak leaching of bases from upper horizons, little or no leaching of silica. Synthesis of kaolinite, illite and montmorillonite. Clay translocation weak or absent. Formation of blocky structure.

Salinization. Chemical weathering of moderate or low intensity. No leaching, except sometimes of soluble salts from upper horizons. Precipitation of soluble salts from inward-flowing water. Often gleying in depth. Flocculation of clays.

Solonization (alkalization). Replacement of exchangeable calcium by sodium. Much clay translocation. Deflocculation of clays and subsequent formation of columnar structure.

Gleying. Chemical weathering, in depth anaerobic, in upper horizons sometimes anaerobic, sometimes alternately anaerobic and aerobic. Release of iron oxides in ferrous form, their local migration and partial oxidation and precipitation in ferric forms as mottles and sometimes concretions. Synthesis of 1:1 and 2:1 lattice clay minerals. Formation of coarse blocky or prismatic structure by swell-shrink on wetting and drying.

CHAPTER 3
SOIL PROPERTIES

Non contemnenda esse ea parva sine quibus magna constare non possent.

St Jerome

The properties held by the horizons of a soil profile are divided into *morphological* and *analytical properties*, the former described in the field and the latter requiring laboratory analysis. Horizon boundaries are determined during field description, and analytical data relates to samples taken from horizons determined on the basis of morphological properties. Soil survey rests very largely on morphological properties; only 1–5 percent of the soil profiles described in the field are subsequently analyzed.

The distinction between morphological and analytical properties is not clear-cut. Thus, texture, organic matter and free carbonates are estimated both in the field and in the laboratory. One well-known tropical soil surveyor only described horizon boundaries, structure and roots in the field, making all other morphological observations on samples collected in partitioned split bamboos. Moreover, the two sets of properties do not entirely differ in their essential nature; morphological properties are those which can be detected by sight and feel, and it is something of an accident that our fingers are not sensitive to pH differences within the range encountered in soils. Analytical data, however, for the most part concern properties at the molecular scale, whilst morphology concerns the soil behaviour at the macroscale. The latter is much affected by the former, and with experience it is possible to predict the analytical results from site data and field morphology.

The analytical properties can be divided broadly into chemical and biological properties. This distinction is still more blurred, since much chemical activity occurs on the surfaces of humus, or clay-humus, molecules, whilst many processes customarily called 'chemical' weathering, involving changes to the mineral fraction of the soil, are brought about by biological agents.

This chapter gives accounts of the main morphological and analytical properties of the mineral soil, with particular reference to features found in the tropics. Properties of the organic fraction are discussed in chapter 4.

Soil properties

Colour

Colour is of no direct agricultural significance and yet it is the most easily observed property in soil survey. Colour as such has no known effect on plant growth; there are red soils and yellow soils indistinguishable in agricultural behaviour. There are few soil surveyors, however, who would have sufficient courage of their convictions to include two such soils in the same series, even though they might readily include two texture classes. Colour is such a prominent and observable feature that it is assumed in the field to be diagnostic of properties of agricultural importance, either through causal connection or non-causal association.

Colour description has been universally standardized through the use of Munsell charts. This system is comprehensive, easy to use, and has the advantage of providing standardized verbal terms in addition to the number and letter codes for hue (shade of red to yellow), value (amount of grey) and chroma (strength of colour). The criteria for description of mottles – abundance, size and contrast – are also standardized; it is normally sufficient to record only the names of mottle colours, giving Munsell notations only for the matrix (e.g. 'dark grey (10YR 4/1) with common medium, distinct yellowish red and reddish brown mottles').

In some tropical soils the only visible differentiation is the darker colour caused by humus in the A horizon. Caution should be exercised in interpreting a clear colour difference as indicative of a high humus content. First, in alkaline soils the humus becomes finely divided and coats the mineral particles; this confers a dark grey colour which may be misleading in terms of organic matter content. Secondly, some soils have a low chroma, i.e. a low intensity of colour, in the mineral fraction, and in these circumstances the grey topsoil greatly differs in colour from the subsoil although the organic matter content may be as low as one percent.

Mottling may be produced by drainage impedance or weathering. Impeded drainage produces ferric iron mottles, with colours varying from red to yellowish brown, on a grey matrix. In the tropics such mottles sometimes develop into concretions, intermediate forms of soft concretions being found. A prominent weathering mottle is a feature of the saprolite in many parts of the humid tropics. In its most marked form, in the rainforest zone, it consists of a matrix of whitish kaolinized material with a very prominent red mottle, graphically described as the 'corned beef horizon'.

The first thing every schoolboy learns about tropical soils is that they are red; red earths of the tropics contrast with brown earths of the temperate zone. Allowing that 'red' includes reddish brown and yellowish red this is an acceptable generalization. Temperate soils can be as red as the 5YR hue, not only when colour is inherited from parent material of tropical origin, as in Old Red Sandstone soils, but also on soils from iron-rich rocks such as basalts. Soils of the humid tropics frequently fall squarely within the 2.5YR hue, although hand-on-heart examples of soils on the 10R page are sufficiently rare to call for some small celebration. At the opposite extreme the 2.5Y hue can occur in the olive brown mottle colour, on a dark grey matrix, of some vertisols.

Red, yellowish red and yellow colours are mainly caused by iron products in the clay fraction. The common ferric oxides in soils are, in order of increasing hydration (Oades, 1963):

Magnetite	Fe_3O_4	Black
Haematite	Fe_2O_3	Red
Amorphous ferric oxides	Fe_2O_3	Red
Goethite	$FeO . OH$	Yellow or yellowish
	$(= Fe_2O_3 . H_2O)$	brown
Lepidocrocite	$FeO . OH$	Orange brown
'Limonite'	$FeO . OH . xH_2O$	Yellow or yellowish
		brown
Hydrated amorphous oxides	$Fe(OH)_3 . H_2O$	Yellowish

Haematite only forms under strongly oxidizing conditions, i.e. when the soil dries out, which accounts for its absence in the temperate zone. Goethite and the variants of it with excess H_2O, known as 'limonite', form where the soil, whilst aerated, remains moist. Soils of rainforest climates are characterized by goethite whilst those of the savannas, which dry out annually, contain haematite, goethite and amorphous iron oxides. Lepidocrocite has been found in the mottles of gleyed soils. Haematite also forms when the rate of release of iron is high by weathering of iron-rich rocks; soils over basic rocks are very red. It has been suggested that control is exercised by the ratio between abundance of organic compounds and the rate of release of iron by weathering, a low ratio being conducive to haematite formation (Schwertmann, 1971). Haematite and goethite can be transformed into one another slowly by hydration and dehydration, with a corresponding change in soil colour.

It is not always the case that the difference between red and yellowish soils corresponds to dominance by haematite and goethite respectively. The tendency in savanna catenas (p. 277) for the soil to become less red and more yellow downslope was formerly attributed to the soils lower on the slope remaining moist for longer periods, with consequent hydration of iron minerals. This has always seemed unlikely to me, as the colour change commences well up on the convexity. Attempts to detect such a difference in iron mineral type have given negative results, showing instead that the redder soils contain a higher percentage of free iron oxides (Webster, 1965). In a sample of latosols studied in thin section, Bennema *et al.* (1970) found that

Soil properties

profiles with an Fe_2O_3 content of less than 10 percent had braunlehm fabrics, those with over 20 percent had rotlehm fabrics, whilst intergrade fabrics were associated with intermediate free iron oxide contents.

Red colours can be caused by quite small quantities of very finely-divided amorphous iron oxides, which coat other minerals, including goethite; this is the process of rubefication (Kubiena, 1956; Segalen, 1969, 1971). Yellow soils are only found where this non-crystalline material is absent. When it comes to colour, a little amorphous iron can go a long way.

Texture

Soil texture may be considered in two respects: the texture of the profile as a whole, and differentiation between horizons.

The texture of the profile as a whole is determined by the grain-size and mineralogy of the parent material, the intensity of chemical weathering, and in the special case of immature soils by the time over which weathering has acted. Weathering intensity decreases with decrease in temperature and with the portion of the year in which the soil is dry. Hence soils of lowland rainforest climates tend to be high in clay, with a general decrease in clay content at higher altitudes and drier climates. However, the passive effects of parent material on texture are greater than the active effects of variation in weathering intensity (p. 17).

A characteristic feature of soils of the humid tropics is a low silt content. Most profiles have less than 15 percent silt, and hence occur along one side of the textural triangle (exceptions are some alluvial soils). Finger texturing in such circumstances becomes largely a matter of determining the relative proportions of sand and clay, along the linear sequence of texture classes, sand—loamy sand—sandy loam—sandy clay loam—sandy clay—clay. After taking account of parent material, the more highly weathered the soil the lower is the silt content; the silt:clay ratio is used as an index of weathering, soils with a ratio of <0.15 being regarded as highly weathered (p. 261).

The Australian terminology is a useful means of describing textural differentiation within the profile (Northcote, 1971). *Uniform texture profiles* have a similar texture at all depths, *gradational texture profiles* show a gradual increase of clay with depth, and *duplex texture profiles* show a relatively abrupt clay increase at some depth. Thus, both gradational and duplex texture profiles contain what is normally called a textural B horizon. The degree of development of a textural B horizon can be compared by means of an *index of textural differentiation*, I_t, defined as

$$I_t = \frac{\text{clay percent in topsoil}}{\text{maximum clay percent}}$$

In the rainforest and rainforest–savanna transition zones there is a common soil type that has a notably uniform texture profile, remaining high in clay right into the topsoil. Such soils are red or dark red and have a 'floury' consistence (p. 82). They have a stable micro-aggregation caused by a high content of free iron oxides, which coat and aggregate the clay minerals and inhibit clay translocation (Townsend and Reed, 1971). It is by no means true, however, that all leached ferrallitic soils of rainforest climates have uniform texture profiles; for example, soils on granite have gradational profiles, with a well-developed and deep textural B horizon. Soils developed from basic rocks are high in clay and have either uniform or gradational texture profiles.

Most soils in the savanna zone have a relatively sandy topsoil over a textural B horizon, sometimes separated by two textural classes, e.g. sandy loam over sandy clay. In some of the poorer weathered ferrallitic soils of plateau sites the topsoil is an almost loose loamy sand and the B horizon a structureless sandy clay loam or sandy clay, firm when moist and hard when dry. There is a tendency for textural profiles to be gradational in actively weathering soils and duplex in highly-weathered soils. Duplex profiles reach an extreme degree of development in solods.

Until recently it was assumed that textural differentiation within the profile was brought about only by clay translocation. This has been challenged, on the grounds that clay skins are not always present in textural B horizons, nor do the latter always show the increase in bulk density that would be expected from clay illuviation (Oertel, 1968; Brewer, 1968). Alternative hypotheses are: (i) differential weathering, with higher clay production in the permanently-moist subsoil; (ii) differential eluviation, with greater clay loss from the eluvial A horizon (but without illuviation in the B horizon). There is micromorphological evidence that illuvial clay skins can be destroyed subsequent to their formation (Nettleton *et al.*, 1969; Eswaran, 1970; Bennema *et al.*, 1970). Micromorphological studies of ferruginous and ferrallitic soils from Gambia, both soil types having an increase in clay below 50–70 cm, showed in each case an increase in the percentage of illuviation cutans within the textural B horizon (Hill, 1970). Clay translocation remains the most likely general explanation of textural B horizons. It is nonetheless unfortunate that the definition of an argilluvic horizon in the FAO soil classification, which is based on that of the US 7th approximation, requires that such an horizon 'contains illuvial layer-lattice clays', since neither the field nor laboratory criteria for distinguishing such clays are free from contention. In this book the term *textural B horizon* (symbol Bt) refers to an horizon with appreciably more clay than some horizon above it, irrespective of the origin of this clay.

Texture is determined in the field by finger texturing and again, for soil series type profiles, in the laboratory by particle-size analysis. It is generally obligatory for soil surveyors to accept the texture as given by applying

89

Soil properties

laboratory data to the US textural triangle, whatever their murmurings about 'field behaviour of the soil' (cf. Childs and Youngs, 1974). Sesqui-oxide-rich tropical soils are difficult to disperse, and it is usual to employ whatever dispersing agent gives the highest value for clay (20 ml sodium hexametaphosphate with 5 ml sodium hydroxide to 25 g soil, made up to 1000 ml, disperses most iron-rich soils). This may result in breaking down strong inter-particle bonds which are very stable in the natural soil, hence the difference between field and laboratory assessments of texture.

As there is no consensus over whether the International Society of Soil Science or US definitions of the silt/clay division should be used, it is customary to determine both. 'International silt' is 2–20 μm diameter, and 'US silt' 2–50 μm. The 20–50 μm fraction is sometimes called coarse silt, and the 2–20 μm fraction fine silt. The proportions of gravel (0.2–7.5 cm diameter) and stones (> 7.5 cm) should not be omitted from analytical data, as they can substantially affect the agricultural behaviour of the soil, notably where nodular laterite or a stone line is present. Stones and gravel are given as percentages by weight of the fine fraction (< 0.2 mm).

Structure

One of the first things to look at in examining a soil profile is the structure of the B horizon. A well-developed blocky structure, with clay skins on the ped surfaces, indicates that the soil is still actively evolving by weathering of primary minerals and translocation of clay. If the B horizon is structure-less, or weakly structured without clay skins, the soil is 'dead': all weather-able minerals have gone and little further profile differentiation is taking place (This does not apply to sandy profiles which are perforce weakly structured.) Well-structured soils generally respond well to cultivation, giving ease of root penetration and free drainage but good moisture retention. Massive soils lack these qualities, and additions of fertilizer can only partly overcome their agricultural disadvantages.

Blocky structures, subangular or angular, are by far the most common in tropical soils. Where the structural grade is moderate to strong, the blocky structure is frequently angular. Topsoils tend to be intermediate between blocky and crumb, rather than having the crumb structure typical of temperate latitudes. Coarse blocky and prismatic structures are found in gleys and vertisols, and the columnar structure is unique to solonetz.

The soils of the rainforest zone with uniform texture profiles (referred to above, p. 89) have a characteristic structure and consistence. This is a weak medium and fine blocky structure breaking down to weak or moderate fine crumb, with very friable consistence; more graphically the consistence can be called 'floury'. On the Butler system (p. 92) the consistence is crumbly, breaking with a small force and having a pulverescence close to

100 percent. These properties are caused by a clay fraction consisting almost entirely of kaolinite and sesquioxides, with the 2:1 lattice clay minerals that normally confer plasticity on clays completely absent. On mechanical analysis such profiles may show over 80 percent clay, and yet still be friable and permeable.

The presence of structure in the pedological sense, rather than 'shadow' structures, is one of the features differentiating soil from weathered rock. Weathered rock and soil may be mineralogically identical, the difference between them lying in the arrangement of particles. The processes by which weathered rock is transformed into soil are not definitely known. They include *pedoplasmation*, repeated swelling and shrinking on wetting and drying which causes a re-arrangement of constituents at the micro-scale (Flach *et al.*, 1968). The activities of microfauna are probably also important.

Soil structure can be modified by management. Structure occupies in this respect a position intermediate between texture, largely unalterable, and nutrient content, readily increased by fertilizers. Topsoil structure depends on organic matter content, and loss of structure and the erosion resistance which it confers is one of the main consequences of over-cropping. The use of soil conditioners to improve structure, even where technically effective, is not likely to be economic. Planted grass leys, although little used in the tropics at present, are probably superior (p. 119).

Experimental studies of tropical red and black earths have shown that potassium has the same unfavourable effect on the stability of structural aggregates as is well known for sodium; the presence of magnesium and calcium in the exchange complex favours structure. Hence heavy additions of potash fertilizers may cause structural deterioration (Ahmed *et al.*, 1969).

Clay skins

Clay skins, also called cutans, are thin coverings of clay particles on the surfaces of peds, stones or cavity walls (stone or root channel). They are visible to the naked eye as smooth, slightly shiny surfaces, contrasting with the rougher surfaces of broken ped interiors. Sometimes there is a slight colour difference between ped surface and interior. Most clay skins appear under the microscope as orientated domains of clay particles, in which case they probably result from clay illuviation. (Note, however, that the converse, that all illuvial clay should appear as orientated domains, in unproven.) Clay skins can also be caused by the rubbing together of adjacent peds whilst they are expanding on wetting, a process which occurs in vertisols.

The FAO system for field description of clay skins is in terms of (i) quantity: patchy, broken or continuous; (ii) thickness: thin, moderately thick or thick; (iii) nature: i.e. composition, as estimated using a hand lens;

(iv) location: on vertical and/or horizontal ped surfaces, on pores or root channels. The first and second of these criteria are not easy to separate, and can alternatively be subsumed as weakly, moderately and strongly developed clay skins. Where clay skins are absent it is important to record this fact.

Clay skins are a very useful diagnostic feature of soil type. For profiles of clay and sandy clay texture, clay skins are absent in weathered ferrallitic soils, absent or weakly developed in leached ferrallitic soils, and moderately developed in ferruginous soils; strongly developed clay skins are one of the main diagnostic features of ferrisols (and are used for their appellation in the FAO system, nitosols, derived from Latin *nitidus* = 'shiny'). Also in the FAO system, the way of estimating in the field whether a soil is a ferralsol or a luvisol is that only in the latter do B horizon peds have clay skins.

Consistence

Consistence describes several aspects of the feel of the soil between the fingers. Unlike structure, it can be at least partially assessed from auger observations. The standard system in the US and FAO soil survey manuals requires the description of four properties: stickiness when wet, plasticity when wet, consistence when moist (on a scale of friable to firm) and consistence when dry (on a scale of soft to hard). The FAO *Guidelines* hints at the inadequacy of some of these terms and definitions, but states that no one has succeeded in improving upon them. This is not true. The less well known Butler system is superior (Butler, 1955). A 2 cm block of moist soil is manipulated for two seconds between thumb and fingers, using just sufficient force to cause rupture or deformation.

Records are made of

1. *Kind of consistence*:
   Plastic;
   Labile: initially friable, but pieces partly coalesce;
   Crumbly: friable, breaking into structural aggregates;
   Fragmentary: friable, breaking into irregular fragments;
   Pulverescent: friable, breaking into ultimate particles.

2. *Force*: the force needed to cause disruption, estimated as nil, very small, small moderate, strong, very strong.

3. *Pulverescence*: the proportion of the original soil block which, after two seconds' manipulation, is in non-coalesced form, recorded as a percentage.

Consistence, taken in conjunction with texture, is the main way of estimating clay mineral type in the field. The presence of expanding-lattice minerals confers stickiness and plasticity. Soils with a clay fraction composed entirely of kaolinite, iron minerals and gibbsite are friable and non-sticky, whilst even small quantities of 2:1 minerals confer a degree of stickiness. Vertisols, with large amounts of montmorillonite, are plastic, needing a strong force to mould and having zero pulverescence.

Minerals

The conventional term *weatherable minerals* is used in field soil survey to refer to the content of minerals, other than quartz and muscovite, as visible with a hand lens. Quartz weathers so slowly that it is regarded as 'unweatherable', whilst if muscovite flakes are present the soil is described as micaceous. The main weatherable minerals are the feldspars and ferromagnesian minerals. Under a hand lens, sand-sized quartz grains are greyish and slightly translucent, feldspar grains are opaque and whiter, and ferromagnesian minerals are black. There is no established terminology for describing abundance of weatherable minerals (the FAO terms, in which 'many' means over 15 percent, are appropriate only for content of rock fragments). Van Wambeke (1962) used 3 percent of the total soil as a critical value to separate highly-weathered materials. The following terms are suitable for describing the abundance of sand-sized grains of weatherable minerals, estimated visually by hand lens:

Absent – None seen

Very few – 1 percent, only occasional grains seen

Few – 1–3 percent

Common – 3–6 percent

Many – 6 percent

The presence or absence of weatherable minerals at the base of the rooting zone, about 150 cm depth, is an important indicator of agricultural potential. Their absence, with all visible sand grains being of quartz, indicates that the profile has reached an advanced stage of weathering and is probably a ferrallitic soil. Conversely the presence of weatherable minerals in the profile provides a reserve of cations that are slowly released by weathering, partially able to renew the losses in cropping; common or many weatherable minerals generally indicates an agriculturally valuable soil.

Concretions are discussed in the context of laterite (p. 154) and calcrete (p. 198).

ANALYTICAL PROPERTIES

The analytical properties can be divided into properties of the exchange complex and the composition of the soil as a whole. This section is concerned with general and pedogenic aspects of the interpretation of analytical data.

Soil properties

Aspects particularly related to agricultural evaluation, including nutrient levels, are considered in chapter 16.

Properties of the exchange complex

Soil reaction is regarded by some as the most useful single analytical test for predicting crop response. It is measured in a soil–water suspension, the dilution of which ranges from a paste to a 1:5 ratio, 1:2.5 being the most common; pH values usually, although not invariably, increase with the dilution of the suspension. Data are sometimes additionally given for a soil suspension in potassium chloride solution, which gives pH readings 0.5–2.0 points lower. Exchange sites in weakly and moderately acid soils are largely occupied by H^+ ions. Below pH 5.2 there is a rapid increase in exchangeable Al^{3+}, possibly exceeding 50 percent near pH 4.0.

There is no standard scale of qualitative terms for pH ranges, but the following are useful for verbal description:

pH:		
	< 4.0	very strongly acid
	4.0–5.0	strongly acid
	5.0–6.0	moderately acid
	6.0–7.0	weakly acid
	(c.6.5–7.5	'neutral')
	7.0–8.0	weakly alkaline
	8.0–9.0	moderately alkaline
	> 9.0	strongly alkaline

Values of pH below 3.5 occur only in acid sulphate soils, and above 9.0 only in solonetz. The strongly and moderately acid ranges are typical of the rainforest and savanna zones respectively. Whereas in temperate latitudes strong acidity is indicative of low soil fertility, this is not so in the rainforest zone where a high proportion of freely-drained soils are strongly acid. A neutral reaction marks the transition from pedalfers to pedocals, the usual situation being weakly acid upper horizons over weakly alkaline lower horizons, as a consequence of greater leaching in the former.

In the humid tropics the topsoil is generally less acid than the B horizon, owing to exchangeable calcium brought up from depth by roots and deposited on the surface as plant litter. This difference is greater in the savanna zone than in more strongly leached rainforest soils which can remain strongly acid up to the surface. In the savannas the difference is largest on soils derived from intermediate to basic rocks which provide a greater supply of cations; pH values of 6.5 in the topsoil falling to 5.5 at 100 cm depth can occur. Acidity rises slowly again in depth as the regolith becomes less strongly leached.

Cation exchange capacity (CEC) is a measure of the negatively-charged sites on the surfaces of clay-humus molecules. *Total exchangeable bases* (TEB) is the sum of exchangeable Ca^{2+}, Mg^{2+}, K^+ and Na^+ ions adsorbed onto these surfaces. In strongly acid soils some of the exchange sites may be occupied by Fe^{3+} ions, which are conventionally excluded from the 'total'. The *base saturation* is the total exchangeable bases divided by the cation exchange capacity, expressed as a percentage. The cation exchange capacity is analogous to the coathooks in a cloakroom, the total exchangeable bases to the number of coats: in a full cloakroom the saturation is 100 percent, in an empty one, zero.

It has been shown that the cation exchange capacity of soils rich in kaolinite, halloysite and iron and aluminium oxides is pH-dependent (Sawney and Norrish, 1971; Barber and Rowell, 1972). The conventional method for determination of cation exchange capacity is by saturation with ammonium acetate at a pH of 7.0. However, as pH is decreased there is a substantial decrease in the negative charge due to proton transfer on the surface of the clays. At a pH of 4.8, cation exchange capacities one half to one quarter those in weakly alkaline conditions have been recorded. Thus, the values given for cation exchange capacities of many tropical soils are substantially above the cation-holding power of the soil under natural conditions. Values for base saturation, the total exchangeable bases as a percentage of the cation exchange capacity, are also affected. In particular, the low apparent values of base saturation on strongly acid soils are in part a consequence of the artificially high values of cation exchange capacity as conventionally determined.

As it is not the practice to determine cation exchange capacity at the natural pH, the alternatives are to use the conventional value or to abandon its use, and that of base saturation, altogether. Since both are useful differentiating parameters they are retained, albeit with a different interpretation from that formerly made. The terms cation exchange capacity and base saturation are therefore used in the text to refer to what are more correctly called *apparent* cation exchange capacity and *apparent* saturation.

The *cation exchange capacity of the clay fraction* gives an indication of the nature of the clay mineral where only standard analytical data is available. It is obtained as CEC of clay = 100 (CEC of soil/percent clay). The cation exchange capacities of common clay minerals are as follows:

Kaolinite	3–15 m.e./100 g clay
Halloysite	5–50
Illite, chlorite	10–40
Montmorillonite	80–150
Vermiculite	100–150
(Organic carbon	c. 100–350)

Soil properties

This is a very approximate but nevertheless useful means of estimating clay mineral type. It is valid only insofar as the cation exchange sites are on the surfaces of the clay fraction; in some soils an appreciable proportion of it may reside with the fine silt (20–50 μm diameter) (Martini, 1970). The index is only valid for subsoil horizons which are low in organic matter.

The cation exchange capacity of organic matter may vary with soil type. Greene (1961) reports the following values (probable equivalent terms in brackets):

Oxysol (leached ferrallitic soil)	127 m.e./100 g carbon
Ochrosol (ferruginous soil?)	213
Basisol (ferrisol)	266

It appears from these data that the richer the weathering complex, the higher the cation exchange capacity of the humus, a result of the highest interest.

The cation exchange capacity of the clay fraction and the percentage base saturation are useful as indications of soil-forming processes and as differentiating characteristics between soil types. The cut-off value for defining highly weathered soils (ferrallitic soils, ferralsols, oxisols) is usually taken as 15, 16 or 20 m.e./100 g clay, indicating dominance by kaolinite, with free sesquioxides present but no 2:1 lattice minerals. Conversely values of over 40 m.e./100 g clay indicate the presence of substantial amounts of 2:1 lattice minerals. Values of cation exchange capacity for the soil as a whole, the analytical figure usually given, reflect clay content as much as clay mineral type.

Percentage base saturation indicates the intensity of present-day leaching, and is a very useful differentiating parameter between soil types of the humid tropics. In the rainforest zone, apparent base saturation is often less than 15 percent, sometimes as low as 5 percent. It progressively rises from the savanna–rainforest transition zone through the moist and dry savannas, reaching 100 percent in the transition to pedocals at the margin of the semi-arid zone. The pH range of 5.0–6.0 corresponds approximately to a saturation range of 25–75 percent. A pH of 7.0 indicates 100 percent saturation, whilst with an alkaline reaction, free calcium carbonate appears. Saturation values are primarily dependent on intensity of leaching, and hence on rainfall, site drainage and permeability; there is a smaller parent material effect, saturation being higher in soils derived from basic rocks and in those with a reserve of weatherable minerals within root range.

Whereas cation exchange capacity and saturation are indicative, respectively, of clay mineral type and leaching intensity, values of total exchangeable bases are of less pedogenetic interest. Notwithstanding the fact that the analytical procedure is to determine total bases and exchange capacity, and then derive saturation by calculation, genetically it is exchange capacity and saturation that are the fundamental determinants, and total exchangeable bases their derivative. The quantities of individual bases are of agricultural importance.

Soil composition

The composition of the inorganic fraction of the soil may be stated in terms of minerals or chemical elements. Properties sometimes determined include mineralogy of the sand fraction, clay minerals, the ratios between silica, iron and alumina in either the whole soil or the clay fraction, and total element analysis of the whole soil. None of these are routine analyses, although determination of silica and the sesquioxides is becoming more frequent.

Information on the *mineralogy of the sand fraction*, usually determined for fine sand, serves both pure and applied purposes. In studies of soil evolution, mineralogy is more useful than data from chemical analysis. The heavy minerals present can indicate the provenance of a soil horizon, in particular whether it could have been derived from the rock which it now overlies; this technique may indicate a former cover of different rocks now removed, or a multi-phase origin to the soil (e.g. Watson, 1964/5; Mabbutt and Scott, 1966). The ratio of weatherable minerals to quartz is the most useful of all indices of weathering; a content of < 15 percent weatherable minerals in the fine sand fraction has been used to indicate highly-weathered soils (p. 261). Other indices of degree of weathering can be obtained from the proportions of more and less resistant minerals, using mineral weathering series (p. 69). The proportion of weatherable minerals is also, as noted above, a valuable indication of the probable long-term response to the soil to intensive cultivation. This index could usefully be determined more often. Whereas mineral identification by thin sectioning is time-consuming, a rapid, if approximate, separation into quartz, iron concretions and 'other minerals' can be made by visual sorting of sand grains under a binocular microscope, and other rapid techniques exist (e.g. Duchaufour, 1960, p. 394).

Clay minerals are estimated by differential thermal analysis or X-ray diffraction; the latter is fairly rapid, and the small clay sample needed can be extracted by pipette during particle-size analysis. Both methods give qualitative results only, although for similar soils, semi-quantitative comparisons of relative abundance can be made. An approximate means of estimating clay mineralogy from standard analytical data is noted above (p. 95), whilst it can also be judged in the field by comparison of texture and consistence.

Determination of the *silica*:*sesquioxide ratio* and related ratios serves pedogenetic purposes rather than being of any direct agricultural significance. The procedure is to determine the total element content of silica, ferric iron and alumina, by digestion in sulphuric or other strong acid, and to obtain molecular ratios. The molecular weights by which the percentages by weight are divided to give molecular ratios are SiO_2 60, Fe_2O_3 160, Al_2O_3 102. The parameters most often quoted are $SiO_2:(Fe_2O_3 + Al_2O_3)$, usually written $SiO_2:R_2O_3$, the silica:sesquioxide ratio, and $SiO_2:Al_2O_3$, the silica:alumina ratio. For convenience they are usually given as values of silica to one unit of the denominator, i.e. 2.0 rather than 2.0:1.

The major process affecting soil composition in humid climates is the differential leaching of silica, preferentially to iron and alumina. This results in a lowering of the silica:sesquioxide ratio from values of 5.0 and above in rocks to about 2.0 ± 1.0 in soils. Differential leaching of silica is fundamentally the same process in temperate and tropical climates, differing only in its greater intensity in the tropics. A general zonal difference is that whereas most freely-drained soils of the humid tropics have silica:sequioxide ratios below 2.0, brown earths tend to have ratios above this value. An exception of interest is the acid brown soils of sloping ground under high rainfall in temperate latitudes, where intense leaching produces not only low silica:sesquioxide ratios but also the yellowish red colours and friable consistence usually associated with latosols (Crompton, 1960).

The silica:alumina ratio is one means of differentiating between more and less strongly leaching latosols. In the intensely leached ferrallitic soils of the rainforest zone the silica:alumina ratio is below 2.0 (and the silica:sesquioxide ratio may fall to 1.5 or 1.0), whereas in less intensely weathered and leached ferruginous soils of the savanna zone the silica:alumina ratio is close to 2.0.

Total element analysis of soils and parent rock can be used to estimate the relative loss, or 'mobility', of different elements. A difficulty is to find a suitable element from which no loss has occurred for use as an index. Two possible bases of calculation are constant–alumina and constant–titanium oxide. The former is present in suitably large quantities, but the assumption that none is lost certainly cannot be made for the more intensely weathered tropical soils. Titanium is less likely to be removed, although if in the form of $Ti(OH)_4$ it may show limited mobility in strongly acid conditions. Unfortunately titanium is only present in quantities of the order of one percent, so a small sampling or experimental error has a large effect on values based upon it.

Total element analysis is also used for agricultural purposes, to indicate the reserves of potassium and phosphorous which can potentially be released by weathering.

Micromorphology

The technique of micromorphology, the study of soils in thin section under the microscope, has proved to be of much value in understanding the properties and genesis of tropical soils. In part, the results confirm what can be seen in the field with a hand lens. Thus, clay skins visible by eye usually coincide with ferri-argillans as seen in thin section, whilst the quantity of weatherable minerals identified in the field can be confirmed by thin-section counts of the ratio between quartz and other minerals. The contrast between blocky aggregates in ferruginous soils and the fine crumb micro-aggregation of some ferrallitic soils can also be confirmed. There are, however, other aspects detectable only under the microscope. Illuviation causes a reduction in the optical porosity of Bt horizons. A possible sequential change in the nature of the plasmic fabric, from sepic to asepic and finally isotic, has been suggested. Used in conjunction with field description, micromorphology appears to have the potential to make a substantial contribution towards the understanding of soil-forming processes (Stoops, 1968; Bennema *et al.*, 1970; Hill, 1970; Eswaran, 1970; Slager *et al.*, 1970).

HORIZON NOMENCLATURE

The letter-symbols applied to soil horizons, called *horizon designations*, are useful both when describing a profile in the field and when interpreting such a description. They are partly descriptive and partly genetic, and in the latter respect an element of subjective interpretation enters. Nevertheless, they provide a valuable *verb. sap.* indication, enabling the main features of a profile to be seen at a glance. The capital letters used for master horizons are less precisely defined, and less useful, than the descriptive suffixes. Unresolved points of disagreement are whether E (eluvial), G (gley) and D or R (unweathered rock) should be recognized as master horizons.

The usual system of horizon designations is based on that of the US *Soil survey manual* (1951, pp. 173–88), and the terms in the FAO *Guidelines for soil description* are identical to these. An extended system of genetic terms was proposed by Whiteside (1959), some of which are incorporated here.

The system of interpretative horizon designations given below, and used in this book, follows the FAO *Guidelines* with modifications. Layers containing more than 30 percent organic matter are O horizons. Master horizons A and B follow the standard (albeit rather meaningless) definitions, with AB as a transitional horizon. The (B) horizon, pronounced B-bracket, is used in the sense of Kubiena (1953) to describe a 'structural B' horizon, one which is distinctive in structure and often also colour from the A horizon but does not have more clay. Number symbols, as B1, B2, are used only where horizons cannot be distinguished by more meaningful letter suffixes. In particular, only A horizons which do not qualify as either Ah or Ae need be called A2 horizons.

Material altered by weathering but retaining signs of rock structure is termed the C horizon, distinguished from the R horizon of unaltered rock.

Soil properties

Where patches of weathering rock are mixed with soil proper it is termed the BC horizon (in preference to B3 of FAO). Many tropical soils, if exposed to their full depth, have a long sequence of horizons of progressively decreasing degrees of weathering which might be distinguished as, for example, BC1, BC2, C1, C2, C3, R.

The following suffixes are used:

l litter; undecomposed plant remains

f 'fermentation'; partly-decomposed plant remains

h as Oh: humus; fully-decomposed organic matter
  as Ah: with humus (A1 in the FAO system)
  as Bh: with illuvial humus

ca accumulation of calcium carbonate

cs accumulation of calcium sulphate (gypsum)

cn concretionary; containing hard sesquioxide concretions or laterite (p. 155)

e eluvial; with less clay and/or of paler colour (through loss of sesquioxides) than horizon below (A2 in the FAO system)

fe iron; applied to horizons with very red colours produced by finely-divided ferric iron (compare ses)

g gleyed; mottled (compare r)

j 'juvenile'; the property which precedes this symbol is only weakly developed

p 'plough'; the cultivated layer, either hoed into cultivation ridges or ploughed

r reduced; fully reduced, uniformly grey with little or no mottling

sa accumulation of readily soluble salts (salts more soluble than calcium sulphate)

ses sesquioxide; accumulation of non-concretionary iron and aluminium

st stone line; with more rock fragments, commonly quartz, than horizon below

t textural; with more clay than horizon above (not necessarily illuvial clay)

Some additional terms are used in the text. *Topsoil* refers to the upper mineral horizon, either the humic A horizon (Ah) or the cultivated layer (Ap), normally 10–30 cm thick. *Subsoil* is used as an alternative to B horizon. The description *in depth* refers to soil in the lower part of a profile pit, at approximately 100–150 cm depth.

CHAPTER 4

ORGANIC MATTER

The difference between weathered rock and soil is largely a biological phenomenon. If the B horizon of a tropical soil is compared with the BC horizon of highly-weathered rock below, it is often found that neither mechanical nor chemical analysis reveals any substantial difference between them. Yet in appearance and still more in feel between the fingers they are clearly distinct. The weathering rock bears traces of rock structure (e.g. gneissic banding) and has a rougher and drier feel. In the true soil, which does not usually reach below 2–3 m, all traces of rock structure have been destroyed, the clay is intimately mixed in, and the components have been homogenized and then re-arranged into units of pedological origin such as peds. These distinctive features of soil in the agricultural sense, which distinguish it from the engineering use of the same term, are brought about through the action of soil organisms and roots.

The agricultural significance of organic matter in tropical soils is greater than that of any other property with the exception of moisture. Its functions are to improve soil structure, and thereby root penetration and erosion resistance; to augment cation exchange capacity; and to act as a store of nutrients, slowly converted to forms available to plants. It has been plausibly argued that soil fertility is largely a biological phenomenon (Jacks, 1963).

Although also a central principle of soil management in the temperate zone, the importance of maintaining organic matter under agricultural use is still greater in the tropics. The greater intensities of both weathering and leaching in the humid tropics lead to an absence or scarcity of weathering minerals and to dominance of the clay fraction by minerals with a low cation exchange capacity. Therefore the nutrient content and cation exchange capacity of most tropical soils is very much more concentrated in the organic complex, largely held within the top 20 cm of mineral soil, than is the case with temperate soils. The cation exchange capacity of soil humus is on average about 350 m.e./100 g, some 30 times that of kaolinite; it may vary from about 100 to 500 m.e./100 g (Greene, 1961). As the soil erosion hazard is greater in the tropics, so also is the importance of keeping a sufficiently well-structured topsoil to resist rainsplash during the period of exposure inevitable in cultivation.

The organic matter content of soil is an unstable property. Under undisturbed natural vegetation it remains relatively constant but under agricultural use it can decline substantially within a few years. This is because of the cyclic nature of organic materials within the plant–soil system, flowing from plant to soil via litter and from soil to plant as root uptake. Agricultural use necessarily cuts off the flow from plant to soil, leading to a decline in the soil store. The ratio of the annual plant-to-soil flow to the amount held in the soil store is greater in the tropics than in the temperate zone, and hence the rate of decline when this flow is cut off is more rapid in the tropics.

UNITS

Humified organic matter in soils has a complex composition and there is no chemical method, suitable for routine analysis, of separating it. Organic matter is therefore usually estimated by determining organic carbon (commonly by some modification of the Walkley–Black procedure) and multiplying by a constant. This constant is usually taken as 1.72, a convention followed in this book, although it depends on the experimental procedure employed and the true value may be nearer to 1.9–2.0 (Broadbent, 1953). Total organic matter can also be determined by ignition, a procedure which slightly over-estimates its value owing to the loss of some combined water; ignition for 16 hours at 375 °C reduces this source of error as compared with a short period at a higher temperature (Ball, 1964).

Dead vegetative matter is determined as oven-dry weight, of which approximately 50 percent may be taken as carbon.

Organic matter and carbon values from soil analysis are given as percentages by horizons, e.g. 2.7 percent within the 0–15 cm horizon. Discussion of the organic matter cycle under natural conditions or agriculture is conveniently carried out in terms of weights of organic carbon per unit area. Assuming the dry bulk density of soil to be 1.5, a layer of soil 1 cm thick covering 1 m^2 weighs 15 kg, so a 1 percent carbon content is equivalent to 150 g/m^2 or 1500 kg/ha of carbon per centimetre of horizon thickness. The humic topsoil (Ah horizon) is usually some 20 cm thick, therefore a 1 percent carbon content within it is equivalent to 3000 g/m^2 or 30 t/ha of carbon.

Much earlier work is given in non-metric units, for which conversions are as follows:

1 lb/acre = 0.112 g/m^2 = 1.12 kg/ha
1 kg/ha = 0.1 g/m^2 = 0.89 lb/acre
1 percent carbon per inch horizon thickness ≡ 3810 kg/ha or 3391 lb/acre
1 percent carbon per centimetre horizon thickness ≡ 1500 kg/ha or 1335 lb/acre

Where only orders of magnitude or approximate estimates are involved, 1 kg/ha may be taken as roughly equal to 1 lb/acre.

Discussion of the organic matter cycle is here confined to the circulation of materials and is not considered in terms of energy flow (cf. Jordan, 1971).

SOIL ORGANIC MATTER AND ENVIRONMENT

The zonal variation in soil organic matter content is from 3–5 percent in lowland rainforest climates, 2–3 percent and 1–2 percent in the moist and dry savanna zones respectively, falling to 0.5–1.0 percent in the semi-arid zone, and dropping to zero in deserts. These are typical values for total organic matter content of the topsoil (Ah horizon, 0–10 to 0–20 cm thick) on freely-drained soils of medium to heavy texture under natural vegetation. There is no wholly organic (Oh) horizon in most lowland tropical soils. Under rainforest there is a leaf litter (Ol) horizon, ranging from a continuous horizon some 5 cm thick to a discontinuous layer of one or two leaves thickness; there may be some 0.5–1.0 cm of highly decomposed leaf litter (Of horizon). In the savanna zone and the dry tropics the Ah horizon is exposed at the ground surface.

In tropical montane conditions there is a substantial increase in the thickness of the Ah horizon and its organic matter content, the latter being typically 5–10 percent at 1500–3000 m altitude. An Oh horizon may occur at high altitudes.

Other factors which influence humus levels are texture, composition of parent material, slope and drainage. Sandy soils hold less organic matter than clays, by a factor of about one half as between a sandy loam or loamy sand and a sandy clay or clay. Soils derived from basic rocks tend to have higher organic contents than those from felsic rocks; whilst in part accounted for by associated texture differences there is also an effect arising from the richer base supply and more luxuriant plant growth on the former. On steep slopes where the mineral soil is thin there is no corresponding reduction in the thickness of the Ah horizon; quite highly humic topsoils, with a dense root mat and a well-developed crumb structure, can occur on 25°–35° slopes. Gleys of valley floors have more humus than soils of adjacent slopes, but peat does not form in the tropics except in permanently flooded swamps or at very high altitudes.

Soil humus content under natural vegetation is the outcome of equilibrium between supply from plant growth and loss by oxidation. Increase in rainfall causes an increase in plant growth. With respect to temperature it is generally held that the optimum for plant growth occurs at 20–25 °C and the optimum for activity of decomposing bacteria at 30–35 °C. Mohr

103

and van Baren (1954) show an excess of growth over decomposition reaching a maximum at 20 °C and falling to zero at 30 °C. This, however, ignores the fact that the rate of decomposition is dependent on the amount of humus present, and hence the number of bacteria. The existence of a negative feedback mechanism between the amount of humus present and the activity of oxidizing bacteria means that there is neither a complete absence of soil humus at very high temperatures nor, under free drainage, to accumulation of organic materials to form peat.

Quantitative studies have shown that soil organic matter varies directly with rainfall and clay content and inversely with temperature. Jenny and Raychaudhuri (1960) obtained carbon contents for 500 soils from India, adopting an adjustment procedure to allow for variations associated with texture. The effect of temperature was greatest in the 750–1000 mm rainfall zone, within which carbon content decreased by half for a temperature rise of 4 °C; the corresponding 'half-temperature' was 7 °C in the 1300–2200 mm rainfall zone and 8 °C for very wet areas, whilst with less than 750 mm rainfall the effect of temperature was only slight. In warmer regions carbon increased logarithmically with rainfall, but near the 14 °C annual isotherm all values were high and there was no correlation with rainfall.

For a large sample of West African soils, Jones (1973) found partial correlation coefficients of soil carbon with clay of +0.52 and with rainfall of +0.50, and a multiple linear regression with these two independent variables explained 46 percent of the observed variation in carbon. The carbon: nitrogen ratio became greater with increase in rainfall. For soils of similar texture, carbon was slightly lower on sandstone parent materials than on Basement Complex rocks, and over twice as high on basalts as on the latter, this difference becoming still more marked at rainfall values above 1800 mm. For 605 well-drained soils under rainfalls of 350–1905 mm, averaging 941 mm, the mean carbon content was 0.68 percent.

For East African soils, Birch and Friend (1956) found an increase of 0.8 percent topsoil organic matter per 300 m increase in altitude (approximately equivalent to 0.5 percent organic matter, or 0.28 percent carbon, per 1 °C fall in temperature). There was also a rise of 0.5 percent organic matter per 100 mm increase in rainfall. Since there is a strong covariation of altitude and rainfall in this area their relative effects could not be statistically separated.

There may be some discontinuity in the rise of organic matter with altitude. In Malaŵi there is an abrupt increase at 1500–1700 m altitude, corresponding to a mean annual temperature of 17–18 °C (Young and Stephen, 1965). A change at a similar temperature occurs in Ruanda Urundi (Denisoff, 1959). There may be some fundamental change in decomposition processes at this altitude and temperature. It marks the division between humic latosols, with topsoil organic matter contents of over 5 percent, and other latosols within which they are mostly below 3 percent.

Annual burning of savannas lowers soil organic matter levels, although the magnitude of this reduction as reported from long-term experiments is not as great as might have been expected (Moore, 1960).

Cultivated soils in the tropics have topsoil organic contents ranging from 30 to 70 percent of the corresponding values under natural vegetation. Jenny and Raychaudhuri (1960) found that the relative nitrogen status (which may be taken as indicative of organic matter) of Indian soils varied from 30 to 60 percent, with lowest values, typically 30–40 percent, in the Deccan. Jones (1973) reports that for similar soils in northern Nigeria, under a rainfall of 1120–1300 mm, mean carbon contents were 1.03 percent for 23 uncultivated sites and 0.58 percent for 19 sites cultivated or under recent fallow; this gives a relative organic matter status for the latter sites of 56 percent. There have been few other systematic studies, but it seems likely that typical values for the relative organic matter status of tropical soils that have been under cultivation for two years or more are 30–60 percent. Values below 50 percent represent an undesirable situation calling for remedial measures.

PROCESSES

The plant–soil system contains four stores of organic materials: living vegetation, dead vegetation, soil humus and soil organisms. Living vegetation comprises above-ground vegetation and roots. The above-ground component of dead vegetation is the litter. The above-ground living vegetation and the litter may each be divided into woody matter and herbaceous matter. Completing the cycle is the reservoir of carbon dioxide in the atmosphere.

As with processes affecting the mineral fraction, there are few biological processes unique to the tropics, with the one (but major) exception of termite activity. Differences between biological processes of tropical and temperate zones are largely those of higher absolute and different relative rates in the tropics.

The processes which affect the organic fraction, and in particular carbon, are listed in table 14. The first group comprises the growth and death of plants, which whilst not soil-forming processes proper provide the primary source of organic compounds. The major input to the carbon cycle is the fixation of atmospheric carbon dioxide by photosynthesis, given, after deducting losses by respiration, as net primary production. Root uptake from the soil brings inorganic materials (nutrients) into the system and is also responsible for a small proportion, possibly 10–15 percent, of plant

TABLE 14   *Processes which affect the organic fraction of soils*

Processes affecting the vegetation
Photosynthesis ⎫
Respiration ⎬ Plant growth
Root uptake ⎬
Within-plant transfers ⎭
Litter fall ⎫ Plant death
Timber fall ⎬
Root exudation ⎭
Rainwash
Burning

Processes affecting the litter (inclusive of dead wood)
Breakdown
Humification
Oxidation
Erosion
Burning

Processes affecting dead roots
Humification
Oxidation

Processes affecting soil humus
Mineralization
Oxidation
Erosion
Leaching

Processes caused or accelerated by man
Burning
Vegetation clearance
Agricultural activities – losses: harvesting of crops
                                                     grazing
                                                     acceleration of rates of oxidation
Agricultural activities – gains: natural fallows
                                                     crop residues
                                                     incorporation of mulches
                                                     farm-yard manure
                                                     compost
                                                     green manure crops
                                                     grass leys

carbon, derived from humus and taken up in solution. The main within-plant transfers are movements of nutrients upwards from roots to above-ground vegetation and of carbon downwards from leaves to woody tissue and roots. The above-ground vegetation passes into litter by the continuous or annual process of litter fall and the longer term process of the death of woody tissue or timber fall. The corresponding process within the soil is root exudation, the death of roots whereupon they become open to attack by soil fauna. In some plant communities, carbon is lost to the atmosphere by burning. There is a substantial transfer of nutrients from leaves to the soil by removal in rainwash (Nye, 1961). From a pedological point of view, plant growth and death may sometimes be treated as one comprehensive process which provides the input to the organic matter cycle. Rates of plant growth and litter production in tropical ecosystems have been summarized above (table 7, p. 43).

The main processes which affect the litter are breakdown and humification, which convert it to humus, and loss to the atmosphere by oxidation. Physical breakdown is accomplished by the soil macrofauna and larger mesofauna. Under rainforest, arthropods can reduce leaf litter to its main veins in 2–3 weeks (Bullock, 1967). Only after it has been finely comminuted is litter open to microbial attack, which ultimately converts it to the amorphous organic substances, of variable composition, termed humus. Only a low proportion of the litter, possibly 10–20 percent, is converted to humus, 80–90 percent being lost to the atmosphere by oxidation (Nye and Greenland, 1960). Litter erosion by surface wash moves leaf litter short distances downslope, but net erosion is small (Kellman, 1969). Litter may also be lost by burning.

Dead roots, the below-ground equivalent of litter, are similarly in part converted to humus and in part lost by oxidation, the proportion converted being larger, possibly 20–50 percent.

Soil humus is lost mainly by mineralization and oxidation. *Mineralization* is the production of inorganic ions by the oxidation of organic compounds. *Oxidation* is the accompanying process by which organic carbon is converted to carbon dioxide and lost to the atmosphere. Mineralization is in part counteracted by *immobilization*, the assimilation of inorganic ions into microbial tissue; if there is a slow but continuous addition of plant residues, mineralization exceeds immobilization resulting in net mineralization, the situation in which ions become available for uptake by roots. Most bacteria, fungi, protozoa and other soil mirco-organisms are chemoheterotrophs, organisms which utilize organic compounds as sources of both energy and carbon (Richards, 1974). The organic substances oxidized to provide energy are the *substrate* for these organisms. The activity of chemoheterotrophic organisms is dependent on the amount of substrate present. This is the principle underlying the observed fact that the greater the amount of humus present, the faster is its rate of loss by oxidation. Thus, decline in soil organic matter is eventually counteracted by a homeostatic process, or negative feedback mechanism, that of the corresponding decline in the rate of oxidation. Provided external conditions remain constant, this leads to a state of equilibrium. Soil humus is also affected by erosion and leaching but losses from these processes are relatively small.

The organic matter cycle under natural conditions thus contains the following flows, or transfers from one state to another:

107

Organic matter

Atmosphere:	to vegetation – photosynthesis
Above-ground vegetation:	to atmosphere – respiration; burning
	to roots
	to litter – litter fall, timber fall
	to soil – rainwash
Roots:	to atmosphere – oxidation
	to above-ground vegetation
	to soil humus – root exudation and humification
Litter:	to atmosphere – oxidation
	to soil humus – breakdown and humification
	to rivers – litter erosion
Soil humus:	to atmosphere – oxidation
	to roots – root uptake
	to groundwater – leaching
	to rivers – erosion

THE CARBON CYCLE

The plant–soil cycle of organic matter, as represented by carbon, for three representative tropical environments under natural vegetation is given in table 15 and fig. 6.

Features of the rainforest cycle (fig. 6(*a*)) are the large size of the vegetation store, mainly as woody tissue, and the high rates of flow. The amount of carbon in above-ground vegetation exceeds that in the soil. Lowland rainforest soils have topsoil carbon contents of 1–3 percent, equivalent to 3000–9000 g/m^2. The annual gains and losses of soil humus are small in comparison, of the order of 200 $g/m^2/yr$. The *turnover period* is the volume of a store divided by its annual inflow and outflow in a steady state. For soil humus under rainforest the turnover period is of the order of 20–50 years. (In terms of energy flow it has been estimated as 25 years: Richards, 1974, p. 126.) The litter store consists largely of wood, i.e. dead tree trunks, which decay much more slowly than leaf litter. The turnover period for dead litter is substantially less than one year, that is, the annual litter fall exceeds the litter present. Expressed in another way, leaf litter on the forest floor decomposes at 1–3 percent per day; once humified, it is lost at 2–4 percent per year (Nye, 1963). A large loss occurs in the rainforest cycle through the low efficiency of conversion of litter to humus.

It may be noted that although the annual volume of timber fall, in terms of oven-dry weight, is greater than that of litter fall, the leaf litter contains considerably higher quantities of plant nutrients than the timber (Nye, 1961).

The cycle in the moist savanna zone is shown in fig. 6(*b*), and values for both moist and dry savanna conditions in table 15. The vegetation store is smaller than in rainforest. The greater part of vegetative carbon is contained in trees and shrubs, and so if these are reduced or removed by an-

108

nual burning, the size of this store is considerably reduced. The litter component is small, there being no continuous cover of dead plant material on the ground surface. The size of the soil store declines with decrease in rainfall, reaching a topsoil carbon content of 0.5 percent, equivalent to 1500 g/m², at the savanna/steppe margin. Flows are also at lower rates than in rainforest, giving a turnover period for soil humus of the order of 40–50 years. There are large differences in carbon flows according to whether the vegetation is regularly burnt. With annual burning, soil humus is acquired largely by root exudation.

The plant–soil carbon cycle is a highly open system, with substantial inflows from and outflows to the atmosphere, by photosynthesis and oxidation respectively. In this respect it contrasts with the cycles of plant nutrients which are less open, cyclic flows exceeding inputs, from rock weathering and from the atmosphere, and leaching losses.

The turnover period for soil humus is considerably less in the tropics than in temperate zone woodlands, where it is typically some 200 years. It is nevertheless still substantial, and would appear to be sufficient to sustain cultivation over a substantial period following vegetation clearance. In these circumstances, however, rates of oxidation loss are increased above their natural levels.

ORGANIC MATTER UNDER SHIFTING CULTIVATION

The maintenance of organic matter levels under agriculture in the tropics has been studied mainly with respect to systems of shifting cultivation (bush fallowing). These employ the natural plant succession as the mechanism for restoration of organic matter and nutrients. Although now a less important question than maintenance of humus levels under continuous annual cropping, data from shifting cultivation has enabled a conceptual model to be constructed which may be of wider applicability. Hence shifting cultivation systems will be considered first, followed by a discussion of organic matter under agricultural systems in which long-term natural fallows are no longer possible.

The Nye–Greenland model

Most information on soil conditions under shifting cultivation rests on the work of Nye and Greenland (1960; Greenland and Nye, 1959). This was based mainly on West Africa. Their studies concerned cycles of nutrients as well as organic matter, but only the carbon cycle will be discussed here. The model is based on empirical studies combined with considerations of the manner of action of the processes involved.

109

TABLE 15 *Estimates of typical values of parameters in the carbon cycle under natural vegetation. Based on data in Nye and Greenland (1960), with some modifications. All values refer to organic carbon, which is assumed by Nye and Greenland to constitute approximately half of organic matter*

A. Stores of carbon (Values in g/m^2)

		Lowland rainforest	Moist savannas	Dry savannas
Living vegetation:	trees	11000	2500	2400
	grasses	0	800	400
	roots	1200	400	200
Dead vegetation:	leaf litter	100	60	30
	wood	3600	200	200
Total living vegetation		12200	3700	2800
Total dead vegetation		3700	260	230
Soil humus		7000	5700	1700
Total in plant-soil system		22900	9660	4530

B. flows of carbon (Values in $g/m^2/yr$)

	Lowland rainforest	Moist savannas Unburnt	Moist savannas Burnt	Dry savannas Unburnt	Dry savannas Burnt
Net intake from atmosphere	850	600		170	
Loss in burning	0	0	450	0	120
Root growth	280	150		50	
Litter and timber fall	570	450	0	120	0
Humification of litter	90	70	0	20	0
Oxidation loss in above	480	380	0	100	0
Root exudation	280	150		50	
Humification of roots	100	50		20	
Oxidation loss in above	180	100		30	
Total additions to soil humus	190	120	50	40	20

Although of some complexity, this model is the main framework available for studies of soil organic matter and will be set out in detail. The following parameters are employed:

C	= humus carbon actually present in the soil	kg/ha
C_E	= humus carbon present in the soil at equilibrium	kg/ha
A	= rate of addition of humus carbon to the soil	kg/ha/yr
B	= rate of loss of humus carbon from the soil	kg/ha/yr
I	= rate of net increase of humus carbon	kg/ha/yr
p	= proportion of equilibrium level reached	fraction
L	= litter fall	kg/ha/yr
R	= root death	kg/ha/yr
f_1	= fraction of litter becoming humus carbon	fraction
f_r	= fraction of roots becoming humus carbon	fraction
k_f	= decomposition constant under fallow	percent/yr
k_c	= decomposition constant under cultivation	percent/yr
T_c	= time in cultivation	years
T_f	= time in bush fallow	years

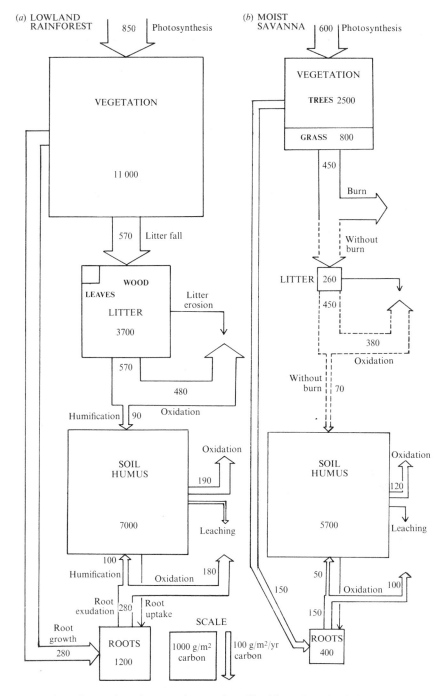

Fig. 6. The carbon cycle under natural vegetation. Cf. table 15. Based mainly on data in Nye and Greenland (1960).

Organic matter

(i) *Rate of increase of soil carbon under fallow*

$$I = A - B \tag{1}$$

$$A = f_l L + f_r R \tag{2}$$

$$B = k_f C \tag{3}$$

The net carbon increase, I, is equal to addition minus loss (1). Addition, A, is from two sources, litter and roots, each multiplied by the fraction which is converted to humus carbon, i.e. which is not lost by oxidation (2). The fraction of litter becoming carbon, f_l, is estimated as between 0.1 and 0.25, and the corresponding fraction of roots, f_r, as 0.2 to 0.5. Equation (3) contains the critical assumption of the model: that the amount of soil carbon lost, by oxidation, is proportional to the amount initially present. This proportion is the *decomposition constant* under fallow, k_f. The existence of this relationship is deduced from the fact that addition and loss are observed to reach an equilibrium, therefore some feedback mechanism must be operating. Its justification in physical terms is that the activity of oxidizing bacteria is dependent on the quantity of the substrate which provides them with energy. Combining (1)–(3) gives the expression for the rate of carbon increase under fallow:

$$I = (f_l L + f_r R) - k_f C \tag{4}$$

Estimates for typical values of these parameters are given in table 16. Substituting these values in (2), and taking both minimum and maximum estimates for f_l and f_r, gives the following estimated rates of annual increase of soil humus carbon:

Lowland rainforest: $A = 960$ to 2400 kg/ha/yr
Moist savanna (burnt): $A = 250$ to 650 kg/ha/yr
Dry savanna (burnt): $A = 80$ to 200 kg/ha/yr

If there was no burning the value for moist savanna would be raised to 650–1625 kg/ha/yr, but this situation could never occur in a shifting cultivation system.

TABLE 16   *Estimates of Nye and Greenland (1960) for decomposition constants under fallow and cultivation. For discussion of these estimates see p. 114.*

	Lowland rainforest	Moist savannas	Dry savannas
Decomposition constant under fallow, k_f, percent/yr	3.0	0.9	0.8
Decomposition constant under cultivation, k_c, percent/yr	3.3	4.5	4.5

Assuming these estimates to be of the correct magnitude, the main feature is the very much lower value under savanna than rainforest. It may be concluded that natural fallows under rainforest are a much more efficient way of building up organic matter levels than fallows under savanna.

(ii) *Variation in rate of increase according to fraction of equilibrium level attained.*
Let (4) be expressed as

$$I = A - k_f C \tag{5}$$

At equilibrium level, $C = C_E$ and $I = 0$, therefore

$$A = k_f C_E \tag{6}$$

Let p represent the fraction of equilibrium level reached, so that $C = pC_E$. Substituting in (5):

$$I = A - k_f p C_E \tag{7}$$

Substituting (6) in (7):

$$I = A - pA = A(1 - p) \tag{8}$$

This demonstrates that the rate of increase in soil carbon is greatest at low levels and falls off as equilibrium is approached. Thus at 50 percent of equilibrium level ($p = 0.5$) the annual increase, I, equals $0.5A$, at 75 percent of equilibrium ($p = 0.75$) I equals $0.25A$ and at 90 percent I equals $0.1A$. This result follows from the critical assumption of the decomposition constant.

(iii) *Evaluation of the effect of a rotation.*
   (a) *Carbon increase under fallow.* Let it be assumed that the soil content follows a cultivation–fallow cycle that fluctuates at around 75 percent of equilibrium level ($p = 0.75$). From (8), the average rate of annual increase, I, is equal to $0.25A$. Substituting the estimates given above yields the following average rates of annual increase:

Lowland rainforest: $I = 240$ to 600 kg/ha/yr
Moist savanna: $I = 60$ to 160 kg/ha/yr
Dry savanna: $I = 20$ to 50 kg/ha/yr

   (b) *Carbon loss under cultivation.* The loss under cultivation is also considered in the model to be proportional to the amount of carbon present, but with higher values of the decomposition constant k_c (table 16). The loss over the cultivation period of a shifting cultivation cycle is given by:

$$\text{Loss in } T_c \text{ years} = k_c C T_c = k_c p C_E T_c \tag{9}$$

Organic matter

Taking the equilibrium values of soil carbon shown in table 16, assuming the cycle oscillates around 75 percent of equilibrium level, and assuming 3 years of cultivation gives the following:

Lowland rainforest: Loss in 3 years $= 0.03 \times 0.75 \times 60000 \times 3 =$
4000 kg/ha

Moist savanna: Loss in 3 years $= 0.04 \times 0.75 \times 50000 \times 3 =$
4500 kg/ha

Dry savanna: Loss in 3 years $= 0.04 \times 0.75 \times 15000 \times 3 =$
1350 kg/ha

The feature of this result is that the organic matter loss under moist savanna conditions is no slower than under rainforest.

(c) *Years of fallow needed to restore cultivation loss.* Summarizing the results from sections (i) and (ii) above gives the figures shown in table 17. Allowing for a short lag for re-establishment, this gives the following ratios of cultivation to fallow necessary for the maintenance of organic matter at 75 percent of equilibrium levels:

Rainforest: 1:3 to 1:6
Moist savanna: 1:9 to 1:21
Dry savanna: 1:9 to 1:22

This shows up clearly the contrast in efficiency of bush fallowing as between the rainforest zone and the savannas. The difference arises in part from the assumption that in the savanna zone the decomposition constant under cultivation is considerably higher than that under fallow, and in part from the loss of above-ground vegetation by savanna burning. There are considerable generalizations and uncertainties in many of the values used, but the final result is in agreement with experience, namely that under rainforest a cycle of 2–3 years of cultivation followed by some 15–20 years of bush fallow maintains the soil in a steady state.

The rate of decline of organic matter under cultivation

The values for the decomposition constant under cultivation estimated by Nye and Greenland (table 16) appear to be substantially too low with respect to the first 1–2 years of cultivation after vegetation clearance. In an experi-

TABLE 17 *Restoration of organic matter under natural fallow*

	Cultivation loss (3 years)	Annual fallow gain	Years of fallow needed to restore loss
Lowland rainforest	4000	250–600	7–16
Moist savanna	4500	70–160	28–64
Dry savanna	1350	20–50	27–67

ment in the semi-deciduous forest zone of Ghana, Cunningham (1963) found a decline, under unshaded cultivation, from 3.7 to 1.6 percent carbon in the top 5 cm and from 1.1 to 0.8 percent carbon in the 5–15 cm layer, decreases of 57 and 30 percent respectively over 3 years. Kellman's (1969) data for montane forest in the Phillipines appear to imply k_c values in excess of 10 percent, although the cultivation periods are not precisely specified. D'Hoore (1968*a*) states that on clearing rainforest in the Congo, topsoil organic matter is halved in six months.

Later during a period of cultivation, the rate of organic matter decline lessens more than would be expected if it were dependent only on the amount present; that is, the decomposition constant becomes lower. Siband (1972) presents data for a red ferrallitic soil in Senegal on which, if the reported dates of clearance are correct, topsoil organic matter decreased from 2.4 percent after 2 years of cultivation to 1.5 percent after 15 years and 1.3 percent after 50 years, results which yield decomposition constants of 2.9 percent for the first interval and 0.4 percent for the second. Data given by Gokhale (1959) refers to continuous tea cultivation in north-east India. Values obtained for nitrogen are in 1930, 0.12 percent, in 1937, 0.094 percent, in 1940, 0.088 percent, and in 1956, 0.074 percent. If a constant 10:1 carbon:nitrogen ratio is assumed, the decomposition constants for the three successive intervals are approximately 3, 2 and 1 percent respectively. The data of Jones (1971) for a ferruginous soil in Northern Nigeria, under continuous cultivation without manure or fertilizer, yield a value for the decomposition constant of 4.7 percent between the tenth and eighteenth years of cultivation; the carbon content prior to the start of cultivation is not given, but estimated at 0.8 percent gives a decomposition constant averaging some 11 percent during the first ten years of cultivation.

It therefore appears that whilst equation (3), which states that the annual loss of humus carbon is in simple proportion to the amount of carbon initially present, may be applicable to periods of fallow under natural vegetation, the same law does not hold for periods of cultivation. The decomposition constant under cultivation seems itself to show an exponential decline, from over 10 percent in the first year after vegetation clearance to less than 1 percent after long periods of cultivation.

MAINTENANCE OF ORGANIC MATTTER UNDER PERMANENT AGRICULTURE

The ecological balance achieved by shifting cultivation under conditions of low population density has been widely rehearsed. There are also descriptions of the adverse consequences that result when population increase compels a shortening of the bush fallow period without any accompanying change in the agricultural system. What is not always appreciated is that there are very large areas of the tropics where there is no question of extended bush fallows since virtually all the cultivable land is cultivated each year. This is the normal situation in the Indian subcontinent and is becoming so over large parts of Africa. Non-irrigated cultivation is becoming increasingly dominated by what may be termed *permanent agriculture*, i.e. continuous arable cropping with only short fallows or none at all. It is on these densely settled areas that the problems of maintaining the soil in good condition are the most acute.

Certain agricultural systems may be excluded from further consideration. Swamp rice cultivation contains its own mechanisms for organic matter renewal and is clearly capable of reaching an equilibrium condition that permits continuous cultivation for indefinite periods. Perennial cropping, the growing of tree and shrub crops such as rubber, oil palm, coconuts and tea, also contains means whereby organic matter can be sustained, including leaf fall, the avoidance of bare soil, and in some crops the maintenance of a grass or leguminous ground cover. This is not to deny that serious soil degradation, sometimes leading to erosion, can occur under such crops as tea if badly managed, but the means for avoiding such degradation exist.

It is the permanent cultivation of annual crops which presents the major problem of organic matter maintenance in the tropics. A distinction can next be made between the situation in the rainforest and savanna zones. In rainforest climates it is only necessary to cultivate annual crops under subsistence farming. Commercial agriculture is more profitable if perennial crops are grown, but these do not supply basic dietary carbohydrate requirements. Hence annuals other than rice need only be grown in the rainforest zone for subsistence purposes and where land suitable for rice is not available. This practice is most widespread in the Amazon Basin and the more sparsely settled islands of the East Indies.

It would appear that there are no practicable means of maintaining organic matter under cultivation of annuals in the rainforest zone other than by extended fallows. Hence bush fallowing is not only, as seen above, more efficient in the rainforest zone than in the savannas, it is also more necessary. An alternative is to make productive use of the vegetative fallow by planting quick-growing softwoods, which can be as effective as natural fallow in restoring soil humus levels (Kellman, 1969; Cornforth, 1970); such a system is certainly worthy of consideration, but suitable economic and social conditions for its practice are unlikely to be widespread. Hence, in the case of subsistence cultivation of annuals under rainforest there may be merit in attempts to adapt and improve the basic ecological mechanism of bush fallowing, as in the corridor system developed in the Congo (Jurion and Henry, 1969).

Within the savanna zone, using this term to include the 'wet-and-dry' tropics of Asia and America as well as Africa, cultivation of annual crops is the means of both subsistence and commercial agriculture. It is clearly apparent, for example from the Indian Deccan, that continuous or nearly continuous cultivation of annuals for indefinite periods is at least possible in this zone, if not necessarily without soil degradation. Further discussion of organic matter maintenance will be confined to this one, but very widespread, system of agriculture.

The problem of maintaining soil organic matter is of a different nature and magnitude to that of providing nutrient supplies. The total weight of nutrients removed in a good crop amounts to no more than a few hundred kilograms per hectare. In contrast, the quantity of organic matter lost from the soil during one year of cultivation in the lowland tropics is of the order of two tonnes per hectare. The traditional method of bush fallowing serves the twofold purpose of restoring both nutrients and organic matter. In the case of nutrients it is practicable to replace this natural method by the use of artificial fertilizers. In the case of soil humus, the physical and economic difficulties of replacing losses are of a different order of magnitude.

The mechanism whereby a certain minimum annual addition of humus carbon reaches the soil is that of root exudation from crops. This 'root carbon' is inadequate to maintain high amounts but will ultimately stabilize soil humus at a lower level. In the temperate zone this may be 50–70 percent of the level under natural vegetation (Allison, 1973), but in the tropics it is likely that it is nearer to 30–40 percent. Thus, if farmers take no measures other than those necessary to prevent physical erosion, root exudation from crops will maintain soil humus in a low but steady state.

It is very desirable that humus should be stabilized at substantially higher levels, possibly 50–75 percent of those under natural vegetation. Methods of augmenting humus supplies, other than by long bush fallows, are as follows:

1. Short fallows
2. Crop residues
3. Incorporation of mulches
4. Farm-yard manure
5. Compost
6. Green manure crops
7. Grass leys

The use of short fallows, leaving the land to self-sown weeds for 1–4 years, is transitional with the long fallows of shifting cultivation. A suitable rotation for some ferruginous soils, for example, is groundnuts–tobacco–maize– maize followed by 4 years fallow. This method requires no capital but its effectiveness, as rate of organic matter increase, is low in the savanna zone and it uses land unproductively.

The incorporation of crop residues, ploughed or hoed into the soil, also requires no capital. It is ruled out for crops for which disease control measures require burning, e.g. cotton, but is common with cereals. Where coarse grains are grown on hoed cultivation ridges it is sometimes the practice to lay the dead plant along the hollow and hoe next year's ridge onto it from both sides.

The return of crop residues involves no extra land or cost and little additional labour, yet the gains can be substantial. A maize or sorghum crop residue is about 5–10 t/ha; assuming that 5–10 percent of the carbon in the residue becomes soil humus carbon, the gain is 125–500 kg/ha carbon, equivalent to 0.004–0.017 percent carbon within the top 20 cm of soil. This could well restore something of the order of half the organic matter lost during one year of cultivation.

The incorporation of mulches, dead grass or other leafy matter carried from elsewhere, is usually neither economic nor practicable in view of the large labour requirement and the need for land as a source of supply. A mulch of groundnut shells laid along the furrows between cultivation ridges has been shown to have a beneficial effect (Jones, 1971). In the case of coffee, mulching is a common practice for moisture conservation, and there can be useful secondary effects on nutrition and soil humus from its subsequent decomposition.

Farm-yard manure has rightly been regarded with some reverence in the farming of temperate lands, and the infrequency with which it is used in the tropics stems in Africa mainly from social and institutional customs which have separated cattle-owning from cultivating peoples, and in parts of Asia from the use of dung as fuel. Its virtues are not unknown in traditional agriculture. There are examples of farming systems in which much trouble is taken to distribute cattle or goat manure from kholas over continuously cultivated 'inner fields' (e.g. Netting, 1968). Village cattle are often turned onto crop residues, although the main benefit from this is to rest valley-floor grasslands. There is a basic difference between the normal situation in Africa, where the keeping of cattle by cultivating peoples is still the exception, and the Indian subcontinent where most farmers have cattle for ploughing and allocate a portion of land to fodder crops. The spreading of farm-yard manure, taken from cattle kholas, over arable land is one of the two best methods of improving the condition of the soil, both with respect to organic matter and the provision of a balanced nutrient supply. The fairly high labour requirement, which now limits this practice in high-income countries, is less of an obstacle in the developing world.

The amount of manure required is more of a problem. Long-term experiments on a ferruginous soil at Samaru, Nigeria, showed that topsoil carbon levels after 20 years of cropping averaged 0.22 percent with no manure, 0.34 percent with 5 t/ha per year (this level remaining constant between the tenth and eighteenth years of cultivation), and 0.82 percent with 12.5 t/ha per year (Jones, 1971). The latter is possibly close to the uncultivated equilibrium level of carbon for the sandy topsoil involved, in which case the 5 t/ha treatment maintained the soil at about 40 percent of equilibrium. Thus, amounts of the order of 10 t/ha per year would seem to be required, and it will not often be possible to obtain quantities sufficient for more than a small proportion of the cultivated area of a farm.

It may be noted that the use of artificial fertilizers has some indirect beneficial effect on soil humus, through permitting a larger volume of growth and hence greater photosynthetic intake of carbon and greater root exudation. Furthermore, the use of fertilizers in conjunction with farm-yard manure may increase the effectiveness of the latter in building up organic matter; there is also a converse effect, in that the use of manure generally increases responses to fertilizers, probably through improving the nutrient balance.

A disadvantage of fresh plant material, such as crop residues or mulches, is its high carbon:nitrogen ratio. Undecomposed plant matter has a carbon:nitrogen ratio exceeding 20:1, and when large amounts of such matter are incorporated into the soil the result is net immobilization of nitrogen. This effect is avoided by the use of compost, in which the carbon:nitrogen ratio is first reduced by partial decomposition under well-aerated conditions. As with farm-yard manure, a limitation is the large weight of matter per unit area needed to produce a substantial increase in soil humus. Its use is more common on special-purpose land, e.g. that used for vegetables.

Green manuring is the growing of a crop, usually a legume, specifically for the purpose of hoeing or ploughing it into the soil. It thus has no more economic benefit than a fallow, and uses labour as well as land. Reports of the effects of green manuring on subsequent crops are very variable, from no improvement at all to a large increase in yields, and no general pattern can be discerned. Where a substantial beneficial effect is empirically found to occur, this can be a useful method of increasing both organic matter and nitrogen levels.

The final method is that which has for a long period been the standard means of improving the soil condition on arable lands of the temperate zone, the planting of a grass ley. The advantages are the addition of organic material through root exudation and the incorporation of the grass at the end of the ley; the beneficial effects on soil structure produced by grass roots; and the fact that economic use is being made of the land at the same time. The inclusion of a herbaceous legume (e.g. *Centrosema pubescens*, *Stylosanthes guyanensis*) with the grass seed gives the added benefit of improvement of nitrogen status (Williams, 1967; Thomas, 1973). It has been shown in Rhodesia that soils under improved and stocked pastures contain up to 100 times as many nitrifying bacteria as those under wild grasses (Meiklejohn, 1968). Plots of Gamba grass (*Andropogon gayanus*) in Northern Nigeria were found to be most effective as a 3-year ley; the annual increase

119

in soil carbon and nitrogen was over twice that of a 2-year ley and higher than one of 6 years. The increase in organic matter from the 3-year ley was equivalent to that from 12 t/ha per year of farm-yard manure (Jones, 1971).

A major advantage of sown grass leys over natural fallows or green manure crops is that the land remains productive, the proviso being that livestock are kept. Grass leys are most effective at moderately high altitudes, where the nutritive value of grasses is higher. However, a number of species grow well in the standard lowland savanna environment and have a good fodder value (e.g. *Chloris gayana*, *Cynodon dactylon* for field grazing, *Panicum maximum*, *Pennisetum purpureum* for cut-and-carry feeding).

To persuade farmers in the tropics, particularly in Africa, that cattle should be integrated with arable farming, and still more that grass can be a productive crop, is clearly a formidable task for extension and education services, and this is not the place to discuss the problems. What may be stressed is that from a pedological point of view, that of keeping the soil in good condition so that sustained high yields can be obtained from arable cropping, there is everything to be gained from encouraging integrated farming systems involving grass leys and cattle kept for productive purposes. The argument that livestock are a biologically inefficient means of converting land potential into human food can be countered firstly, by the important nutritional role of animal protein and the frequent situation of protein deficiency in diets; and secondly, by the fact that productive use is made of land at the same time as the soil is being improved preparatory to the succeeding arable crop. Among the various means of counteracting the decline of organic matter under cultivation in the savanna zone, farming systems which combine arable cropping with productive livestock, hence permitting use of farm-yard manure, composts and sown grass–legume leys, offer the best opportunities.

SOIL DEGRADATION

Soil cultivated continuously does not attain or even closely approach a zero humus and nutrient content. This applies even to the most soil-exhausting cropping system, such as 20 years or more of maize monoculture. But under such treatment *soil degradation* occurs; this is a decline in organic matter content accompanied by a deterioration of soil structure and a lowering of fertility.

One consequence may be that soil erosion occurs, leading to permanent destruction of resource potential. Much more frequently, however, the physical conservation measures necessary to prevent the more spectacular manifestations of erosion are taken, such as terracing or simply the hoeing of cultivation ridges along the contour. Where such physical conservation is the only type of agriculture improvement practised, the soil reaches a state of *low-level equilibrium*, in which humus and nutrient levels remain constant and crop yields are stabilized at a low level. This condition is probably widespread in India and is becoming increasingly common in Africa. Ruthenburg (1971, p. 123) describes an example from the Indian Deccan where wheat yields have fallen to the extremely low level of 350 kg/ha. After 18 years' cultivation the topsoil carbon content of a soil at Samaru, Nigeria, fell from over 0.5 percent to less than 0.15 percent and was still dropping (Jones, 1971).

The hazard of soil erosion is widely recognized. Most government departments of agriculture now have a division of soil conservation, or land husbandry, and there are well-established conservation techniques (Hudson, 1971). This is very far from saying that erosion has been eliminated, but at least the problem is appreciated. The less spectacular but more widespread problem of soil degradation is far less generally recognized. In many parts of the tropics farmers are living on soil capital, using up the fertility stored from the past without restoration. This leads to a vicious circle, since with lowered yields still more intensive cropping becomes necessary.

There is a danger that well-meaning attempts at improving productivity through intensive multiple cropping may exacerbate this problem. Systems are being advocated whereby, through overlap of harvesting and sowing dates coupled with heavy use of artificial fertilizers and sometimes irrigation, up to five crops may be harvested in one year from the same land (e.g. Bradfield, 1974). Experimental work on such systems should include a programme of soil monitoring, with particular attention to organic matter levels and structure.

The incidence of soil degradation has received little attention from governments or systematic study by research organizations. As population increase leads to land pressure it is becoming a problem of considerable and growing magnitude.

TERMITES

The extent of profile modification by the action of termites is one of the major unanswered questions of tropical soil formation. Termites transfer, vertically and laterally, large volumes of both mineral and organic matter. One theory of the origin of stone lines holds that the whole of the overlying soil has been brought there by termites, without which the stone line would form the ground surface (p. 169). However, termites are neither more nor less abundant where there are stone lines than where there are not. So it could be that the uppermost metre of many tropical soils is radically modified by termite action, a possibility that is usually ignored when discussing horizon differentiation and its causes.

Termites, commonly called (biologically inaccurately) 'white ants',

121

constitute the order Isoptera. They quantitatively dominate the macro-fauna of tropical and subtropical soils, extending to latitudes 45° N. and S., in the way that earthworms dominate temperate soils. The Isoptera contain six families, five of which, the Mastotermitidae, Kalotermitidae, Hodotermitidae, Rhinotermitidae and Serritermitidae are known as the lower termites and the sixth, the Termitidae, as the higher termites. This distinction is biological, and does not correspond with either habits or distribution. The Hodotermitidae, Rhinotermitidae and Termitidae are divided into sub-families. The Termitidae contain 75 percent of known species, and are divided into the sub-families Amitermitinae, Termitinae, Macrotermitinae and Nasutitermitinae.

Termites are social insects which construct subterranean *nests*, the *termitaria*, containing chambers used to house and protect the colonies, store food, and in some species as fungus gardens. There is a distinction between mound-building species, in which the nests are contained in earth casings built up above the ground surface, and subterranean nesting species, whose presence is not visibly apparent. A peculiarity of the South American continent is that several species share the same mound. Radiating from the nests are subterranean *galleries*, which may reach 30 m or more in length, and are commonly the diameter of a forearm; these are used as routeways to transport food, in safety, to the nests. On digging a soil profile pit, as often as not you will cut across a gallery. The mounds are composed of a fabric of re-packed soil particles, often mixed with excreta, and the galleries may be similarly lined. The feeding habits vary between species, and food may include living and dead wood, living and dead herbaceous vegetable matter, and humus. They are best known, however, for the destruction caused by their habit of feeding on dead wood, being biologically distinctive in their ability to digest polysaccharides, including celluslose, and in some species lignin. Man's augmentation of the natural supply of fallen tree trunks by timber buildings has been adopted with enthusiasm, and spectacular results can ensue from their practice of consuming a chair leg leaving only a paper-thin outer casing intact.

Mounds vary in height from less than 1 m up to a maximum, in *Macrotermes* spp., of 9 m. Approximately circular in plan, they vary in shape between species, and the conflict between destruction by rainsplash and the reconstructive efforts of the termites can result in bizarre forms. Mounds can be constructed in a few months. They typically last 10–50 years before being abandoned, but on archaelogical evidence an age of 700 years has been claimed (Watson, 1967). Mound frequencies vary widely, from less than one to over 500 per hectare, and may occupy over 5 percent of the ground surface. Where they are abundant, a strikingly regular pattern is exhibited on air photographs which can often be diagnostic of a particular soil type.

The depth of galleries and colour comparison shows that most soil used in mounds is brought from depths up to 1.5 m. Rates of mound growth have been observed, and assuming that the soil in mounds is cyclically redistributed evenly across the ground surface, average rates of surface soil accumulation are 0.01–0.1 mm/yr (Lee and Wood, 1971, p. 154; Williams, 1968a; Watson, 1974a). This is equivalent to one centimetre of soil in 100–1000 years.

The soil within a termite mound shows substantial differences from the surrounds, although the nature and extent of these differences vary owing to the habits of different species. There is generally rather more clay and less coarse sand within the mound, and the largest particle that can be carried sets a size cut-off, commonly at 2–3 mm. Structural aggregation is greater. The pH is nearly always higher within the mound, often by one unit or more, and there is usually a higher content of carbon, nitrogen and exchangeable bases, especially calcium. Thus, a disproportionate amount of the total soil nutrients is contained within the mounds. A peculiarity of *Macrotermes* mounds in some parts of Africa is the presence of calcium carbonate in concentrations of several percent, forming concretions (Sys, 1955; Watson, 1962c, 1974b). There is less downward water movement within mounds than in the surrounding soil (Watson, 1969), which explains why the carbonates are not leached away, but not their origin. Hesse (1955) suggested periodic inundation with calcium-charged groundwater. Some method of selective accumulation by the termites themselves, perhaps of material originating as plant debris, seems more probable.

Termite mounds have a marked effect on vegetation patterns. Often they are sites of tree growth on grassy plains with predominantly hydromorphic soils.

It is uncertain whether the net effects of termites on the soil profile are agriculturally beneficial or not. On the credit side would be the creation of a gravel- and stone-free surface soil, if it could be shown that this would not be there in their absence. The constant mixing of the upper 1.5 m must tend to counteract profile differentiation, and thus favour root growth. On the other hand, the abundance and voracious feeding habits of termites not only speeds the breakdown of plant litter but selectively concentrates it in nest sites; there is a transference of organic matter and nutrients from the surrounds to the nests, which might otherwise be more slowly and evenly incorporated into the topsoil. In some areas of ferrallitic soils in Africa, maize is grown only on the richer soil of degraded mounds. Termites can cause damage to crops, since some species feed on living plant matter, and compete with grazing animals for food (Pullan, 1974). Once farming becomes mechanized the mounds become an obstacle to ploughing and are sometimes bulldozed away.

There has been speculation that termite mounds might be used for geochemical prospecting in regions where the presence of a thick regolith makes access to rock difficult. The idea is that the termites bring up material from depth. Anomalous concentrations of gold and zinc have been detected in mounds over Kalahari Sand (Watson, 1974*a*).

What would tropical soil profiles look like in the absence of termite activity? Possibilities are that profile differentiation, as between eluvial A and textural B horizons, would be greatly increased, and that laterite layers and stone lines would be widely exposed at the surface. The answers to these speculations may never be known.

WORMS

Earthworms, which play such an important part in brown earths of temperate regions, occur also in some tropical soils, although not in comparable numbers or biomass to termites. They ingest soil, excluding gravel and much of the coarse sand fraction, and deposit it as casts on the surface. The content of organic matter and exchangeable bases is higher in casts than in the soil below. In a dry forest climate (rainfall 1200 mm) in Ghana, Nye (1955) estimated the rate of worm casting to be considerably greater than that of surface soil accumulation by termites, although the soil is carried a much shorter distance upwards. At the end of the rainy season a humic fine sandy loam derived from worm casts had formed a surface horizon of irregular 0–2 cm thickness.

SELECTED REFERENCES

Soil organic matter. Greenland and Nye (1959), Nye and Greenland (1960), Cunningham (1963), Ignatieff and Lemos (1963), Siband (1972), Abu-Zeid (1973), Jones (1971, 1973).

Termites. Lee and Wood (1971), Watson (1962*c*, 1974*a*).

SOILS OF THE TROPICS

CHAPTER 5
SOIL TYPES

Part Two of this book describes the main soil types that occur in the tropics and discusses their origin and agricultural properties. Soils cannot be described without a system of classification, but conversely it is difficult to discuss existing classification systems without a basis of established soil types for reference. The procedure adopted is to give in this chapter an outline of the classification and nomenclature used throughout the book, and to compare classification systems in chapter 13.

SOIL TYPES AND NOMENCLATURE

In giving a general account of soil types, natural units are more satisfactory than the use of one particular formal classification system (cf. p. 257). The nomenclature employed here is eclectic. Where possible it is based on established names in common use. In the case of the various types of 'red and yellow' soils, here grouped as latosols, the older names (e.g. 'lateritic soil') are unsatisfactory. The terminology used for these soils, together with the classification in modified form, is that of the CCTA *Soil map of Africa* (D'Hoore, 1964). One term (arenosols) is taken from the FAO classification (FAO–Unesco, 1974). In conformity with the practice advocated below, approximate equivalents on the FAO system are given for each soil type discussed; in many cases the established names are either identical with FAO classes (e.g. vertisol, solonchak) or the customary term has an easily identifiable FAO equivalent (e.g. FAO gleysols for gleys, fluvisols for alluvial soils).

The nomenclature used is given in table 18, together with approximate FAO and CCTA equivalents. This is not intended as a formal classification system but as a means of identifying the natural soil types that exist in the tropics.

The primary distinction is into four groups: pedalfers, pedocals, a group of hydromorphic and alluvial soils, and a group of shallow, stony or immature soils. *Pedalfers* include all mature soils of the humid tropics that have free or slightly impeded drainage; they have an acid reaction and are leached of exchangeable bases, contain neither carbonates nor salts, and lack strong development of features caused by gleying. *Pedocals* have an alkaline reaction and contain free carbonates or salts: they occur mainly,

TABLE 18 *A classification of tropical soils. Approximate equivalents are given for the classifications of the FAO–Unesco Soil Map of the World (FAO–Unesco, 1974) and the CCTA Soil Map of Africa (D'Hoore, 1964). The two classes given in quotation marks cut across other types. Soils containing an horizon of laterite are distinguished as 'with laterite', e.g. ferruginous soil with laterite. The term latosol is used in its broader sense (p. 132)*

Present nomenclature		Approximate equivalent, FAO	Approximate equivalent, CCTA
PEDALFERS: LATOSOLS:	FERRUGINOUS SOILS	Ferric luvisols	Ferruginous tropical soils
	LEACHED FERRALLITIC SOILS	Ferralsols p.p., ferric acrisols p.p.	Ferrallitic soils p.p.
	WEATHERED FERRALLITIC SOILS	Ferralsols p.p., ferric acrisols p.p.	Ferrallitic soils p.p.
	FERRISOLS	Dystric nitosols	Ferrisols (excluding humic ferrisols)
	EUTROPHIC BROWN SOILS	Eutric nitosols	Eutrophic brown soils (excluding those on volcanic ash)
	HUMIC FERRALLITIC SOILS	Humic ferralsols	Humic ferrallitic soils
	HUMIC FERRISOLS	Humic nitosols	Humic ferrisols
OTHER PEDALFERS:	LOWLAND TROPICAL PODZOLS	Albic arenosols	Podzolic soils p.p.
	MONTANE TROPICAL PODZOLS	Orthic and humic podzols	Podzolic soils p.p.
	ANDOSOLS	Andosols	Juvenile soils on volcanic ash, eutrophic brown soils on volcanic ash
	ARENOSOLS	Arenosols (mainly ferralic)	Ferruginous tropical soils on sandy parent materials, ferrallitic soils on loose sandy sediments

PEDOCALS:			
CALCIMORPHIC SOILS:	VERTISOLS	Vertisols	Vertisols
	RENDZINAS	Rendzinas	Rendzinas
	BROWN CALCIMORPHIC SOILS	Calcic luvisols, xerosols p.p.	Brown soils of arid and semi-arid tropical regions
	SIEROZEMS	Xerosols p.p.	Sub-desert soils p.p.
	GREY AND RED DESERT SOILS	Yermosols	Sub-desert soils p.p.
	DESERT DETRITUS	Lithosols p.p.	Desert detritus
	'CALCAREOUS SOILS'	'Calcic' prefix	Soils with calcareous pans
	SOILS WITH GYPSUM	Gypsic xerosols and yermosols	Soils containing more than 15 percent gypsum
HALOMORPHIC SOILS:	SOLONCHAKS	Solonchaks	Saline soils, saline alkali soils
	SOLONETZ	Solonetz	Solonetz, alkali soils
	SOLODS	Solodic planosols	Solodized solonetz
HYDROMORPHIC AND	GLEYS	Gleysols	Mineral hydromorphic soils
ALLUVIAL SOILS:	ORGANIC SOILS	Histosols	Organic hydromorphic soils, organic non-hydromorphic soils
	ALLUVIAL SOILS	Fluvisols	Juvenile soils on riverine, lacustrine and fluvio-marine alluvium
	ACID SULPHATE SOILS	Thionic fluvisols	Not differentiated; juvenile soils on fluvio-marine alluvium p.p.
	'PADI SOILS'	—	—
SHALLOW, STONY OR	LITHOSOLS	Lithosols	Rock and rock debris, lithosols and lithic soils
IMMATURE SOILS:	REGOSOLS	Regosols	Juvenile soils on wind-borne sands

but not exclusively, in semi-arid and arid regions. *Hydromorphic soils* are poorly-drained profiles, dominated by mottling, dark grey colours or peat accumulation; they are grouped with *alluvial soils* on the basis that both occur on flat and low-lying sites and many alluvial soils are also hydromorphic. The fourth group consists largely of *lithosols*, shallow or stony soils; their separation at a high category is justified by the fact that this is probably the most extensive soil type in the tropics. In respect of properties of the fine fraction, lithosols may be pedalfers or pedocals, but their morphology is dominated by rock, stones or concretionary material. *Regosols*, immature soils on sands, are included in this group for convenience.

This grouping corresponds, with some exceptions, to distinctions between dominant soil-forming processes. Pedalfers experience a downward movement of water through the profile, caused by an excess of rainfall over evapotranspiration for at least part of the year. This causes leaching of salts, carbonates, exchangeable bases and some silica. In most pedocals, potential evapotranspiration exceeds rainfall for much of the year and there is no net leaching, at least in the lower horizons. Rendzinas, in which the calcium carbonate is inherited from the parent material, are an exception. The dominant features of hydromorphic and alluvial soils result from permanent or seasonal waterlogging, sedimentation, or both of these processes. The fourth group contains what in terms of origin are two distinct subgroups. Firstly, shallow or stony soils which are mature in the sense that they are unchanging with time (p. 263), but in which profile development is inhibited by rapid surface denudation on steep slopes or by the high resistance to weathering of the parent rock; secondly, soils which are immature in the sense that insufficient time has elapsed for profile differentiation to occur.

Soils intermediate between these classes occur. Many pedalfers show some gleying in depth; they become hydromorphic soils where there is at least a faint mottling in the topsoil, or where marked features of gleying occur at a depth of less than 50 cm. In topographic depressions of semi-arid regions the distinction between pedocals and hydromorphic soils becomes blurred. Both pedalfers and pedocals pass into lithosols at whatever depth is arbitrarily adopted in the definition of the latter, here taken as 25 cm. Between pedalfers and pedocals there is no soil type with a neutral reaction, since on proceeding across the boundary between sub-humid and semi-arid climates a weakly acid topsoil persists whilst carbonates accumulate in depth.

Two of these four main groups are subdivided: the pedalfers into latosols and other pedalfers, and the pedocals into calcimorphic and halomorphic soils. The term *latosol* is used in its wider sense, to cover all of what are loosely called 'red and yellow' soils of the tropics (p. 132). It includes all tropical pedalfers except those with certain distinctive characteristics. The latter include andosols, developed from recent volcanic ash, tropical podzols and very sandy soils (arenosols), which are grouped for convenience as *other pedalfers*. *Calcimorphic soils* are characterized by the presence of free calcium carbonate but do not contain large quantities of soluble salts. *Halomorphic soils* have properties dominated by the presence of readily soluble salts; they may or may not also contain free calcium carbonate. Soils of extreme desert regions, although they may sometimes contain neither accumulations of carbonates nor salts, are for convenience included with the calcimorphic soils.

The 29 soil types described in the following chapters fall into these six groups: latosols, other pedalfers, calcimorphic soils, halomorphic soils, hydromorphic and alluvial soils, and a group consisting largely of lithosols. As well as acting as a framework for discussion, these groups form a useful basis for the initial division of the soils of any particular tropical region. Thus, in the humid tropics most areas can be divided into latosols, hydromorphic and alluvial soils, and lithosols; other pedalfers are less commonly represented. In semi-arid and arid regions the groups present will usually be calcimorphic soils, halomorphic soils, hydromorphic and alluvial soils, and lithosols.

The subdivision of the latosols is discussed in the following chapter. For the other groups, the soil types described are well-established natural classes. Two special classes, not soil types in the normal sense, are also discussed, namely calcareous soils and padi soils.

CLASSIFICATION SYSTEMS

In the descriptions of soil types, reference is made to their equivalents in other major classification schemes. To avoid repeated citation of references these are referred to by the following short titles, discussion of which will be found in chapter 13 on the pages indicated:

Short title:	FAO	Year:	1974	See page:	240
	CCTA		1964		236
	ORSTOM		–		248
	7th approximation		1960/67		249
	US 1938		1938		251
	INEAC		–		252
	Ghana		–		252

The terminology used to describe the morphological and analytical characteristics of the soil types, together with the horizon nomenclature employed, is given in chapter 4.

CHAPTER 6
LATOSOLS

Latosol was defined by Kellogg (1949) as a term 'to comprehend all the zonal soils in tropical and equatorial regions having their dominant characteristics associated with low silica–sesquioxide ratios of clay fractions, low base-exchange capacities, low activities of the clay, low content of most primary minerals, low content of soluble constituents, a high degree of aggregate stability, and (perhaps) some red colour . . . It is a collective term for those zonal soils previously called 'lateritic soils' . . . as contrasted to character-istics associated with the terms podzolic, chernozemic, or desertic.' Kellogg specifically excluded the American class of 'red-yellow podzolic soils', found in warm temperate latitudes as well as in the tropics, from the latosol sub-order. Unfortunately his use of the adjective 'low' led to subsequent writers interpreting the term in two different ways: *sensu stricto* for soils where the properties specified reach their lowest, and *sensu lato* to include also a much wider range of reddish to yellowish leached, acid soils. For the former sense there are other terms: ferrallitic soil, ferralsol or oxisol. For the broader concept the only alternative is kaolisol, from the INEAC classification system, meaning a soil in which kaolinite dominates the clay fraction.

It is frequently convenient to be able to refer to the broader class, com-prising all 'red and yellow' soils of the tropics. *Latosol* is used here with this meaning. It is a class that covers all tropical pedalfers except those with unusual and distinctive characteristics, namely andosols, tropical podzols and arenosols.

Within the latosols as here defined there are three major groups: fer-ruginous soils, ferrallitic soils, and a group of soils derived from basic rocks, collectively termed *basisols*. *Ferruginous soils* and *ferrallitic soils* (in part, see p. 138) are the zonal soils of the savanna and rainforest zones respectively. The distinctive properties of the third group are caused by parent material; the basisols of the rainforest and transitional rainforest–savanna zones are *ferrisols*, and those of the drier savanna zone *eutrophic brown soils*. There exists also a group of ferrallitic soils which owe their properties not to climate but to the highly-weathered nature of the parent material. These latter are termed *weathered ferrallitic soils*, the zonal soils of the rainforest being distinguished as *leached ferrallitic soils*. There is a further group of *humic latosols*, occurring at high altitudes.

The groups and sub-groups recognized are as follows (tables 19 and 20):

TABLE 19   *Limiting values of properties of the main types of latosols. Structural grade refers to the B horizon. Saturation = apparent base saturation of the B horizon. CEC of clay = apparent cation exchange capacity of the clay fraction in the B horizon*

LATOSOLS	Depth to rock >25 cm, clay >15 percent, not gleyed above 50 cm, no free carbonates; pH <7.0, usually <6.0, saturation <100 percent.
FERRUGINOUS SOILS	Structural grade at least weak, usually moderate, clay skins present, some weatherable minerals in depth; pH >5.0, usually >5.5, saturation >40 percent; silt:clay ratio >0.15, silica:sesquioxide ratio ≪2.0.
LEACHED FERRALLITIC SOILS	Structural grade weak or structureless, clay skins absent or very weakly developed, no or very few weatherable minerals in depth; pH <5.5, usually <5.0, saturation <40 percent, CEC of clay <20 m.e./100 g; silt:clay ratio <0.25, silica:sesquioxide ratio <2.0.
WEATHERED FERRALLITIC SOILS	Structural grade structureless or very weak, no clay skins, no or very few weatherable minerals in depth; weathered to >2 m, usually >3 m, depth; CEC of clay <20 m.e./100 g.
FERRISOLS	Structural grade at least moderate, usually strong, clear or prominent clay skins, common or many weatherable minerals in depth; CEC of clay >15 m.e./100 g, silt:clay ratio >0.15; saturation <50 percent.
EUTROPHIC BROWN SOILS	Properties as ferrisols but saturation >50 percent.
HUMIC FERRALLITIC SOILS	Topsoil thickness (cm) multiplied by organic matter percent >50; structural grade weak or structureless, clay skins absent or very weakly developed.
HUMIC FERRISOLS	Topsoil thickness (cm) multiplied by organic matter percent >50; structural grade moderate or strong, clay skins present.

1. *Ferruginous soils.* The zonal soils of the savannas, on rocks of felsic to intermediate composition. Sub-group:

   1b. *Leached ferruginous soils.* Ae horizon paler than B horizon.

2. *Leached ferrallitic soils.* The zonal soils of the rainforest lands, on rocks of felsic to intermediate composition.

3. *Weathered ferrallitic soils.* Soils on level to gently-sloping sites in the savanna zone, on rocks of felsic composition which are deeply and highly weathered. Sub-groups:

   3b. *Pallid soils.* With imperfect drainage.

   3c. *Weakly ferrallitic soils.* Intermediate between ferruginous soils and weathered ferrallitic or, possibly, leached ferrallitic soils.

4. *Ferrisols.* Soils on rocks of basic composition in the rainforest and intermediate rainforest–savanna zones. Sub-group:

   4b. *Strongly leached ferrisols.* In permanently-humid rainforest climates.

133

5. *Eutrophic brown soils.* Soils on rocks of basic composition in the dry savanna zone.

6. *Humic latosols.* Soils developed in the cooler climates of high-altitude tropical regions. Subdivided into:

    6a. *Humic ferrallitic soils.* On felsic to intermediate rocks.

    6b. *Humic ferrisols.* On basic rocks.

The following adjectival terms are used as qualifications to these groups:

Sandy. Profiles with no horizon of clay or sandy clay, and with at least 65 percent sand in the topsoil (the latter criterion defines *coarse textured* in the FAO classification).

Gleyed. Imperfectly drained profiles, with hydromorphic features in the lower horizons.

With laterite. Profiles containing an horizon at least 25 cm thick, within 100 cm of the surface, consisting of at least 40 percent by volume of hard laterite, massive or nodular (see p. 155). This includes both the *petric* and *petroferric phases* of the FAO classification.

FERRUGINOUS SOILS

(Profile 1, p. 425)*

CCTA: Ferruginous soils
FAO: Ferric luvisols

TABLE 20 *Relations of types of latosols to environmental factors. Sub-groups shown in lower-case lettering Cf fig. 4(b) (p. 75). After Young (1974b)*

Climatic zone	Parent material			
	Basic	Felsic to intermediate	Felsic, highly weathered*	
Rainforest	Strongly leached ferrisols	LEACHED FERRALLITIC SOILS		
Savanna	BASISOLS → FERRISOLS / EUTROPHIC BROWN SOILS	FERRUGINOUS SOILS	Weakly ferrallitic soils	WEATHERED FERRALLITIC SOILS
Tropical high-altitude	HUMIC FERRISOLS	HUMIC FERRALLITIC SOILS		
	HUMIC LATOSOLS			

*Landform factor important; gently undulating erosion surfaces.

* Representative profiles of some major soil types are given in the Appendix, p. 425.

Ferruginous soils are characterised by a B horizon with a well-developed fine or medium blocky structure. Where the texture is clayey the structural grade is moderate or strong and clay skins are often present. There is usually a textural B horizon. Depths to weathered rock are moderate, not usually exceeding 2.5 m, and there is commonly a reserve of weatherable minerals within the profile. Colours are normally red, reddish brown or yellowish red, caused by the separation and dehydration of iron compounds. Part of the free iron may be deposited at the upper margin of weathered rock, giving reddish mottling or staining.

Sandy ferruginous soils, with sandy clay loam or sandy loam textures, also have a blocky structure in the B horizon, but the structural grade may only be weak.

Iron concretions, either soft or hard, occur in many ferruginous soils. A gravel horizon or stone line may occur, formed of iron concretions, quartz fragments or a mixture of these components; this feature seems especially common in West Africa. Soils containing massive laterite, however, do not usually possess the well-developed structure and other properties of ferruginous soils.

In some profiles there is an A2 horizon of paler colour than the underlying textural B horizon. This variant is termed *leached ferruginous soils*. It was through the existence of this paler horizon that the epithet 'podzolic' came to be applied to ferruginous soils in the US 1938 classification, as in 'red yellow podzolic soils'.

Ferruginous soils have a base saturation in the B horizon of more than 40 percent, sometimes as high as 80–90 percent. They are weakly to moderately acid (pH > 5.5). The silt:clay ratio, whilst low, is generally above 0.15. Clay minerals are predominantly kaolinite and free iron oxides, but small amounts of 2:1 lattice clays are present. Hence the cation exchange capacity of the clay fraction is above 15 m.e./100 g, never reaching the very low values found in ferrallitic soils. The silica:sesquioxide ratio of the clay fraction is generally below 2, but the silica:alumina ratio exceeds 2.

Viewed in thin-section, ferruginous soils are characterized by the presence of illuvial ferri-argillans (the micromorphological equivalent of clay skins) in the Bt horizon; there is a marked increase in their proportion between the A and B horizons. The plasma consists of clay stained with amorphous iron hydroxides and there is often a sepic plasmic fabric. Due to illuviation, the optical porosity of the Bt horizon is some 6–8 percent lower than that of ferrallitic soils of comparable clay content (Hill, 1970; Bennema *et al.*, 1970).

These are the zonal soils of the savanna climates, with a winter dry season (Köppen Aw). They are also common in savanna climates with two wet seasons (Aw''). At the drier margin, the transition to pedocals is reached at a rainfall of 500–600 mm, falling in a wet season of 3–4 months. The upper rainfall limit is commonly at about 1200 mm rainfall, above which the soil passes into the weakly ferrallitic class; but ferruginous soils can sometimes occur with 1500 mm or more rainfall, for example in the 'Dry Zone' of Ceylon. The leached ferruginous type is characteristic of warm temperate humid climates (Cfa) of continental east coasts. A savanna vegetation is typical, but the soil class extends beyond both the drier limit of this formation into moderately xeromorphic vegetation types, and the wetter limit into semi-deciduous forest.

There is a wide range of parent materials, including crystalline rocks of felsic to intermediate composition, arenaceous and argillaceous sedimentaries, and limestones. Sandstones and unconsolidated sands are the commonest parent materials of the sandy ferruginous type; for example, near the savanna–steppe transition in West Africa the 'Continental Terminal' sandstones, together with extensive sandy drifts, give rise to an extensive area of sandy ferruginous soils and arenosols. Site drainage is free to imperfect. Slopes may be of any angle. Steep slopes, where the profile is not thin enough to be classed as a lithosol, often carry ferruginous soils. A common feature is for weathered ferrallitic soils developed on a gently undulating plateau to give place to ferruginous soils on the steeper slopes of the dissected plateau margin; this may be accompanied by a change in vegetation.

Weathering is moderately intense, releasing free iron oxides but not alumina. The process of iron oxide release and dehydration, or rubefication (p. 88), imparts the red or dark red colour to the soil. Natural erosion removes the more highly-weathered materials from the surface. The occurrence of clay skins in conjunction with a textural B horizon probably indicates clay translocation, although other explanations are possible. Carbonates are removed from the profile, there is weak to moderate leaching of cations, and differential removal of silica.

Essentially, ferruginous soils are relatively young latosols, neither very highly weathered nor very strongly leached. Both weathering and translocation processes are still active. The reserve of weatherable minerals from the parent rock is being fed into the soil; the removal of material from the surface maintains this condition, preventing the development of a deep, highly altered layer. The moisture content is close to field capacity during the greater part of the wet season, and surplus rain therefore drains rapidly downward through the profile. In the dry season leaching ceases, the upper horizons dry out, and precipitation of ferric iron occurs. The colour range from yellow to red corresponds to increasing states of dehydration of these ferric compounds.

The definition of ferruginous soils given here largely follows that in the CCTA classification. They correspond to the *sols ferrugineux* (*lessivés* and *non-lessivés*) of the ORSTOM classification, although this latter also includes many profiles here termed arenosols. Ferruginous soils are the *savanna ochrosols* of the Ghana scheme, and in the INEAC classification includes both *hydro-xerokaolisols* and *xerokaolisols*. Ferruginous soils fall mainly into the class of *ferric luvisols* in the FAO classification, and into the *udult* suborder of the *ultisols* in the 7th approximation, although in neither case is there exact correspondence. The 'red loams' of India and 'reddish-brown earths' of Ceylon are ferruginous soils.

Ferruginous soils are widely distributed in the savanna zone. They are a major soil type of the Deccan of India and the 'Dry Zone' of Ceylon, and are widely distributed in both West and East Africa. Frequently they occur on undulating land surfaces in association with weathered ferrallitic soils, the ferruginous soils occurring on areas of steeper slope or less acid parent rock. On the *Soil Map of Africa* they are shown as occupying 11 percent of the continent, but this includes substantial areas of very sandy soils that are here termed arenosols. Moreover, the extent of all zonal soils is greatly exaggerated on small-scale, reconnaissance maps; detailed surveys show that even under typical climatic conditions for the zone, ferruginous soils frequently occupy less than half the total surface area.

Agricultural properties

Ferruginous soils are suited to annual crops, for example maize, tobacco and cotton. Yams are grown towards the wetter limits. Moisture retention in clayey profiles is good owing to the heavy-textured B horizon, whilst the well-developed structure of this horizon assists root penetration. Sandy ferruginous soils have lower moisture retention, and under dry savanna climates this limits cultivation to annuals with a short growing season, for example sorghum, bulrush millet, groundnuts and some varieties of cotton. Gravel horizons cause lower moisture retention and fertility but do not completely prevent root penetration, and soils with these horizons may have a moderate cropping potential.

In terms of fertility, ferruginous soils are intermediate between the more fertile basisols and the less fertile weathered ferrallitic soils. Organic matter levels under natural vegetation are 1–2 percent, and the aim of cultivation must be to maintain this level, by farm-yard manure, short fallows or grass leys, green manure or other means (p. 117). Nitrogen levels are moderate following bush fallowing but fall rapidly during cultivation. Potassium levels are generally adequate where there is a reserve of weathering minerals in the profile. Phosphate is often deficient, for although the pH range is favourable the high free iron content may cause fixation. Ferruginous soils usually give a good response to fertilizers. Sandy ferruginous soils have lower organic matter and fertility levels than clayey types, but respond well to fertilizers and are suited to groundnut cultivation.

There is a substantial erosion hazard on ferruginous soils, particularly because they often occur on slopes. Conservation works and the mainten-

137

ance of a good organic matter status are necessary. The hazard of sheet erosion is particularly great where it could lead to a thinning of the soil above a gravel horizon.

FERRALLITIC SOILS

The term ferrallitic soil and its equivalents (ferralsol, oxisol, 'lateritic soil') has been applied to two distinct soil types: the zonal soils of rainforest climates, characterized by strong acidity and very low base saturation, and structureless soils of plateau regions in the savannas. Both these types have in common a high degree of weathering. In the former case this is due to the high intensity of present-day weathering; in the latter it has been caused by the continuance of weathering over a long period, without the corresponding removal of weathered products by erosion. The former type, developed under a rainforest environment, is here described as *leached ferrallitic soils*. Latosols low in weatherable minerals but occurring under savanna conditions are termed *weathered ferrallitic soils*.

LEACHED FERRALLITIC SOILS

(Profile 2, p. 427)

CCTA: Ferrallitic soils p.p.
FAO: Ferralsols p.p., ferric acrisols p.p.

Leached ferrallitic soils are deeply weathered and have clayey and stone-less textures. The horizon boundaries, other than that of the Ah horizon, are merging. On crystalline rocks there is a deep layer of highly-weathered material *in situ*, often extending to over 10 m and sometimes to 50–100 m. Sedimentary rocks may not be so deeply altered. Few or no weatherable minerals remain within the upper 2 m. Textures are usually clay or, where the parent material yields much quartz, sandy clay. Textural profiles may be uniform or gradational. In some cases, particularly on fine-textured red soils, the clay content varies by less than 5 percent throughout the profile. Where a textural B is present its boundaries are gradational. On granite, which yields sandy clay textures, the clay content is typically at a minimum in the upper layer, increases gradually through a transitional horizon and reaches a maximum at a depth of 50–100 cm.

The structural grade is generally weak but not completely structureless. Commonly there is a weak, fine blocky structure which under gentle pressure breaks down into fine crumb aggregates. The consistence is friable, often with a characteristic soft feel described as 'floury'. A feature of this soil type is that even clays remain friable, or at least labile, owing to the absence of 2:1 lattice minerals; soils occur that are friable when moist but which on analysis show clay contents of 80–90 percent.

138

Colours may be red, dark red, reddish yellow or yellow and, apart from the darker organic topsoil, remain constant down to the C horizon. In the latter there is characteristically a prominent weathering mottle; often this is white and red, and may be described graphically as the 'corned beef' horizon. There is a strong concentration of feeder roots in the uppermost 20 cm. Under natural rainforest there is a 1–5 cm layer of leaf litter, but owing to rapid decomposition the partly decomposed (Of) horizon is very thin and often discontinuous, and there is no humus layer. Mixing by termites may cause the topsoil to have an admixture of brightly-coloured material carried up from lower horizons. Scattered iron concretions may occur.

Under the intense weathering conditions found in the rainforest environment, sandy textures are less common than in ferruginous soils; thus, granites and some sandstones weather to a sandy clay. However, where the parent material contains abundant quartz grains this can outweigh the climatic influence, giving *sandy leached ferrallitic soils*, generally with a textural B horizon and extremely low silt values.

Horizontal sheets of thick, massive laterite can occur with this soil type, for example in Malacca District of Malaysia (Eyles, 1970). Also common is a concretionary laterite horizon 20–50 cm thick with its upper boundary at 20–100 cm depth; the thickness of overlying soil may vary substantially over short distances.

Leached ferrallitic soils are characterized by extreme values of the three main analytical parameters associated with the exchange complex: they are strongly acid (pH < 5.5, usually <5.0), have low cation exchange capacities (< 20 m.e./100 g of clay), and have low to very low apparent base saturation values (< 40 percent, usually <20 percent). The silt: clay ratio is normally low. Kaolinite is the dominant clay mineral, 2:1 lattice minerals are absent, gibbsite is frequently present and free sesquioxides are common. The silica: alumina ratio of the clay fraction is usually less than 2, whilst the silica: sesquioxide ratio is well below 2.

The micromorphology of these soils is characterized by the absence of illuvial ferri-argillans, except sometimes in depth in the BC horizon. The skeleton consists almost entirely of quartz grains, set in a plasma of kaolinitic clay; the plasmic fabric may be sepic, insepic or isotic (Stoops, 1968; Bennema *et al.*, 1970).

These are the zonal soils of the permanently humid tropics with evergreen rainforest (Köppen Af) and the wetter parts of the semi-deciduous rainforest zone (Am). The lower limit for annual rainfall lies between 1250 and 1750 mm. Above 1750 mm all the ferrallitic properties are well developed. Between these two rainfall values, base saturation rises to about 20–40 percent and other ferrallitic properties are present in less extreme form; some soils in this rainfall zone still have the properties of the leached ferrallitic

139

class whilst others are better grouped with the intermediate class of weakly ferrallitic soils.

The parent material may be any rock type other than basic igneous. Because of the close dissection characteristic of rainforest relief, moderate and steep slopes are common. Many tea-growing areas have deep dissection, steep slopes, and deep, red ferrallitic soils. Another common type of landscape is gently to moderately undulating relief, with slopes of 5°–20° and a relative relief of 20–100 m. Site drainage is free to imperfect. If subsurface drainage is slow, for example over shales, in conjunction with the high rainfall this leads to imperfectly drained profiles with mottling in depth.

Intense weathering, rapid leaching, and the nature of the clay complex account for the properties of this soil type. Weathering under hot and permanently humid conditions rapidly reduces all but quartz grains to clay size, causing the low silt content. The complete breakdown of primary minerals releases into the weathering complex not only bases and silica but also free iron and alumina. Downward movement of water through the profile for all or much of the year causes strong leaching of bases giving the low saturation values and strong acidity. Silicic acid is also removed in solution. Clay mineral synthesis under conditions of intense leaching is predominantly of kaolinite and gibbsite, the latter being formed partly by direct synthesis and partly by transformation. Some iron is removed in solution but much remains, together with free alumina, to form part of the clay complex. The low exchange capacities of kaolinite, gibbsite and the free sesquioxide account for the low exchange capacities of the soil despite its high clay content.

Where there is much free iron the clay forms stable aggregates, inhibiting clay translocation and giving uniform texture profiles. On iron-poor, quartz-rich parent materials, including granites and some sandstones and shales, this stabilizing effect does not occur and there is clay eluviation from the topsoil.

A leached ferrallitic soil is thus one in which the processes of latosol formation operate at a high intensity. It need not be an old soil, in the sense that the time elapsed since the material now forming the profile existed as unaltered rock is not necessarily great. Indeed, recent evidence suggests that the rate of removal of material by surface wash under rainforest is substantial; and since many ferrallitic soils occur on steep slopes, they may be relatively young in a time sense. The advanced stage of weathering attained is a consequence of the rapidity of weathering at the present time rather than its continuance over a long period in the past.

All tropical soil classifications have recognized a class equivalent to ferrallitic soils. Minor correlation problems arise with respect to the limits at which the defining criteria are placed. Thus, the *oxisols* of the 7th approximation and the *ferralsols* of the FAO system must possess an oxic B horizon, which in the former case is rather narrowly defined. A more serious confusion is that some systems have grouped together, in the same high-category class, the two soil types here defined as leached and weathered ferrallitic soils. This applies to the *ferrallitic soils* in the CCTA and ORSTOM systems. Leached ferrallitic soils are the *hygro-kaolisols* of the INEAC classification, the *oxysols* of Ghana, and in early classifications were termed *red earths*, *lateritic soils* or 'laterite soils'.

Leached ferrallitic soils are widely distributed in the rainforest regions, for example in the Amazon and Congo Basins, Indonesia and Malaysia, Assam, the Western Ghats of India, and the 'Wet Zone' of Ceylon. The broader class termed ferrallitic on the *Soil Map of Africa* is mapped as occupying 17 percent of the continent, of which more than half is probably the present class of weathered ferrallitic soils. They occur in catenary association with valley-floor swamps; the change from freely-drained to strongly hydromorphic profiles is often abrupt (p. 278).

Agricultural properties

Most leached ferrallitic soils have good physical properties. The stable, fine structure permits easy root penetration, and gives good moisture retention but free drainage. Exceptions are some soils derived from iron-poor shales, which have compact subsoils that restrict root penetration and cause drainage impedance. The absence, or shortness, of a dry season, combined with the high field capacity of the more clayey soils, permits cultivation of perennials without the need for precautions against drought. On sandy profiles periodic moisture stress may depress yields, and there is a slight risk of crop loss from exceptional dry spells.

The fertility is quite high under natural rainforest but falls rapidly in the first few years after clearance. The natural organic matter level is usually 2–5 percent, the carbon:nitrogen ratio 10:12, and the nitrogen level is maintained at 0.1–0.3 percent. Much of the phosphorous and the cations are also derived from the organic fraction. The plant and soil stores of organic matter undergo a high rate of cycling. On clearance, the input of plant litter is terminated by clearance whilst the losses from the soil, by oxidation and leaching, continue or are augmented. Hence the soil store is rapidly depleted: since the nutrient content of the mineral horizons is low, the nutrient status of the soil as a whole falls correspondingly. Deep weathering, and the paucity of minerals within the profile, means that the quantity of nutrients released to the soil by weathering is small.

All of the major nutrients may become deficient. Nitrogen in its available form of nitrate is subject to rapid leaching, and is therefore dependent on a continued supply of organic matter. Phosphorous is subject to fixation in unavailable forms, owing both to the acidity and the presence of free iron oxides. The rapid leaching may cause potassium and magnesium deficiencies.

This situation has consequences for both annual and perennial crops. Under annuals, yields fall rapidly after the second or third year of cultivation. Under a rainforest climate there is no economic way of restoring the organic matter and fertility levels, other than bush fallowing. A fallow of more than 10 years, or a fallow cultivation ratio of at least 3 : 1, is necessary for the long-term maintenance of fertility (p. 114). The traditional system of shifting cultivation is better adapted on this soil type than in other environments; if annual crops are to be grown at all, then there is an argument for retaining the ecological feature of bush fallowing, in some modernized form such as the 'corridor' system (Jurion and Henry, 1969). Any attempt to cultivate annuals continuously or with short fallows leads to rapid soil degradation and ultimately to erosion.

The growth of perennial crops overcomes two hazards of the soil type: organic matter depletion and erosion. Provided that a cover crop is established, the organic matter status can be maintained, and a leguminous cover crop (e.g. *Pueraria*) augments atmospheric nitrogen fixation. But the loss of nutrients by leaching is not overcome. Crops grown under traditional conditions, in mixed stands, may be able to maintain unchanged but low yields. But for sustained high yields, in pure stands, heavy additions of artificial fertilizers are necessary. Oil palm is one of the most demanding crops in this respect.

Because of their good physical condition these soils are resistant to erosion on first clearance, even on steep slopes. With the depletion of organic matter this resistance is lost after 1–3 years, after which they must under annual cropping be returned to a fallow, either of natural recolonization or planted softwoods. Under perennial cultivation a cover crop should be established as soon as possible after clearance. On slopes of more than 12° terracing is advisable.

The management properties of ferrallitic soils, including the effects of fallows, have been reviewed by Van Wambeke (1974).

WEATHERED FERRALLITIC SOILS

(Profile 3, p. 428)

CCTA: Ferrallitic soils p.p.
FAO: Ferralsols p.p., ferric acrisols p.p.

Weathered ferrallitic soils are structureless or weakly structured, have low cation exchange capacities but do not necessarily have low base saturation values or strong acidity. There is a deeply-weathered regolith and very few weatherable minerals remain within the soil profile.

It has been a major source of confusion that the weathered ferrallitic soils of the savanna zone and the leached ferrallitic soils of the rainforest zone have been grouped together. Whilst similar in having a low content of primary minerals and a low cation exchange capacity, the two soil types otherwise differ morphologically, chemically and in respect of agricultural properties. The weathered ferrallitic soils are more weakly structured than the leached ferrallitic group, are not rich in free sesquioxides, and include profiles with over 35 percent base saturation. They were earlier known as 'plateau soils' (Milne, 1936; Trapnell *et al.*, 1948/50). Explicit recognition of the distinction between the leached and weathered types of ferrallitic soils (oxisols) was made by Botelho da Costa and Cardoso Franco (1965).

The upper horizons of weathered ferrallitic soils are sandy, typically sandy loams. A humus-stained topsoil overlies a structureless A2 horizon. The mineral soil is often pale in colour, and organic matter makes the Ah horizon appear markedly darker; this is deceptive, since organic matter contents are usually below 2.5 percent, sometimes as low as 0.5–1.0 percent. The B horizon is compact, firm, and at the end of the dry season becomes very hard. It is either structureless or has a fine–medium blocky structure of no more than weak grade; clay skins are absent. Often there is a duplex textural profile, with a sandy clay textural B horizon commencing fairly abruptly and quite close to the surface (30–50 cm). Uniform textural profiles also occur. The B horizon passes transitionally downwards into a thick highly-weathered layer, either of uniform colour or mottled; although *in situ*, the rock structure has been largely or entirely destroyed by weathering, and the transition to weathered rock is not reached until considerable depth. Colours are often pale, especially near the surface, but this is not a defining property and red profiles can occur.

A sub-group of this soil type are the *pallid soils* (Watson, 1962a), which have seasonally restricted subsoil drainage. The upper part of the profile is pale in colour (often in the 10YR hue) and there is reddish mottling or iron concretions in depth. They occur on gentle slopes, including pediments, often in catena below freely-drained, redder weathered ferrallitic soils.

A laterite horizon is common, both in the normal and pallid types. Many of the soils that overlie thick, sub-horizontal sheets of massive laterite exhibit the properties of the weathered ferrallitic class to an extreme degree, being pallid, structureless and compact. Thin horizons of laterite are also frequent; typically a vesicular or concretionary laterite horizon 20–50 cm thick commences at a depth of 20–100 cm. The laterite may include a small or large admixture of quartz fragments. Stone lines consisting largely or entirely of quartzite are also common, especially on slopes. Another common feature is for iron concretions to occur in the transitional BC horizon, passing gradually downwards into highly-weathered, partially ferruginized rock with variegated red, yellow and black colouring.

Of the analytical properties associated with the exchange complex, two are constant and two variable. There is always a low cation exchange capacity of the clay fraction, and values of total exchangeable bases are low or very low. Base saturation values, however, may vary from very low to 80–90 percent, and reaction may be weakly to strongly acid. The silt:clay ratio is low. Kaolinite is the dominant clay mineral, with 2:1 lattice minerals rare or absent; in contrast to ferruginous soils, the profile is not rich in free sesquioxides.

Weathered ferrallitic soils are found on gently-undulating plateau surfaces under a savanna climate. They are the typical 'plateau soils' of Africa and other shield areas. In the conventional geomorphological terminology they are associated with the 'Gondwana' and 'African' erosion surfaces. Be that as it may, the essential feature is that slopes are gentle; maximum valley-side slopes are generally less than 5 ° and sometimes as low as 2°. The regolith is deeply and intensely weathered; total chemical analyses show that the weathered rock of the C horizon differs greatly in composition from the fresh rock beneath; most of the bases and much silica have been lost and the silica:sesquioxide ratio is greatly reduced.

Felsic igneous rocks of the Basement Complex are the most common parent material; the soil is also found in siliceous sedimentaries but not on rocks of basic or calcareous composition. It occurs over the whole of the savanna zone, from about 500 to 1200 mm rainfall. This is the typical soil type under *Brachystegia* savanna (*miombo*) in Africa, and the corresponding *cerrado* formation in South America. Above 1000–1200 mm base saturation decreases and there is a transition to the leached ferrallitic type.

There are two theories of the origin of this soil type: that it is produced by prolonged weathering in conjunction with a slow rate of removal by surface denudation, resulting in the progressive accumulation of weathered material; and alternatively, that the deep and intense weathering took place during a previous cycle of erosion. These hypotheses are discussed below (p. 265). In either event, the outcome is that current pedogenetic processes are acting upon highly-altered proximate parent material. Since few weatherable minerals remain, there is little production of clay by weathering, accounting for the absence of clay skins. The intensity of current leaching varies with the rainfall. Profiles low in exchangeable bases occur under as little as 500–600 mm rainfall; it is possible that this feature is relict from a past period of more humid climate.

Essentially these are 'dead' soils, which are being very little altered by current pedogenetic processes. Every change that processes can bring about, in weathering, translocation or clay mineral synthesis, has already occurred. The profile is old, in the sense that a long period of time has elapsed since its constituents existed as unaltered rock.

Weathered ferrallitic soils form part of the ferrallitic class in the CCTA and ORSTOM classifications. In the Ghana system they include both the *savanna oxysols* and the 'groundwater laterites'. They are unsatisfactorily categorized by the FAO and 7th approximation classifications since they sometimes do not possess an oxic horizon, and are thus not ferralsols or oxisols respectively, but there is no other class defined so as to include their characteristic assemblage of properties.

These soils are of widespread occurrence where shield areas lie under a savanna climate – the Guyana and Mato Grosso shields of South America, the Deccan of India, and large parts of the African continent. They are particularly extensive in East and Central Africa, for example on the extensive plateau areas at 1000–1500 m in Tanzania, Zambia and Malawi. Where the plateaux are dissected they give place to ferruginous soils or lithosols (Webster, 1960).

Three common catenary associations are as follows (fig. 7):

(*a*) On gently-undulating surfaces without inselbergs. Crests and upper slopes have reddish or reddish yellow weathered ferrallitic soils, becoming yellower and paler on middle to lower slopes; this is succeeded by mottled sandy soils on the concavity and dark grey clays in the valley floor (p. 277).

(*b*) On surfaces with inselbergs and pediments. Lithosols or bare rock on the inselberg slopes give place to ferruginous soils, with a high content of weatherable minerals, on the upper fringe of the pediment. The greater part of the pediment is then occupied by weathered ferrallitic soils, becoming pallid towards the base, and succeeded by gleys in the valley floor.

(*c*) On surfaces with level crest areas. The crests carry weathered ferrallitic soils, often with a laterite horizon; this gives place to ferruginous or weakly ferrallitic soils, without laterite, on the valley sides.

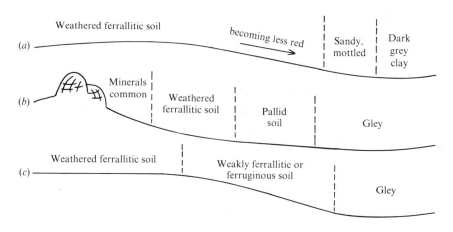

Fig. 7. Catenary associations of weathered ferrallitic soils.

Latosols

Weakly ferrallitic soils

Soil types occur in the savanna zone which do not have the modal properties of either ferruginous or weathered ferrallitic soils; they tend towards the latter, but not in an extreme form. It is useful in mapping to classify soil series of this nature as forming an intermediate subgroup of *weakly ferrallitic soils*, distinct from soils of lower agricultural potential in which ferrallitic properties are strongly expressed.

Another intermediate situation occurs between ferruginous soils and leached ferrallitic soils, in the zone climatically intermediate between savanna and rainforest. The former view that ferrisols were the zonal type of this climate is incorrect (Maignen, 1961). It is possible that weakly ferrallitic soils could also occur under these conditions, and also leached ferruginous soils. However, the nature of the soil climosequence in this zone, for a constant parent material, has not been established.

Agricultural properties

Both physically and in respect of nutrients, weathered ferrallitic soils are poor for agriculture. The compact B horizon inhibits root penetration; where there is a duplex textural profile the abrupt change in the root environment is deleterious to plant growth. Moisture retention is relatively low, particularly in the sandy topsoil; this can lead to crop failure where early rains, causing germination, are followed by a dry spell before the seedlings have rooted deeply. Where there is a dry season of six months or more the soil moisture drops to wilting point to a depth of over a metre.

Both the natural fertility status and the capacity to retain added nutrients are poor. The organic matter content is low. Nitrogen levels are generally 0.1–0.2 percent, and sometimes as low as 0.05 percent. Savanna grasses of the Andropogonae family appear to inhibit nitrification (Nye and Greenland, 1960, p. 96). Sulphur deficiency is sometimes present. There is no reserve of weatherable minerals to replenish the soil.

The vegetation gives some indication of the poverty of this soil type. There is frequently a contrast between more open short-tree savanna on plateaux with weathered ferrallitic soils and taller or denser savanna woodland on adjacent ferruginous soils or even on steep slopes with lithosols.

The response to the soil conditions under primitive technology is shifting cultivation, for example the *citemene* system of Zambia (Allan, 1965). Finger millet (*Eleusine coracana*), which yields satisfactorily on poor soils, is often grown. However, this system is ecologically less well adapted to the savannas than to the rainforest environment, owing to the damaging effects of burning and the need for a higher fallow: cultivation ratio.

Satisfactory yields of maize and other grains, groundnuts and tobacco

146

can be obtained with improved technology. The low nutrient status is in some respects favourable to high-quality tobacco, provided fertilizers are added at the stage of early growth. Profiles with laterite horizons or with sandy or duplex textures are the poorest. Short fallowing (with a fallow: cultivation ratio of at least 1:1) is necessary, together with green manuring, a rotation that includes legumes, and application of manure or fertilizers. Weathered ferrallitic soils demand a high standard of skill if they are to be farmed continuously.

BASISOLS

Two terms that have been used to denote more fertile types of latosols derived from rocks of basic composition are ferrisols and basisols. In the CCTA classification ferrisol, originating from the INEAC classification, is applied to the redder types occurring under more humid conditions, whilst the corresponding soils under drier climates are called eutrophic brown soils. Basisol is derived from the Ghana classification, in which it refers to all latosols derived from basic rocks, the basisols being divided into forest and savanna rubrisols (red) and brunosols (brown).

Although the Ghana terms for these soils are both simple and descriptive, the CCTA nomenclature is preferred here, in conformity with its use for other latosol types. The term *basisol* is retained not for a soil type but to refer to the class of soils comprising all those derived from basic rocks.

Ferrisol is used here for the basisols developed under more humid climates, including both rainforest and the transitional rainforest–savanna zones; these are usually but not necessarily reddish. *Eutrophic brown soil* is applied to the corresponding soils developed under savanna climates, which may be red or brown. The ferrisols partly correspond to the FAO class of *nitosols*, and cover the *forest basisols* of the Ghana classification. *Eutrophic brown soils* include the *savanna rubrisols* and *brunosols* of the Ghana system.

FERRISOLS

(Profile 4, p. 430)

CCTA: Ferrisols
FAO: Dystric nitosols

Ferrisols have a deep, clayey B horizon with a moderately to strongly developed fine or medium blocky structure. Clay skins covering the ped surfaces, clearly visible to the naked eye, are a characteristic feature. The B horizon is usually a clay or sandy clay, and its upper and lower boundaries are merging. The topsoil texture may be a clay, sandy clay or sandy clay loam. The textural profile is either gradational or uniform, but never duplex. The moist consistency is friable to labile. At 1.5–2.0 m depth small to moderate quantities of weatherable minerals can be distinguished with a hand lens. The colour may be dark red, dark reddish brown, red or yellowish red, often becoming gradually brighter with depth. Scattered iron concentrations may occur, and less commonly a horizon of laterite gravel occurs in depth. Massive laterite is not common.

The clay fraction is dominated by kaolinite and free iron oxides, but also includes small amounts of 2:1 lattice minerals. The base saturation of the B horizon is below 50 percent. The cation exchange capacity of the clay fraction is generally 15–25 m.e./100 g, exceeding that of leached ferrallitic soils. Topsoil organic matter content is 2–5 percent.

In thin section, ferrisols show abundant illuvial ferri-argillans. The plasma, consisting of iron-stained clay, is dominant over the skeleton, and possesses a sepic plasmic fabric (Stoops, 1968; Eswaran, 1970).

The comparatively rare ultra-basic rocks also give rise to ferrisols. Profiles on such rocks are dark red to dark reddish brown clays, lack quartz sand grains, possess a strong blocky structure and clay skins, and have indistinct horizons. They may contain comparatively high quantities of minerals such as nickel, copper, cobalt or lead (Fox and Hing, 1971).

Where the rainfall is 1200–1800 mm, with one or two short dry seasons (Köppen Am or Aw″) the base saturation is 10–50 percent and acidity is moderate, sometimes rising to an almost neutral reaction in the topsoil. This is the environment in which the characteristic properties of ferrisols are most clearly developed.

Under extreme rainforest conditions, with a rainfall exceeding 2000 mm and no dry season, a soil type develops which has similar morphological properties but which is strongly acid and may have a base saturation as low as 10 percent. This variant is termed *strongly leached ferrisols* (FAO rhodic ferralsols). Examples occur over andesites in Malaysia, where such soils have excellent physical properties for agriculture. On basalts in Samoa, Wright (1963) records cases of this soil type in which the value obtained for total exchangeable bases was 0.0.

Most ferrisols are derived from rocks of basic composition. The Ghana classification explicitly recognizes this. It is significant that the CCTA system includes a sub-group of 'ferrisols on rocks rich in ferromagnesian minerals' and notes: 'It is on this parent material that the most typical ferrisols are found', whilst the FAO classification comments on 'their favourable physical properties and often their higher fertility, especially when derived from basic rocks' (Dudal, 1968a, p. 9). The well-developed structure and clay skins appear to be associated with continuous renewal of the soil by the weathering of iron and base rich minerals. Many ferrisols occur on moderate or steep slopes, and it has been suggested that the removal of material by surface denudation is an additional factor inhibiting the formation of a highly-weathered regolith. However, they also occur over basic rocks on level ground. In the absence of quartz grains, removal in solution in groundwater may be more important than surface wash in causing ground loss and thus renewal of the regolith. Ferrisols derived from non-basic rocks have been described (e.g. D'Hoore, 1964, Profile 25, p. 163) but are uncommon. Thus, on freely-drained sites within the climatic zone of 1200 mm or more rainfall, basic rock is a sufficient but possibly not a necessary condition for ferrisol formation.

Ferrisols occupy relatively small, scattered areas. On the *Soil Map of Africa* they are shown as covering 3 percent of the continent, but this figure is exaggerated by areas mapped in the Congo where the class is more widely interpreted. These fertile soils probably occupy no more than 1–2 percent of the tropics, or 3–6 percent of the area covered by latosols.

EUTROPHIC BROWN SOILS

CCTA: Eutrophic brown soils p.p.
FAO: Eutric nitosols

The term eutrophic brown soil is here applied to both brown and reddish brown soils developed from basic rocks under a dry savanna climate. They are the *eutrophic brown soils* of the CCTA system except that in the latter, andosols are included as a subdivision of this class. The eutrophic brown soils as here defined include the *savanna brunosols* and *savanna rubrisols* of the Ghana classification.

Eutrophic brown soils are dark brown to dark reddish soils of moderate depth. The topsoil is dark, with a good organic matter content. There is usually a textural and always a structural B horizon with a moderate or strong angular blocky structure and clay skins. The consistence is plastic to labile and sticky. There is a relatively high content of weatherable minerals in the profile; sometimes rock fragments are common.

The clay minerals include substantial amounts of montmorillonite, giving a high cation exchange capacity. Base saturation values are 50–100 percent, and the content of exchangeable bases is high. The reaction is neutral to weakly acid. There is a high content of exchangeable calcium, but free carbonates are not present.

Eutrophic brown soils occur under a rainfall of 700–1500 mm. Where the rainfall is below about 700 mm on basic rocks, free calcium carbonate appears in the profile giving pedocals. At about 1200–1500 mm there is a gradual transition from basisols to ferrisols, with a greater depth of weathering, stronger acidity and lower base saturation; the boundary, necessarily arbitrary, may be taken at a B horizon base saturation of 50 percent.

The genesis of eutrophic brown soils is associated with the same main cause as that of ferrisols: continuous renewal of the mineral soil by the weathering of basic rock. Leaching is less than in ferrisols, giving a more base-rich weathering complex and a greater tendency to the formation of 2:1 lattice minerals. The swelling and shrinkage on wetting and drying of montmorillonitic clays accounts in part for the blocky structure.

Agricultural properties of basisols

Ferrisols and eutrophic brown soils are among the most productive soils of the tropics. They allow easy root penetration, and combine good moisture retention with free drainage; ferrisols in addition have deep profiles. Although frequently occurring on steep slopes, their good structure and permeability confer a relatively high resistance to erosion.

Allied to their favourable physical properties, these soils are able to sustain continuous cropping. The detailed mechanism of nutrient cycling that enables this to occur has not been worked out. Undoubtedly the main cause is the continuous input of elements released by rock weathering; higher biological activity may be a contributory cause. In ferrisols the content of plant nutrients is only moderate, although greater than in leached ferrallitic soils under corresponding climates. Topsoil organic matter contents are higher than in soils on felsic rocks. Basisols have moderate nitrogen contents, but generally high levels of the other main nutrients. In ferrisols the nutrient levels are only moderate, but are greater than in leached ferrallitic soils. What is exceptional in both ferrisols and basisols is the capacity to maintain fertility under cultivation without recourse to fallowing.

Good soils are capable of giving some return to poor agricultural methods. Thus, practices such as maize monoculture or the growing of unimproved varieties of perennials are found on some basisols and ferrisols. This is a misuse of their high resource potential. Ferrisols are suitable for most perennial crops, and are particularly valuable for those with more exacting requirements; they include, for example, the best cocoa soils in West Africa. Oil palm grows well on strongly-leached ferrisols, e.g. on adesite-derived soils in Malaya, provided its high nutrient demands are supplied by fertilizers. Coffee is widely grown; where there is a long dry season mulching, in conjunction with the high moisture retention, enables the soil to remain moist in depth. Near their humid limit, eutrophic brown soils are used for cocoa, coffee and bananas; where there is a long dry season they yield well under annual crops. With good agricultural technology, ferrisols and eutrophic brown soils are among the most productive soils of the tropics.

HUMIC LATOSOLS

CCTA: Humic ferrallitic soils and humic ferrisols
FAO: Humic ferralsols and humic nitosols

Humic latosols are freely to imperfectly drained acid soils, relatively rich in organic matter, which occur at high altitudes. A working definition is that the thickness of the Ah horizon in centimetres multiplied by its percentage of organic matter must exceed 50, e.g. a horizon 10 cm thick with over 5 percent organic matter.

The cause of the humic topsoil is that with falling temperatures the rate of loss of organic matter decreases faster than does its production, i.e. the rate of plant growth (p. 103). Humic latosols occur at altitudes above about 1500 m. For a given organic matter content a decrease in altitude with increase in latitude might be expected, but such a relation has not been positively demonstrated.

Humic ferrallitic soils are the more highly-weathered type of humic latosols. They occur on rocks of felsic to intermediate composition under moderate or high rainfall. *Humic ferrisols* are less weathered and more strongly structured; they occur on basic to intermediate crystalline rocks, generally under moderate rainfall. Highly basic parent material, however, can outweigh the influence of climate in respect of profile morphology (although not of analytical properties) and give humic ferrisols with high rainfall totals.

Humic ferrallitic soils are weakly structured, and do not have visible clay skins. There are few weatherable minerals in the upper horizons, but the profile is not deep, highly-weathered rock sometimes commencing at about 1 m. There is often a textural B horizon of sandy clay, sometimes with quartz gravel. Core-stones, residual boulders with concentric weathering, frequently occur in depth. There is a dark brown humus-rich topsoil, often with abundant roots. Below this the colour is variable, from red or reddish brown to yellow, yellowish brown or brown. Bauxitic variants, high in free alumina, occur.

A variant called humic ferrallitic soils 'with dark horizon' has occasionally been reported, for example from Zaire (Ruhe and Cady 1954; Sys, 1960), and Sri Lanka (de Alwis and Panabokke, 1972–3, p. 88). It occurs at altitudes above 1400 m, and is recognized as a separated sub-group in the CCTA classification. At a depth of 100 cm or more the B horizon changes from dark red to dark reddish brown accompanied by a slight rise in organic matter content: alternatively, the organic matter level remains at about 1.0 percent throughout the lower horizons. It would appear that this is a weakly developed Bh horizon as found in podzols, but there is no suggestion of an associated sesquioxide B horizon.

In humic ferrisols the lower horizons are red or reddish brown. The B horizon has a well-developed fine to medium blocky structure with clearly

151

visible clay skins. Profiles are deeper than in humic ferrallitic soils, commonly over 2 m to weathered rock; weatherable minerals occur in the lower subsoil.

In respect of analytical properties both types of humic latosol are strongly acid and highly leached with a low base saturation. These characteristics are most strongly developed in the humic ferrallitic type, which frequently has a pH of 4.0–4.5 and a saturation below 10 percent. By definition, organic carbon levels are high.

The high-altitude areas at which humic latosols occur normally have rainfalls in excess of 1250 mm. In conjunction with the lower temperatures, low evapotranspiration and free drainage, this results in a large volume of water passing through the profile, and hence the acidity and low saturation resulting from strong leaching. Although commonly found on plateaux originating as uplifted erosion surfaces, these plateaux are usually dissected, with moderate slopes. Surface denudation therefore occurs at moderate rates; together with the lower rates of weathering this accounts for the relatively shallow profiles. Associations with lithosols occur, and sometimes rock outcrops. Intrusive granitic massifs are a common environment of humic ferrallitic soils. Basic rocks are associated with the better structure of the humic ferrisols; acidity and saturation, however, are dominated by the strong leaching, and basic rock only raises their values slightly, in the case of pH by the order of half a unit.

Humic latosols are of small extent and scattered distribution. On the *Soil Map of Africa* they cover one percent of the continent of which the greater part is a lithosol–humic ferrisol association on the basalt lavas of the Ethiopian Plateau. Other occurrences are on isolated high plateaux ("Gondwana surface", see p. 28), for example in Malaŵi (Young and Stephen, 1965). Sequences from low-altitude latosols to humic latosols have been described in Indonesia and Malaysia (Mohr and van Baren, 1954). Other substantial areas occur in the Andes and in the Western Ghats of India.

Agricultural properties

The main agricultural problems of the humic latosols arise from their shallowness, acidity and strong present-day leaching. Depths within a given area are variable, hence patches too shallow for perennial crops may occur. Nitrogen levels are high, but leaching results in potash deficiency. Phosphates suffer from fixation under the strongly acid conditions. Nutrient levels are substantially reduced where, as frequently, the original montane rainforest has been replaced by grassland.

The shallower and most strongly acid areas, principally of humic ferrallitic soils, are frequently uncultivated. Such plateau areas, however, have an increasing value as game reserves. Where depth is sufficient, and particularly but not exclusively on the humic ferrisols, the environment is suitable for perennials tolerant of acidity but requiring lower temperatures and free drainage, such as tea, coffee and tung.

Terracing is necessary on the steep slopes that are common, and severe sheet erosion can occur in its absence, as for example on some tea smallholdings in Ceylon. The high productivity of perennials can, however, economically sustain conservation works. If cultivation is by smallholders, capital must be made available for such works, and some measure of legal enforcement is desirable. Still more essential is control of stock levels where montane grasslands are used for cattle (e.g. Bawden and Tuley, 1967, Appendix I). Overstocking can lead to severe sheet and gully erosion, as has occurred, for example, in Lesotho.

SELECTED REFERENCES

Prescott and Pendleton (1952), Maignen (1961), Watson (1962*a*), Bennema (1963), Botelho da Costa and Cardoso Franco (1965), Delvigne (1965), Haantjens (1965*a*), Sys (1967), Van Wambeke (1967, 1974).

CHAPTER 7

LATERITE

The term *laterite* is used in this book to refer to a hard material, rich in secondary forms of iron. The hardness exceeds that of a dry clay soil, and is retained if the laterite is immersed in water; a hammer or pick-axe is needed to break it. The iron occurs mainly as goethite, haematite and amorphous ferric oxides. Synonyms for laterite are ironstone, ferricrete and murram.

Accessory characteristics are that laterite is usually reddish brown (typically 5YR 5/6); has a moderately high density, 2.5–3.6; usually contains secondary aluminium; may contain quartz and kaolinite but is low in other forms of silica; and exchangeable bases and humus are almost completely absent.

There has also been described an iron-rich, mottled clay which hardens on exposure to air, or to repeated wetting and drying. This material, which is rarely encountered, will be called *soft laterite*.

A soil profile that contains a horizon of laterite will be referred to as *with laterite*. The term 'lateritic' is not used here.

HISTORICAL NOTE

There is rarely any doubt about the identification of laterite when it is encountered in the field. In the languages of indigenous tropical peoples there is frequently a word for it, different from that for rock. It is all the more remarkable, therefore, that published work on laterite should have been constantly afflicted by confusion over terminology.

The first cause of this arose from the original use of the term, by Buchanan (1807). Buchanan described a material that was soft enough to be cut when *in situ*, but hardened on exposure; it was being quarried to make bricks, and he derived the term from the latin *later*, a brick. Adherence to the rule of scientific priority would confine the term of such soft material. Unfortunately soft laterite proved to be rare, although the myth has long been perpetuated. A solution would have been to call the soft material 'Buchanan laterite' had not Stephens (1961) revisited Buchanan's type site to find that, far from being cut with a knife, rock-like ironstone was being hacked out with pickaxes.

The second cause of confusion arose from the fact that laterite has a low silica: sesquioxide ratio, a characteristic which it shares with many freely-drained soils of the tropics; and one theory of the origin of laterite was that it is an extreme form of the same desilication process that produced these soils.

154

Owing to the ignorance about tropical soils in America and Europe up to about 1950, such soils came to be called 'lateritic' and 'laterite soils'. both terms being used in the US 1938 Classification which influenced many text-books.

The problem could have been solved if the US Soil Survey, in devising the 7th approximation, had chosen a term to represent hard laterite; the rest of the world would gladly have adopted it. Instead, they defined a new word *plinthite* as a sesquioxide-rich clay which either 'changes irreversibly to hard-pans . . . on repeated wetting and drying' or 'is the hardened relics of the soft red mottles'. The FAO classification adopts the above definition for iron-rich clay which can be 'cut with a spade' but uses *ironstone* for irreversibly hardened material (FAO–Unesco, 1974, p. 30).

There has probably been more written about laterite than any other aspect of tropical soils. A recent review lists 360 references, said to have been selected from over 2000. The main articles of historical importance are by Buchanan (1807), containing the first use of the term; Lake (1890), a review of early ideas together with a discussion of the physiographic signif-icance of laterite; Walther (1915), who first described the mottled and pallid zones; Campbell (1917), the originator of the groundwater fluctuation hypo-thesis of origin; and Woolnough (1918, 1930), who discussed relations to landforms and described other types of indurated horizons. The most impor-tant research reviews are those of Prescott and Pendleton (1952), D'Hoore (1954*a*, 1954*b*), Sivarajasingham *et al.* (1962), Aubert (1963) and Maignen (1966).

CLASSIFICATION

Laterite may be classified in terms of its morphology, composition, and posi-tion of occurrence in relation to landforms and soil profiles. There have also been genetic classifications (D'Hoore, 1954*b*, 1955; Maignen, 1959; Aubert, 1963) but in view of the uncertainty about its origin these are unsatisfactory.

On the basis of morphology, the following main types and subdivisions are distinguished (based partly on Pullan, 1967):

(1) *Massive laterite.* Possesses a continuous hard fabric. Occasionally completely massive but usually contains cavities, empty or filled with soft earth. Frequently incorporates quartz grains and sometimes other rock material, but rock structure not visible. Subdivided into:

    (1*a*) *Cellular laterite.* Cavities are approximately rounded, and either unconnected or partly connected.

    (1*b*) *Vesicular laterite.* Cavities are predominantly tubular.

(2) *Nodular laterite.* Consists of individual, approximately rounded, con-cretions. (Also called pisolithic laterite, 'pea-iron'.) Subdivided into:

(2*a*) *Cemented nodular laterite.* Individual concretions can be seen, but are strongly joined together by the same ironstone material.

(2*b*) *Partly-cemented nodular laterite.*

(2*c*) *Non-cemented nodular laterite.* Concretions are largely or entirely uncemented, but are adjacent. Concretions form over 60 percent by weight of the total soil. (Also called packed nodular laterite.)

(2*d*) *Iron concretions.* Concretions are separated by soil, and from less than 60 percent by weight of the total horizon. (Also called spaced nodular laterite.)

(3) *Recemented laterite.* Contains fragments of massive laterite or ferruginized rock, broken and wholly or partly cemented.

(4) *Ferruginized rock.* Rock structure is still visible, but with substantial isomorphous replacement by iron.

(5) *Soft laterite.* Mottled iron-rich clay which hardens irreversibly on exposure to air or to repeated wetting and drying.

COMPOSITION AND PROPERTIES

The main constituents of laterite are iron and aluminium. Proportions vary from 80–90 percent iron oxides with 5–10 percent alumina to 40–50 percent of each. Where alumina exceeds about 60 percent the material is a bauxite. Up to 30 percent of silica may occur, as quartz and combined in kaolinite. Small amounts of titanium oxide, typically 1–5 percent, are usually present. Combined water occurs mainly in combination with alumina, the amount varying with the latter. Calcium, magnesium, potassium, sodium and manganese vary from nil to <1 percent.

The iron is entirely in ferric forms, mainly goethite $(FeO.OH)$, haematite (Fe_2O_3) and amorphous oxides. Alumina occurs as gibbsite $(Al_2O_3.3H_2O)$ and boehmite $(Al_2O_3.H_2O)$. Free silica, where present, is in the form of quartz, inherited from the parent rock; feldspars and most other primary minerals are absent. There are often substantial amounts of kaolinite, but 2:1 lattice clay minerals are absent.

Many soils, neither actually nor potentially indurated, have similar chemical and mineralogical composition to laterites. The features which distinguish laterite, and account for its hardness, are a greater degree of crystallinity and a greater continuity of the crystalline phase (Alexander and Cady, 1962). Common micro-structures are: (i) pseudomorphs of rock minerals, formed of goethite, gibbsite or kaolin; (ii) frameworks of oriented materials, particularly goethite and kaolin stained or impregnated with iron oxides, having a high degree of continuity within the mass; (iii) small spherical bodies, or micro-nodules, rich in iron, and sometimes with concentric shells of crystalline goethite; (iv) cracks and pores filled with gibbsite or boehmite, or lined with oriented films of iron-impregnated kaolin or goethite.

These micromorphological features, particularly the frameworks and micro-nodules, imply considerable movement and recrystallization of sesquioxides. Iron in the soil solution first impregnates primary minerals, a feature commonly seen on a macro-scale in the staining of quartz grains and weathered rock fragments. Subsequently, iron replaces rock minerals, either isomorphously or by forming crystalline frameworks in the cavities left by their removal. Inherited quartz grains remain. Iron is absorbed and immobilized by kaolin, but may be released again if the kaolin is destroyed by further weathering. Further mobile iron is added as films and micro-nodules, filling cavities and so increasing the density. Alumina may also be deposited. It is, however, the framework of recrystallized iron that appears to be the essential feature conferring hardness on the material.

CLIMATE AND PARENT MATERIAL

Laterites occur in all humid tropical climates, ranging from rainforest to dry savannas. They extend into the semi-arid zone, below 600 mm rainfall, for example in West Africa, and into middle latitudes, for example in New South Wales (Hallsworth and Costin, 1953), Western Australia (Mulcahy, 1960) and Natal (Maud, 1965). Most examples from semi-arid and from subtropical to temperate climates, however, are probably relict forms, originating from wetter or warmer climates in the Tertiary. The latitudinal extremes are 50° N. and 43° S., which is consonant with the extent in the late Teriary of the 18 °C annual isotherm, as reconstructed from palaeoclimatic evidence (Dury, 1971), albeit probably with an element of circular argument.

It is not true, as is sometimes held, that a savanna climate is necessary for laterite formation. Whilst some laterites of the rainforest zone may be relict, others are forming at the present.

Laterite is found over a wide range of rock types, igneous, metamorphic and sedimentary, with basic to felsic composition. In a given region they are more likely to be present, or are thicker, over basic than felsic rocks, but they are known over many types of felsic siliceous rocks, including granites, felsic gneisses, shales and sandstones. Whilst it is possible in some such cases that the laterite has been derived from a more iron-rich overlying rock, now removed, the occurrence over felsic rocks is too widespread for this explanation to be generally true. Basalts tend to give laterites with a high aluminium content.

PEDOLOGICAL OCCURRENCE

In considering relations with soils and relief, it is convenient to distinguish between that laterite which occurs as relatively thin soil horizons, less than 2 m thick, and that which forms thick, massive crusts.

157

Thin laterite horizons usually commence in the soil at depths from 20 to 200 cm, and range in thickness from 10 to 200 cm. A typical occurrence of non-cemented or partly-cemented nodular laterite might extend from 50 to 70 cm depth. Often the depth is not constant. Road cuttings frequently show nodular laterite horizons varying irregularly between 30 and 60 cm depth, whilst retaining approximately the same thickness. Massive or cemented nodular laterite more often occurs as horizons 1–2 m thick; the upper-boundary may again vary in depth over short distances in an apparently irregular manner, occasionally by as much as 60 cm within a soil pit. Horizons such as these do not usually outcrop at the surface, unless exposed by accelerated erosion.

Spaced nodular laterite, i.e. individual iron concretions, generally occurs in the B or BC horizon of soils, often near the junction between soil and weathered rock. It is common to find iron staining with occasional soft to moderately hard concretions where the C horizon of weathered rock commences.

Individual concretions frequently occur above or below the main laterite layer. Composite laterite horizons are common, for example adjacent layers of massive and cemented nodular, or of cemented and non-cemented nodular material. Multiple laterites, with two laterite horizons separated by soil, have also been described but are rare.

The soil overlying laterite horizons often has ferrallitic properties, being massive to weakly-structured, low in cations and with no weatherable minerals. Below the laterite there is often an horizon of compact soil, slightly to strongly variegated in colour and showing traces of rock structure. Quartzite veins may extend through this underlying soil, showing that it is *in situ*. Sometimes they extend through the laterite itself, either *in situ* or showing evidence of downslope movement.

Thick laterite crusts have something of the nature of a geological formation rather than a soil horizon. Thicknesses range from 3 to 30 m, exceptionally more, and typically about 10 m. They are usually formed of massive laterite, less frequently of ferruginized rock or cemented nodular laterite.

Two soil horizons commonly associated with thick laterite layers are the mottled and pallid zones. A typical profile is shown in fig. 8(*a*). A thin layer of soil overlies massive laterite or ferruginized rock. Below this is a *mottled zone*, which varies from soft clay to a firm earthy material that can be crushed with a boot, and which is prominently mottled red, orange, or purple or white. The proportion of white material increases downwards, merging into the *pallid zone*, consisting of white, kaolinitic clay or rock crushable in the hand. This passes downwards into weathered rock. There are wide variations in the dimensions of such profiles, from some in which the complete profile occupies 3 m to others in which the massive and mottled zones may each be 5–10 m and the pallid zone over 50 m.

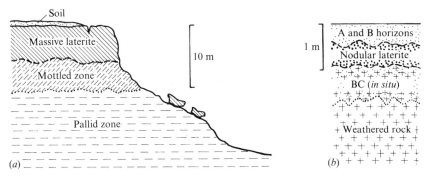

Fig. 8. Profiles containing horizons of laterite: (*a*) the classic profile with a thick laterite layer, shown as a breakaway; (*b*) typical situation of thin laterite horizon in a soil profile.

Where such profiles are dissected the laterite forms scarps, and is undercut by erosion of the softer material beneath, to give a landform graphically described by the Australian term *breakaway*.

MORPHOLOGICAL OCCURRENCE

Thick crusts usually occur as *plateau laterite*, capping erosion surface remnants (fig. 9, *A* and *B*). Whilst appearing quasi-horizontal, the laterite is usually found on survey to possess a gentle relief, up to 2° (Brosh, 1970; McFarlane, 1970). Such crusts act morphologically as a stratum of resistant rock, giving a characteristic landscape of plateau remnants, mesas and buttes, bounded by breakaways, as for example in the Buganda region of Uganda (Pallister, 1956). There may be two distinct surface levels each with a crust, with or without nodular or recemented laterite on the slope between

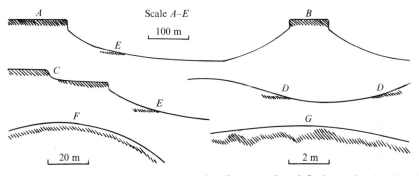

Fig. 9. Occurrence of laterite in relation to landforms. *A*, *B* and *C*: plateau laterite. *D* and *E*: valley laterite. *F* and *G*: laterite as a soil horizon following the relief.

159

them (fig. 9, *C*) (Lake, 1890; McFarlane, 1970; Sombroek, 1971). Thin laterite horizons can also occur in plateau situations, ending at the margins of valley sides.

Valley laterite occurs at valley-floor margins, and on pediments below a scarp (fig. 9, *D* and *E*). It is common in the broadly concave valleys of gently undulating plains in the savanna zone, becoming well-known when driving across them on earth roads. These sites are seepage zones, where soil water draining from the upper parts of the slope encounters the water table (cf. Watson, 1964/5, fig. 1). The laterite sheets do not extend far into the valley side.

Laterite horizons are also common neither in plateau nor valley-floor margin sites, but as soil horizons following the relief. (fig. 9, *F* and *G*). There is no corresponding term for such occurrences. They are particularly common with nodular horizons of 20–100 cm thickness (fig. 8(*b*)). The laterite frequently marks the boundary between soil and highly-weathered saprolite. Exceptionally, it can extend across valley floors.

ORIGIN

To explain the formation of laterite, it is necessary to account for the concentration of iron-rich material, and for its presence in forms in which hardening takes place. For some thin horizons, and for spaced nodular laterite, explanation in terms of processes within the soil profile is possible. But in most cases, including all thick crusts and many thin, this is too short a time scale, and the mechanisms of formation must be related to landform evolution.

Iron-rich material may originate by relative or absolute accumulation. In *relative accumulation*, concentration is the result of selective removal of silica and other constituents. In *absolute accumulation*, iron is brought to the site in solution, and precipitated.

Relative accumulation of sesquioxides takes place in all latosols, as indicated by low silica:sesquioxide ratios. It was formerly supposed that laterite was the product of a continuation of this desilication process. Evidence against this hypothesis is first, the existence of soils with similar chemical composition to laterites but no hardening properties; and secondly, that the micromorphological features of laterite clearly indicate addition of iron. Desilication is a contributory factor in producing a sesquioxide-rich soil as a proximate parent material for laterite.

An early stage in one type of absolute accumulation is seen in soils in which iron staining, mottling and concretions are found at the rock–soil transition. From the field evidence it is clear that iron is taken into solution by water draining through the soil, and some of it precipitated where this water meets weathering rock. Possible reasons are a rise in pH in this horizon, or reduced permeability with temporary drainage impedance. By continued weathering

160

of the rock within the horizon of iron deposition, and renewal of the overlying soil by termite action, this process could continue, the laterite being lowered together with the ground surface and weathering zone. This is one possible mechanism for the formation of nodular laterite horizons which follow the relief.

The most widespread mechanism of laterite formation is precipitation of iron within the zone of groundwater fluctuation. The savanna zone, with a regular seasonal rise and fall of the water table, provides the most favourable conditions, but the process can also occur in rainforest climates, as a result of short dry seasons of dry spells. During the period of high water table, iron is held in solution in ferrous forms; it may originate either by water draining from the soil above, or by anaerobic weathering within the saturated zone. When the water table falls this iron is precipated and oxidized. The resulting ferric compounds are relatively insoluble, and therefore remain through the next period of high water table, to be added to in the following dry period. There is clearly some cause of preferential deposition on existing concretions, giving the possibility of a progressive development from isolated concretions through non-cemented and cemented nodular to massive laterite. Quartz grains and fragments become incorporated in the concretionary material. Ferruginized rock results where the zone of groundwater fluctuation is within the bedrock.

This mechanism for the solution, separation and concentration of iron can readily be reproduced on a miniature scale. Half fill a glass beaker with red or yellowish red tropical soil, and fill to the brim with distilled water in which a small amount of glucose has been dissolved. Stand on a radiator, cover to prevent evaporation, and keep the soil waterlogged for a few weeks. Then remove the cover and allow the soil to dry out completely. Fill again with water, allow to dry, and repeat this cycle a number of times. The soil surface and the sides of the beaker become coated with ferric iron, whilst the soil is progressively bleached. Some concentration of iron towards the upper part of the soil may be visible. This initial addition of glucose is needed as a food for bacteria; without it, the effect is very much slower.

It is commonly supposed that iron deposition and hardening are successive processes well separated in time; this idea derives particularly from plateau laterite, on the assumption that iron concentration took place on plains, and hardening only followed uplift, dissection, and consequent lowering of the water table. There is increasing evidence that this is not the case; in particular, the rarity of soft laterite at the present day, combined with the frequency of hard forms, conflicts with the principle of uniformitarianism. It is more probable that concentration and hardening are simultaneous or separated by a relatively short period of time.

An unusual example which clearly illustrates contemporary laterite formation occurs in a part of the Mato Grosso, Brazil (fig. 10). The landscape

161

Laterite

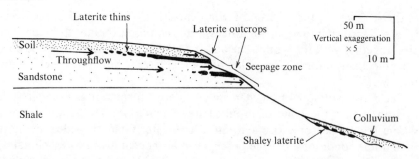

Fig. 10. An example of contemporary laterite formation. Based in part on Townshend (1970).

consists of a gently-undulating plateau formed in horizontally-bedded Palaeozoic sandstones, dissected by valleys. Near the upper margin of the valley sides a step and bench occurs, on which there is an outcrop of cemented nodular laterite. Much seepage takes place on these steps, as a result of less permeable strata below. This laterite has the appearance of a continuous stratum, outcropping over a large area of such valleys. On tracing it back into the plateau by auger, however, it is found to extend only about 20 m, first giving place to an iron-rich mottled clay and then becoming unidentifiable. This is a clear example of a laterite formed by seepage. As the valley side retreats it is eroded, but is maintained by concurrent extension back towards the interfluves (Townshend, 1970).

LATERITE AND LANDFORM EVOLUTION

In considering the origin of thick crusts of plateau laterite, there is a problem of the source of so much iron-rich material. This may be illustrated by calculations for the Buganda laterite of Uganda (Trendall, 1962). This laterite is 10 m thick and overlies granite. The iron content of the granite is 600 kg/m³ and that of the granite 42 kg/m³. Thus, 14 m of granite would be required to produce 1 m of laterite. Allowing for a 30 percent loss by solution, some 200 m of rock must have been needed to produce the observed crust. This calculation neglects the possibility of derivation from a former more iron-rich overlying rock. Whilst such may apply in a few cases, the phenomenon is too widespread for it to be a general explanation.

There is thus a problem that the thickness of rock necessary to provide the iron (and aluminium) present in laterites usually much exceeds the depth of soil. In some cases it also exceeds the thickness of the pallid zone. The following are possible explanations:

(i) The iron originates by downward leaching, from soil above the laterite; the water table originally lay at a considerable depth within bedrock or deeply-weathered saprolite (fig. 11(*a*)).

162

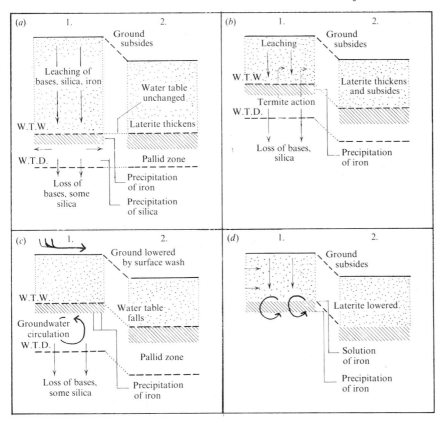

Fig. 11. Theories of the origin of laterite. W.T.W. and W.T.D. = wet season and dry season water-table positions. For explanation, see text.

(ii) The iron originates by downward leaching from the soil above. The soil at any given time is relatively thin, but is constantly renewed by termites bringing up material from beneath the laterite; the laterite subsides as this material is removed (fig. 11(*b*)).

(iii) The iron originates from below the laterite in particular from the pallid zone where present; there is also solution loss, to groundwater, from this zone, and the laterite subsides as material is removed (Trendall, 1962) (fig. 11(*c*)).

(iv) The iron originates from any of three sources: the soil above, the pallid zone below, and lateral movement of iron-charged groundwater; lowering of the laterite is not a physical process of sagging, but is due to continuous solution and redeposition of iron (De Swart, 1964) (fig. 11(*d*)).

The first hypothesis cannot be applied to thick laterite sheets, as it is unlikely either that the water table would be initially located at the necessary

163

depth or that it would remain unchanged in height, with respect to bedrock, for a sufficient period of time. The second involves attributing a dominant role to termites. The second, third and fourth are not mutually exclusive. It has not been established which of these hypotheses is the most widely applicable.

It was formerly considered that the existence of a nearly level erosion surface predated, and was a necessary precondition for, the formation of a laterite sheet. It is more likely, however, that the formation of the surface and of the laterite were simultaneous. The starting-point for models representing laterite formation should therefore be an irregular land surface, of substantial relief, and not a level plain. To make such an assumption removes some of the difficulties in constructing an internally consistent scheme of evolution.

One such model is based on the most detailed study of the field relations of laterite that has yet appeared, by McFarlane (1970). In Kyagwe, Uganda, reconstructed contours of hill summits capped by plateau laterite reveal a gently undulating surface, with a relief of 200 m. On this surface there is a sequence, proceeding downslope, of massive vesicular, spaced nodular and non-cemented nodular laterite. McFarlane suggests that the spaced nodular laterite forms first, as a soil horizon on moderately sloping relief (fig. 12). It is converted to non-cemented (packed) nodular by loss of the intervening soil; the massive laterite results from solution and re-cementation of the nodular. Both these processes accompany a reduction in relief and a lowering of the laterite together with the ground surface, the massive laterite developing simultaneously with the gently undulating surface. The situation is complicated by the presence of two successive erosion cycles.

The origin of the pallid zone is not determined. Du Bois and Jeffrey (1955) considered it was produced by silica deposition, simultaneously with iron deposition in the laterite horizon. In Trendall's hypothesis of laterite formation it is a residual horizon, originating by removal of sesquioxides upwards to the laterite. McFarlane (1970) views it as a weathering zone formed after, and as a result of protection from erosion by, the laterite capping.

Despite the continuing problems concerning fundamental questions of origin, one general aspect has become clear. This is that once formed, laterite is capable of persisting in the soil for long periods, without losing its identity as a continuous horizon; and that as a consequence, much present-day laterite is wholly or in part relict, dating from Pleistocene and Tertiary times. This is not to say that contemporary formation of laterite is absent; some thinner horizons can be explained without recourse to climatic change and accompanying phases of erosion and weathering, whilst the contemporary process is most clearly seen in some seepage laterites. But some thin

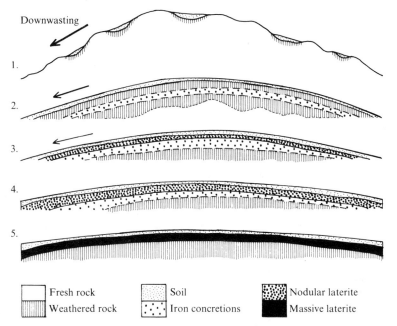

Fig. 12. Theory of laterite evolution and landform downwasting. After McFarlane (1970).

horizons, and all thick crusts, have evolved over the order of time required for landform evolution, rather than the shorter period required for soil profile development.

LATERITE, AGRICULTURE AND SOIL SURVEY

The agricultural problems set by the presence of laterite are considerable, but not complex. Massive or cemented nodular types act as a limiting horizon, preventing root penetration. Non-cemented nodular types do not prevent some roots from reaching within and beneath the laterite, but severely restrict their growth. A guide as to the likelihood of roots being able to penetrate an horizon is whether this can be done with a screw auger. The high proportion of stones and gravel reduces the clay content, which lowers both the chemical activity and the available water capacity of the soil.

There is little that can be done to ameliorate a laterite horizon. For high-value tree crops it may be worth costing the digging of individual holes, refilling with soil and using the laterite for terrace banks. Other than in such special circumstances, the main need is to conserve the overlying soil by preventing sheet erosion.

The agricultural value of soils with laterite depends largely on the thickness of the overlying soil. If it averages less than 50 cm it is usually preferable, where controlled land use planning is possible, to allocate the land to wet-season grazing (thereby providing a resting period for regeneration of valley-floor pastures used in the dry season). Under cereals or root crops yields are low, and the land will give a poor response to inputs, whereas satisfactory growth of grasses can be obtained. If the grass is to fulfil its further purpose of controlling erosion, limits on grazing intensity are necessary.

With laterite at more than about 50 cm depth, soils can be moderately productive. An example is the Malacca Series of Malaya, which has a thick horizon of massive laterite and yet is extensively used for rubber.

The variation in depth and thickness of laterite horizons presents special problems in soil survey. The first is the identification of soil series. It is both unnecessary and highly inconvenient to have a series distinguished by a criterion such as, for example, whether the laterite is more than one metre thick. Once it has been established by augering that there is massive laterite or at least 30 cm of nodular material, the only agriculturally relevant facts concern the thickness and properties of the overlying soil. To dig deeper is a pedological refinement, justifiable only for a few type pits.

Irregular variations in the depth, or presence, of laterite call for special soil survey design, purpose orientated, statistically controlled, and costed. The first aim is to establish, by sample surveys, the proportion of the mapping unit covered by laterite at various depths. Only subsequently, if purpose and costs justify it, is detailed mapping attempted.

A compensation for the agricultural limitations of laterite is its merit in road construction. It provides an excellent source of durable material for road foundations and surfacing, the more sought after since fresh rock is more expensive and unavailable, owing to deep weathering, over large areas. Nodular laterite is the most desirable, and local roadmen develop an ability to select quarry sites that will yield an abundant supply. Occasionally it is possible on plateaux to use undisturbed laterite as a natural foundation, although this is frequently prevented by dissection. (See e.g. Clare and Beaven, 1962, 1965; Dowling, 1968.)

Laterite was formerly used as a building stone, notably in the temples of Angkor Wat, Cambodia (see illustrations in Prescott and Pendleton, 1952). It is still widely used in Northern Nigeria. Another example is Portuguese buildings in Malacca, Malaya. It is both highly durable and visually attractive. We do not know whether such stones were initially, as is commonly supposed, soft laterite, or whether they were shaped from hard, massive material. It is reported that a little soft laterite occurs low in the catena in Malacca, and 'small quantities are still quarried at the present day' (Panton, 1956, p. 419). Had soft laterite been common it would have been an excellent resource for building stones. The difficulty and cost of shaping hard forms prevents its normal use today, although it could form an attractive ornamental architectural component.

DURICRUSTS

Laterites belong to a larger class of indurated regolith materials called *duricrusts*. The types of duricrust are:

Ferricrete: formed mainly of iron; synonymous with massive laterite.

Bauxite: formed mainly of alumina (called *alcrete* by Goudie, 1973*a*).

Silcrete: formed mainly of silica.

Calcrete: formed mainly of calcium carbonate; occurs only in semi-arid and arid regions.

Calcrete and silcrete are found in semi-arid to arid tropical and subtropical climates. Bauxite and laterite are found in the humid tropics.

Dury (1969) proposed a triangular diagram based on the relative percentages of Fe_2O_3, Al_2O_3 and SiO_2. The corner members are ferritic, allitic and silitic duricrusts, and the sides define ferrallitic, fersilitic and (hypothetical?) siallitic types, whilst materials of balanced composition are fersiallitic duricrusts. This scheme is logically tidy and draws attention to the existence of a continuum of variation in composition; it is of little use in the field as, apart from the extreme aluminous and siliceous types, composition cannot be identified visually. In normal usage the silitic duricrusts are silcretes, the allitic duricrusts bauxite, and the four types that include 'fer-' are laterites.

Calcretes are layers of secondary calcium carbonate. They occur in the dry tropics and are discussed in the context of calcimorphic soils below (p. 198).

Silcretes are indurated layers consisting predominantly of secondary silica. They are not very widespread, being reported mainly from Australia and southern Africa. They are hard and can form breakaways. They appear to be associated with semi-arid climates (rainfall about 500 mm or less), but like laterite may frequently be relict forms. The conditions which lead to the accumulation and induration of secondary silica have not been determined.

Bauxites are morphologically akin to laterites but with alumina predominant over iron oxides. The major countries where they are exploited for alumina are Jamaica, Guyana, Surinam, Guinea, and in the temperate zone in southern France, Hungary, and parts of the USSR and the USA. This appears to indicate a strongly localized distribution (although the large investment required for exploitation is no doubt in part responsible). Bauxite appears to be associated with very humid climates (present or past), and to be most readily formed over basic rocks, the latter being one reason for its localization. The origin is apparently some process of leaching so intense that much iron as well as silica is removed leaving alumina, together with greater or lesser amounts of iron, as a residual accumulation.

STONE LINES

Stone lines are soil horizons which have a greater proportion of stones than the material above and below. They range from horizons consisting almost entirely of stones and gravel to isolated stones concentrated at a certain depth. Typically the stone line is 10–100 cm thick with 50–200 cm of overlying soil. The depth often varies over short distances, giving an irregular wave-like appearance in road cuttings. Stone layers mantling the surface and stony horizons that pass downward into less fragmented rock are excluded from the definition.

In the tropics, stone lines usually consist mainly of quartz fragments or a mixture of quartz and iron concretions. A small percentage of other rock fragments may be present. Horizons of nodular laterite usually contain quartz fragments, and hence the distinction between these and stone lines is an arbitrary matter of nomenclature; there is no standard practice, and a division at 50 percent laterite (compared to all other material) is suggested.

Stone lines occur under climates ranging from semi-arid to rainforest, but are most common in the savanna zone. They appear from published descriptions to be especially common in West Africa, but this may be an academic breeder-reaction. They also occur in temperate regions, including the US, where they were first defined (by Sharpe, 1938). But in temperate latitudes they are exceptional, whereas in the tropics they are a common, almost normal feature of the soil profile. Basement Complex rocks are the most common parent material, since these frequently yield quartz fragments.

They may occur on level and gently-sloping ground, but are particularly characteristic of moderate to steep slopes. They also occur on pediments, including those below retreating breakaways where laterite blocks are incorporated in the upslope parts of the stone line. Double stone lines are occasionally found, especially on pediments, and instances of one line crossing and truncating another below have been seen.

Both the upper and lower boundaries are well defined, passing from stony to non-stony material in a few centimetres. The lower boundary of the stone line is very often the upper limit of weathered rock *in situ*, as indicated by undisturbed quartz veins and other rock structures. Within the stone line itself, rock structures have been destroyed.

There are two groups of explanations for the origin of stone lines: those which call upon current processes (steady-state explanations), and those which invoke past climatic change. In the first group, three mechanisms have been suggested:

(i) Soil creep. Quartz and other fragments enter a moving creep layer at its base, and remain there. As weathering, erosion and slope retreat progress, more rock fragments are taken into the creep layer and redistributed as an horizon at its base (Ireland *et al.*, 1939). Clear evidence that this is at least a partial explanation is provided by sections in hillslopes where quartz veins 'feed into' stone lines, sometimes with a short transition zone of outcrop curvature. It helps to account for the greater frequency of stone lines of slopes. But irrespective of whether movement in creep approximates to laminar or turbulent flow, stones would become distributed throughout the moving soil unless there were some further mechanism operating.

(ii) As a residual weathering product. As weathering works downwards into the rock, other material is lost in solution whilst quartz resists weathering and accumulates. This is a possible contributory mechanism, and can apply on level ground as well as on slopes. But like the preceding explanation it is incomplete as a model, since its logical result would be an accumulation of stone on the ground surface.

(iii) Termite action. Weathering differentiates the soil into coarse fragments and the fine fraction. Part of the fine fraction is lost in solution, but part is selectively removed into the horizon above by termites. Not only is this the most logically satisfactory of the steady state explanations, but it accords with what is known of termite action. Stone lines occur at approximately the depth commonly reached by termite channels. Nye (1955) found that the maximum particle size in the soil overlying the stone line (4 mm) was the same as that in termite mounds. By extrapolation from the present-day rate of termite casting Williams (1968*a*) calculated that the soil above stone lines in part of Northern Territory, Australia, could have been wholly deposited by termites in 12000–18000 years; the estimate of casting rate by Nye (1955) in West Africa gives a similar order of time. This explanation incorporates that of residual weathering and can, but need not necessarily, take place in conjunction with downslope movement of the soil by creep. The strongest evidence in favour of termites as the main mechanism is the restriction of stone lines as a frequent feature to the tropics.

Opposed to steady state explanations are those involving climatic change. During an 'unstable' phase of the landscape, with a drier climate than the present, scarps retreat leaving pediments, and a lag gravel is left on the pediment as fine material is removed by wash and coarse material remains. Such layers are known in semi-arid regions. The climate becomes more moist, giving a 'stable' period of landscape development, and finer material is deposited above the stony layer. In examples described by Fölster (1969) an horizon of fine gravel, interpreted as a deposit laid down prior to the fine-grained hillwash, overlies the stone line proper. Termite action during the humid period then further concentrates fine material toward the surface.

A further possibility is to attribute the origin of the stony layer to a former dry period, and that of the soil above not to deposition but solely to termite action.

169

Laterite

There is no doubt that both types of explanation are locally applicable. It is clear from many descriptions of complex layering that soils in West Africa have been subject to variations in erosional intensity caused by climatic change. Complications such as the morphological relations with laterite, and the inclusion of rocks or minerals that are not present in the bedrock beneath make stone lines a fine subject for academic study. The interest lies not only in the origin of the stone line itself, but in what it tells of the evolution of the overlying soil.

With respect to agriculture, stone lines have similar effects to an horizon of non-cemented nodular laterite: to restrict root growth and lower the clay content, with consequences for nutrients and moisture retention. It is possible that stone lines have a greater world extent than laterite. They are, however, less of an agricultural problem because the gravel content is usually lower, they are non-cemented, and, unlike laterite, are more frequent on sloping than on level land.

SELECTED REFERENCES

Laterite. Prescott and Pendleton (1952), D'Hoore (1954*a*, 1954*b*), Sivarajasingham *et al.* (1962), Trendall (1962), Aubert (1963), Maignen (1966), Pullan (1967), McFarlane (1970).

Duricrusts. Dury (1969), Stephens (1971), Goudie (1973*a*, 1973*b*).

Stone lines. Ruhe (1959), Fölster (1969), Riquier (1969).

CHAPTER 8

OTHER PEDALFERS

Grouped together under the heading of other pedalfers are those freely-drained soils of the humid tropics which have special characteristics that set them apart from the latosols. These characteristics are the presence of a bleached Ae horizon (tropical podzols), a high proportion of glassy material of volcanic origin (andosols), or a high sand content (arenosols). Lithosols and regosols, whilst found in both the humid and dry tropics, are for convenience included in this chapter.

TROPICAL PODZOLS

CCTA: Podzolic soils
FAO: Orthic and humic podzols; albic arenosols

Soils having a bleached Ae horizon and some accumulation of secondary humus or iron occur in two situations in the tropics: in humid lowlands on very sandy parent materials, and on mountains.

Lowland tropical podzols, sometimes referred to as 'bleached sands', occur mainly in tropical evergreen and monsoon rainforest climates, and less commonly in the moist savanna zone. Rainfall is most frequently 1200–1700 mm, may be higher, and there is usually a substantial moisture surplus for 6–12 months (Klinge, 1969). The parent material is invariably sandy, sometimes over 95 percent sand, and very permeable. Raised beach sands, alluvial terraces, and colluvial sands of lower valley sides are common materials. There is often a distinctive edaphic climax vegetation, called 'heath forest'. They have been described particularly from Borneo, Malacca, northern Australia, the Guianas and Amazonia, but are uncommon in Africa. Their total extent has been estimated at 4 000 000 ha, or about 0.1 percent of the tropical zone, plus a similar area in subtropical Australia and Florida (Klinge, 1966).

The profile has a surface accumulation of raw humus, poorly decomposed, acid and with a high carbon : nitrogen ratio, below which is a humus-stained sandy horizon. There is then a light grey to white Ae horizon, usually a sand, the thickness of which can vary from 20 to 50 cm to over 2 m ('giant podzols'). The illuvial horizons are variable. A dark brown Bh horizon is common, often patchy or irregular in depth, and relatively thin. Less frequently there is some secondary concentration of iron or aluminium, often as a mottle or soft concretions. They may be slightly more clay in the illuvial horizons, but they are still very sandy. Sometimes a hardpan occurs, possibly of recemented silica. The brightly-coloured sesquioxidic B horizons of temperate podzols do not occur. The profile is structureless, and the reaction moderately to very strongly acid.

The probable origin is that the presence of permeable sands under a high rainfall leads to strong leaching of any bases that may be present initially, and the growth of an acidophyllous vegetation. This produces raw humus, intensifying the leaching and assisting in the removal of iron oxides released on weathering. Andreisse (1969/70) has described a chronosequence in which rubefication, through the release of iron by weathering of ilmenite, precedes the leaching of iron. Deposition of humus, and sesquioxides if present, occurs either where there is a textural change (of sedimentary origin) or at the water table. The reason that iron B horizons are infrequent is that this type of development sequence only occurs on highly siliceous parent materials. Thus, lowland tropical podzols are an intrazonal soil type, caused by the presence of highly permeable parent material under a humid climate.

Montane tropical podzols occur on some mountains with a high rainfall. Askew (1964) describes their occurrence as an altitudinal zone in Borneo. The parent rocks are shales and sandstones. At lower altitudes there are pale yellow ferrallitic soils under dipterocarp forest. This passes into micropodzols at 1200 m, the podzol zone at 1450–1700 m, and peaty gleys under moss forest above this. These zones are accompanied by changes in vegetation, the podzols coinciding with the presence of conifers, e.g. *Agathis*. There is an accumulation of raw, mor-type humus. The profile is less than 1 m deep, and consists mainly of a white Ae horizon. Organic matter and iron oxide deposition occurs patchily on weathering rock fragments. A pH of 3.0 is reported. This soil type appears to be the result of a climatically-induced vegetation change, in association with a siliceous, but not excessively permeable, parent rock.

Most lowland tropical podzols are classed as *albic arenosols* on the FAO system, only falling into the podzol class if there is a sesquioxide B horizon within 125 cm, which is not usually the case. Tropical montane podzols qualify for the podzol group, there being no specific soil unit for tropical profiles.

Agricultural properties

Lowland tropical podzols have a very low agricultural value, and are generally uncultivated. Cassava can be grown, with periodic bush fallowing.

A better form of land use, if these soils are to be developed at all, is for forestry, possibly using quick-growing softwoods or species that produce straight poles.

Podzolization and 'laterization'

From a pedogenetic viewpoint, the interest of tropical podzols is that leaching of sesquioxides takes place, contrary to their tendency to accumulate within the latosols. It has been suggested that there is only one fundamental process in humid-climate soils, that of podzolization, and that red tropical soils represent an early stage of this (Carter and Pendleton, 1956). Andreisse (1971) described soil types on five different rocks under the same climate, in Sarawak, each of which showed some of the features of podzols; he argued from this that podzolization could follow laterization, as the soil becomes progressively more acid.

This view is incorrect. Strongly leached ferrallitic soils, very low in cations and with a pH of 4 or less, show no tendency to develop an Ae horizon; the ferric iron is present in highly stable forms. The bleached horizons of lowland tropical podzols result when virtually all constituents – including non-quartz silica and such iron and aluminium oxides as may be present – are removed by intense leaching, leaving behind only a skeleton of the quartz sand grains. If the parent material is less siliceous or less highly permeable, secondary minerals are formed, the soil provides a richer medium for plants with consequent recycling of nutrients, and podzolization does not occur. The occurrence of montane podzols where the acid litter-producing vegetation is in part climatically induced is consistent with this explanation.

ANDOSOLS

CCTA: Juvenile soils on volcanic ash, eutrophic brown soils on volcanic ash
FAO: Andosols

Andosols are genetically defined as soils derived from volcanic ash, containing a high proportion of vitric (glassy) material. The term is derived from Japanese *an do*, 'dark soil', referring to the characteristic thick blackish A horizon. Other terms are volcanic ash soils and humic allophane soils. Andosols are azonal soils, with properties still dominated by characteristics derived from the parent material. Old volcanic ash deposits may have acquired the properties of zonal soils. Andosols occur from polar to equatorial latitudes, being particularly widespread in Japan. In the tropics they cover substantial areas in Java, Sumatra, Hawaii and Colombia, and occur also in other islands of the East Indies, and the Pacific, Ecuador, El Salvador, and scattered localities in East Africa.

Andosols have a thick (e.g. 50 cm) dark greyish brown to black friable A horizon. Typically they have a high silt content, although profiles dominated by fine sand also occur. There is an open granular structure, high porosity and unusually low bulk density. The consistency is greasy but neither sticky nor plastic. This dark layer often passes directly into a yellowish brown C horizon, in which some measure of depositional layering may be present. There may be a transitional brownish (B) horizon, but there is no appreciable clay translocation.

The dark horizon has a high, well-distributed organic matter content of 5–20 percent. The pH varies from 5.0 to 7.0, cation exchange capacity is moderately high (20–50 m.e./100 g) and base saturation varies from high to medium.

The unusual properties are mainly associated with a high proportion of amorphous mineral colloids. This material, amorphous to X-rays, is commonly referred to as 'allophane' although there are problems in the precise definition of this term (Swindale, 1965). In the silt fraction, opaline silica may be present (Shoji and Masui, 1971). Allophane has a large active surface, which can combine with organic matter and absorb much water. On weathering, allophane is transformed into metahalloysite. Soils formed from feldspathic volcanic ash, without allophane, have been recorded in the Sudan (White, 1967).

An horizon of silcrete, a hard, silica-cemented pan at about 100 cm depth, has been reported from Indonesia. Silica is dissolved in the topsoil and precipitated below. The silcrete is relatively impervious, giving drainage impedance and possibly causing precipitation of iron in the soil above. Mohr and Van Baren (1954, pp. 300–4) have described an elegant model of successive stages of soil formation on basic volcanic ash under a rainforest climate, based on the secondary effects of a silicified tuff horizon (p. 54).

Andosols are of particular pedogenetic interest in that they provide opportunity to study successive stages in soil formation commencing with a known parent material. The transformation can be quite rapid, fresh ash being converted into deep, productive andosols in as little as ten years (Colmet-Daage, 1967). Under a rainforest climate the chronosequence is generally fresh ash—andosols—eutrophic brown soils—(dark red) ferrisols or strongly leached ferrisols, possibly continuing towards leached ferrallitic soils under very high rainfall. Allophane is transformed simultaneously to gibbsite and kaolinite under extremely humid conditions, and to halloysite with gibbsite where there is a dry season; the halloysite subsequently evolves by dehydration to metahalloysite (Dudal and Soepraptohardjo, 1960; Sieffermann, 1973).

Agricultural properties

Andosols derived from basic (andesitic) ash have a high agricultural potential. They have a high available water capacity and, except when silcrete is present, are freely drained. The natural fertility is high, and is renewed by weathering of minerals. There is a problem of phosphate fixation. The combination of a fine texture, high available water capacity, good drainage and satisfactory nutrient status makes these soils particularly favourable for sugar cane. Tobacco, sweet potatoes and padi are also suitable, and the soils have a good repayment potential for irrigation. Coarse-textured andosols, those derived from felsic material and those with a silcrete horizon are less productive.

ARENOSOLS

CCTA: Ferruginous tropical soils on sandy parent materials, ferrallitic soils on loose sandy sediments
FAO: Arenosols (mainly ferralic)

Where the parent material is extremely sandy the morphological properties of the resulting soil are dominated by its texture. Very sandy soils can occur over coarse-grained sandstones but are more common on drift deposits, e.g. marine, wind-blown or alluvial sands. Such soils are distinguished as *arenosols*. Their definition is that all horizons must have less than 15 percent clay, with the additional requirement (to eliminate highly silty soils) that there should be over 50 percent sand. Thus, all sands and loamy sands are included, together with some sandy loams. The name arenosol is taken from the FAO classification. It is useful to cover soils which are not regosols, since they have a mature profile, but which do not have the usual properties of latosol classes owing to the predominance of sand.

Morphological properties consequent upon the sandy texture are that the profiles are structureless (single-grain to very weakly massive) and have a loose to highly friable, fragmentary consistency. Colours are usually yellow to yellowish red, beneath a yellowish brown topsoil. Horizon boundaries are transitional. Because such clay as is present moves easily between the sand grains there is commonly an increase in clay with depth, i.e. the topsoil is still more sandy than the lower horizons.

Arensols may be subdivided according to the latosol class to which they would have been assigned had the texture been less sandy. *Ferruginous arenosols* have a subsoil base saturation of more than 40 percent, and are weakly to moderately acid; *ferrallitic arenosols* have below 40 percent saturation and are strongly acid. The climatically-induced transition between these subclasses takes place at a rainfall of about 1000 mm. Clay mineral composition varies in accordance with that of ferruginious and ferrallitic soils respectively, i.e. small amounts of 2:1 lattice minerals may be present in the former. Organic matter levels are less than those of more clayey profiles in corresponding environments.

Many soil maps do not recognize very sandy profiles as a class. It is nevertheless clear that these soils are of considerable extent. On the *Soil Map of Africa* the subclasses 'ferruginous soils on sandy parent materials' and 'ferrallitic soils on loose sandy sediments' occupy respectively 4 and 5 percent of the continent. They are particularly widespread around the equatorward margins of the subtropical deserts; during drier phases of the Pleistocene, vast sand deposits were spread across areas that are now relatively humid, as evidenced both by fixed dunes and more evenly-spread sand sheets. Thus in West Africa, some areas mapped on the French classification as 'sols ferrugineux tropicaux' have over 85 percent of sand, sometimes as high as 98 percent (White, 1971). In Australia, 'sand soils' are recognized as a widely-distributed profile type. Marine sands, either on raised beaches or on sand bars created by marine progradation, are another distinctive environment; miniature catenas develop on successive bars and depressions, conditioned by the high, sometimes saline, water-table in the latter. Arenosols on both raised beaches and sand bars also occur around lakes, e.g. Lake Nyasa.

REGOSOLS

CCTA: Juvenile soils on recent deposits (mainly on wind-borne sands)
FAO: Regosols

Regosols are soils developed from unconsolidated materials, usually sands, that possess little or no profile development. The distinction from arenosols depends on whether there is an identifiable B horizon. Regosols may possess a weakly developed Ah horizon with less than 1 percent organic matter. Coastal dune sands and wind-blown sands of desert margins are the two main environments.

Because of the sandy texture, regosols have a low available water capacity and a low nutrient content. Other properties vary with climate and the groundwater table. In humid regions they are strongly acid but under dry climates they may contain secondary calcium carbonate, or become saline, in depth.

Agricultural properties of arenosols and regosols

The dominant agricultural consequences of the sandy texture of these two soil classes are ease of cultivation and rooting, but low moisture retention

176

and nutrient content. The field capacity may be as low as 10 percent of moisture, tending to give physiological drought in areas which climatically are moderately humid. Also because of low clay content, the cation exchange capacity and ability to retain added nutrients are low. Cassava, with its tolerance of low nutrient levels, is a common subsistence crop, and groundnuts can also be grown on some arenosols. Under high rainfalls the very severe leaching of ferrallitic arenosols prohibits the attainment of high yields from perennial crops. Regosols in coastal situations are most suitably kept under their traditional use, coconuts. An alternative productive use is that of softwood plantations.

Where arenosols and regosols occur under a 500–1000 mm rainfall, as is common in the steppe zone, they give rise to an appearance of semi-aridity for which irrigation appears an obvious remedy. The low moisture-retaining capacity requires, however, that water be provided in small, frequent applications; moreover there is often a micro-relief of dunes, sand bars or relict channels. Sprinkler or trickle irrigation solves these technical problems, but with the infertility of the soil it is difficult to obtain the high productivity necessary to justify this on economic grounds. Fine sands are less ill-suited to irrigation than coarse, and detailed special-purpose soil survey may be desirable to identify such differences. With a rainfall over 750 mm the entire profile is raised to its (low) field capacity quite early in the wet season, and improvement in methods of dry farming may be a better form of investment. It could be said that these soils need clay more than water.

LITHOSOLS AND LITHIC SOILS

CCTA: Rock and rock debris, lithosols and lithic soils
FAO: Lithosols and the lithic phase of other soil types

Lithosols may be described in broad terms as soils which are too shallow or stony for agriculture. Usually there is a limiting horizon of consolidated rock or massive laterite close to the surface, the overlying soil sometimes also being stony. Less commonly the profile is deeper but dominated by gravel, stones or boulders. They are the most extensive soil unit in the tropics, and indeed the world. The *Soil Map of Africa* maps 19 percent of the continent as lithosols and lithic soils, plus a further 6 percent as desert pavements. On the FAO map of South America, 11 percent is occupied by soil associations dominated by lithosols.

The cause of lithosols may be climate, relief, or a combination of these factors. They are most extensive in deserts, in association with regosols, and mountainous regions. In the semi-arid and savanna zones, most slopes of over 25° and many of 10°–25° are covered by lithosols, commonly with stones or boulders. Quartzites produce shallow and stony profiles even on gentle slopes. In the rainforest zone, slopes of up to 40° may carry a soil cover of adequate depth for cultivation and lithosols are confined to quartzites and the near-vertical slopes of limestone inselbergs. A further common situation is that of lithosols on almost horizontal surfaces produced by a plateau capping of massive laterite.

A distinction is commonly made between lithosols *sensu stricto* and lithic soils, the former being very shallow whilst the latter have sufficient profile depth to be recognizable as shallow phase of other soil types; deep but very stony profiles form a third sub-group. It is also possible to sub-divide according to whether the limiting horizon, or stones, consist of rock, massive laterite or concretionary laterite. Depth limits are necessarily arbitrary but the following, based on agricultural limitations, offers a convenient standard:

Consolidated rock (including ferruginized rock) or massive laterite at <25 cm, irrespective of stoniness of overlying soil (very severe agricultural limitation): *lithosol.*

Consolidated rock commencing at 25–50 cm, irrespective of stoniness of overlying soil (severe agricultural limitation): *lithic phase.*

Massive laterite commencing at 25–50 cm, irrespective of stoniness of overlying soil (severe agricultural limitation): *petroferric phase.*

Consolidated rock or massive laterite at 50–100 cm (moderate agricultural limitation): *shallow phase* of other soil types.

Depth to rock >50 cm, but upper 50 cm dominated (>40 percent by volume) by stones, boulders or gravel, consisting mainly of rock fragments (mechanized cultivation impossible): *stony phase.*

Depth to rock >50 cm, but upper 50 cm dominated (>50 percent by volume) by hard laterite concretions: *petric phase.*

These definitions are in general accord with those of the FAO classification, except that in the latter the boundary between lithosols and the lithic or petroferric phases is placed at a depth of only 10 cm, and the shallow phase is not recognized. In most circumstances, 25 cm is a more suitable depth than 10 cm at which to deem a soil uncultivable.

If such subdivisions are made, it frequently becomes necessary in soil survey to map compound units, e.g. 'lithosols and lithic soils', 'series X including concretionary phase'. The problems of mapping laterite were discussed above (p. 165). With respect to soils limited by depth to rock, there is frequently the further agricultural limitation of steep slopes, and in soil survey such areas are normally dismissed on casual inspection. More careful survey becomes necessary where there may be potential agricultural land but soil depth is marginal. It is necessary to establish the proportions of the area in various depth classes and state these, erring on the cautious side with respect to confidence limits, in the map legend. This is a circumstance in which survey by statistically-controlled areal sampling is called for.

In normal circumstances lithosols are uncultivated. They have a resource potential for wet-season grazing, as forest reserves (for indigenous woodland rather than planted species) and as water catchment areas. The unsuitability for cultivation is due to the depth limitation and erosion hazard, for the level of natural fertility of hillslope soils is often higher than on gently sloping land in the same climate, owing to the input of weathering minerals. An unfortunate consequence is that one or a few good crops can be obtained from a steeply-sloping hillside, followed by a severe danger of soil erosion; there is a temptation for this to happen where population pressure presents the need and laws forbidding it are not enforced.

By intensive use of hard labour, steep slopes with thin and stony soils can be transformed into cultivable land, through terracing, the removal of stones by hand and their use as terrace fronts. Such practices are found for example in the Himalayan foothill zone of Pakistan. An unusual situation occurs on the margins of the Jos Plateau in Nigeria. Former slave raiding by Fulani horsemen led to the retreat of subject peoples into the hills. A tradition was established of cultivating very stony land, which persists to this day. Stones and small boulders are piled into conical heaps, the larger boulders left, and good crops of sorghum sown individually in the soil between. The local belief that the sorghum 'feeds on the stones' shows an appreciation, unusual among farmers, of the value of weathering minerals in the soil.

SELECTED REFERENCES

Tropical podzols. Andriesse (1968/9, 1969/70), Askew (1964), Bleackley and Khan (1963), Klinge (1965), Richards (1941).

Andosols. FAO (1965a), Sieffermann (1973).

CHAPTER 9

VERTISOLS

Vertisols are *sui generis*, in that whilst they frequently contain both free calcium carbonate and features associated with drainage impedance they do not necessarily possess either. In table 18 they are grouped with the calcimorphic soil types, but they may alternatively be regarded as intra-zonal.

The last part of this chapter is concerned with soils on limestones in the humid tropics, not a single soil type but a range from rendzinas, with marked calcimorphism, to acid latosols. Zonal calcimorphic soils, those restricted to semi-arid and arid climates, are treated in chapter 10.

VERTISOLS

(Profile 5, p. 432)

CCTA: Vertisols
FAO: Vertisols

Vertisols are dark-coloured clays which expand and contract markedly with changes in moisture content, and develop deep vertical drying cracks. These distinctive features have led to their recognition by a large number of local names, vernacular and otherwise, of which the more widely used are black earths, black cracking clays, black cotton soils, *regur* (India) and *gilgai* soils (Australia). Of the scientific names, vertisol (from Latin *vertere*, to turn), indicative of their property of self mixing or 'turning over', has gained preference over the earlier US term *grumusol*.

These soils are of special interest from the points of view of both origin and agriculture. Pedologically, the fact that they are dominated by mont-morillonitic clays contrasts with almost all other soils of the tropics. Agri-culturally, they are intensively cultivated and highly productive in some areas, yet in others remain under grazing.

Vertisols cover a world area estimated at 257 million hectares (Dudal, 1965). Their full extent is to latitude 45° but some 230 million hectares occur between latitudes 30° North and South. The largest areas are in east-central Australia, the north-west part of the Indian Deccan, the Sudan, Ethiopia and Chad. Extensions poleward of 30° latitude occur in the three southern hemisphere continents, in the US and around the Black Sea.

Properties

There is a deep and uniform A horizon, 30–100 cm thick, consisting of a dark grey or very dark greyish brown clay, sometimes with a faint, diffuse olive-brown mottle. This possesses a very strongly developed, very coarse prismatic or blocky structure, with the main peds of the order of 15 × 15 × 30 cm. Usually these larger peds are subdivided by less marked planes of weakness into smaller blocky aggregates. When the soil is dry there is a polygonal pattern of deep vertical cracks on the surface. Peds have polished and grooved surfaces. These are not illuvial clay skins, but are produced when peds surfaces slide across each other during expansion; hence the geological term for abrasions along a fault plane, slickensides, is applied to them. The consistency of the A horizon is hard to very hard when dry, very firm when moist, and very plastic, stiff and sticky when wet. There may be a few calcium carbonate concretions in this horizon. During the dry season the upper 5–10 cm on some profiles develops an open granular structure, the peds becoming very loosely aggregated.

The dark colour of the A horizon does not indicate a high organic matter content. It is caused by the presence of diffused humus in an alkaline to neutral environment, and possibly also by reduced iron compounds. The organic matter is typically 1–3 percent and is very well distributed through the horizon.

There is commonly a transitional horizon with merging boundaries and properties intermediate between the A and C horizons. As it does not have the characteristics of either a textural B or a structural (B) it is called an AC horizon. Below this the C horizon may be either of hard rock or of unconsolidated sedimentary materials. There may be a Bca or Cca horizon of calcium carbonate concretions or less commonly calcrete.

A peculiar type of wedge-shaped structure, not known in any other soil type, has been recorded. It resembles platy structure in that the peds have greater horizontal dimensions than vertical, but the upper and lower ped surfaces instead of being parallel are inclined away from each other at 20°–30°, forming wedges. These structures occur at some preferred depth in the A horizon and cut across prismatic peds. Their mechanism of origin is not known, but may be related to differential horizontal stresses caused by the wetting front advancing parallel to the ground surface (de Vos and Virgo, 1969).

Reaction varies from weakly acid to weakly alkaline, typically pH 6.0–7.5 in the A horizon, rising slightly with depth. Base saturation varies correspondingly from 70 to 100 percent. Cation exchange capacity is unusually high, 30–60 m.e./100 g of total soil or over 50 m.e./100 g for the clay fraction alone. Toward the drier limits of occurrence of vertisols moderate levels of salinity can occur, and the clay complex often contains appreciable amounts of exchangeable sodium.

Bulk density is unusually high, typically 1.8. Permeability is very low, either remaining uniform throughout the A horizon or decreasing to a minimum value in its lowest part. There can be a great deal of standing water on the surface in the wet season, and yet it is possible to dig a pit within a few yards of water and it remain open for a day or more.

The distinctive morphological and physical properties are caused by the presence of substantial amounts of expanding lattice minerals in conjunction with a high total clay content. The texture is usually clay, occasionally silty or sandy clay, and always with a clay content exceeding 35 percent. Montmorillonite is usually the most common clay mineral, but mixed-layer expanding lattice clays as well as illite and kaolinite may also occur. The montmorillonite occurs within the 'fine clay' fraction (<0.2 μm nominal diameter); in one study it was found to constitute over 90 percent of this fraction (Fadl, 1971).

Dominance by montmorillonite causes the stickiness, high cation exchange capacity, and marked swelling when wet and shrinkage when dry. Shrinkage produces drying cracks, the high cohesion accounting for the large size of peds. In the dry season some of the loose material from the surface falls into the cracks, or is washed in at the start of the rains. When the soil is again moistened the peds expand and press against each other, producing the slickenside surfaces. The seasonally repeated infilling of cracks produces a self-mixing effect to the depth to which they penetrate, which gives the soil its name. The high bulk density and low permeability may be accounted for partly by the very fine pores and partly by compaction as a result of lateral pressure on swelling.

The *regur* or 'black soils' of the Indian Deccan are vertisols developed on basaltic lavas and alluvium derived from them. The climate is monsoonal in origin but is effectively a one-wet-season or savanna regime, with 500–1000 mm rainfall mainly between June and September. Colours are typically very dark greyish brown, and clay content 40–60 percent. Reaction is weakly alkaline, pH 7.0–8.5, and calcium carbonate concretions occur in depth. Classes of shallow, medium and deep black soils are recognized, the first two mainly on basalt and the last on alluvium. The shallow black soils may alternatively be classed as lithosols, and grade into very shallow ones and rocky outcrops on steeper slopes, giving a vertisol–lithosol association on the solid basalt topography (Simonson, 1954).

In the Gezira of the Sudan there is an extensive plain of alluvial and lacustrine clays under a semi-arid climate (300–600 mm rainfall). The soils developed on this, known as 'Gezira clay', have properties of vertisols rather than alluvial soils (Jewitt, 1955; de Vos and Virgo, 1969). They are browner than most vertisols, a dark brown A horizon overlying yellowish brown parent material. Clay content ranges from 50 to 80 percent. There is usually a loose surface layer in the dry season, less commonly a thin, hard brittle crust. Reaction is alkaline, pH 8.5–9.5, and there are many calcium carbonate concretions. Soluble salts are commonly present in depth. Another alluvial vertisol is the Kano Clay, the most extensive soil type on the Kano Plain east of Lake Victoria, Kenya (D'Costa and Ominde, 1973).

Two subdivisions of the vertisol class can be recognized, the extreme members of which are distinct although there are many intermediate profiles. At one extreme are soils which are very dark in colour, often mottled, have large and hard peds, and are transitional to non-vertisolic gleys. At the opposite extreme are profiles which are browner in colour with smaller structural aggregates and a less hard consistency; the *makande* soils of Malawi, developed on basalt-derived alluvium, belong to this category. Commonly (but not invariably) the former, darker, class is developed under imperfect drainage. The FAO classification distinguishes between darker *pellic vertisols* (moist chromas < 1.5) and browner *chromic vertisols*. The CCTA system makes the distinction on genetic grounds, between *vertisols of topographic depressions* and *vertisols of lithomorphic origin*.

It is incorrect to include with vertisols the black clays of valley floors in savanna climates, except where these are derived from basic rocks. Although black, with coarse blocky structure and some montmorillonite, these do not have the distinctive cracking and other characteristics of vertisols, and are correctly classed as gleys (p. 213).

Environment and genesis

As the distinctive features of vertisols result from dominance by montmorillonite, their genesis takes place in conditions that favour its synthesis. This requires a profile environment rich in magnesium and calcium; magnesium occurs in the structure of montmorillonite, whilst calcium maintains a pH favourable to its formation. Three factors favour such a base-rich environment: basic parent material, impeded drainage and a semi-arid climate.

Because of the interaction between these different factors, ranges for individual environmental conditions are wide. Two invariable features are that the parent material must be fine-textured or produce a fine-textured weathering product, and the slope gentle, usually less than 3°. Basic igneous rocks, limestones, and river or lacustrine alluvium derived from these are the most common parent materials, but non-basic shales and unconsolidated sediments can also develop into vertisols if leaching is weak and the base supply is augmented by seepage or flooding. There is a complete range of drainage classes from free to poor. The most favourable climate is semi-arid to dry savanna, tropical to subtropical, with a rainfall of 500–1000 mm; the full range of rainfall is 250–1500 mm. There is invariably a dry season of at least four months (Dudal, 1965).

Working inductively from observed conditions under which vertisols occur, the factors governing their formation can be stated in terms of necessary and sufficient conditions. The conditions are:

(1) A parent material yielding a fine-textured weathering product
(2) A slope of $<3°$
(3) A mean annual rainfall of 250–1500 mm
(4) A dry season of four months or more
(5) Parent material of basic composition
(6) Impeded drainage
(7) A mean annual rainfall of 500–1000 mm

Conditions (1), (2), (3) and (4) are necessary. Even where all other factors are favourable, sandy soils do not develop vertisolic properties, whilst more sloping sites give latosols or lithosols. Condition (5) is not necessary, since vertisols can develop on non-basic rocks where all other factors are favourable. Given (1)–(4), then no single one out of (5), (6) and (7) is either necessary or sufficient. The co-existence of (1), (2) and (4)–(7) is certainly sufficient for the invariable development of vertisols (six conditions only need be specified, as (3) necessarily follows from (7)). (1)–(6) are probably also sufficient. (1), (2), (4), (5) and (7) with free drainage would probably yield vertisols with a semi-arid climate but not with rainfall exceeding 750 mm. (1)–(4) with both (6) and (7) are not invariably sufficient.

Transitions occur between vertisols and lithosols, gleys and ferrisols. Those between shallow vertisols and lithosols and between poorly-drained vertisols and non-vertisolic gleys have already been noted. The most interesting feature is the existence in humid savanna climates of profiles intermediate between vertisols and ferrisols. In the Shemankar Valley of Nigeria, a tributary of the Benue south of Jos Plateau, there are basalt flows under a rainfall of 1400 mm. Where the water table is deep, extremely red ferrisols are developed; where it lies close to the surface in the wet season, dark grey vertisols occur. Profiles can be found on intermediate sites which have reddish and friable clays in the upper horizon, becoming black and plastic in depth. Presumably this is related to dominance by kaolinitic clay minerals above and montmorillonite synthesis in depth. 'Mixed red and black soils' on transported material have been reported from India (FAO, 1965*b*, p. 36). On basic rocks at about 600–1100 mm rainfall drainage is the critical factor tipping the balance towards or away from vertisol formation. Ferrisols and vertisols ('red soils' and 'black soils') occur in close topographic association, the former dominated by kaolinite and the latter by montmorillonite. On basic rocks montmorillonite is usually the first weathering product; in topographic depressions this persists, whereas on freely-drained sites it is converted to kaolinite as a result of progressive lowering of the silica and magnesium concentrations in the soil solution (Kantor and Schwertmann, 1974; Beckmann *et al.*, 1974).

Agricultural properties

The type of present use and land productivity of vertisols varies widely. Some are under extensive grazing, not only in semi-arid areas where there is no alternative, but also in savanna regions climatically suited to annual cropping. Others, especially in India, are under fairly intensive non-irrigated cultivation, particularly of cereals (maize, sorghum and millet), cotton and tobacco, together with rice under natural flooding or irrigation. It is not clear to what extent the contrast between intensity of use in India and frequent non-cultivation elsewhere (for example in West Africa) is associated with a more favourable range of soil properties, and to what extent it results solely from land pressure. Vertisols are quite commonly irrigated, notably the 740000 ha Gezira/Manaqil scheme in the Sudan on which cotton is the principal cash crop. Sugar cane is another common crop under irrigation.

The principal reason why vertisols in savanna climates are commonly left uncultivated under traditional farming is the difficulty of tillage with hand implements; where choice is available, latosols of lower fertility but greater ease of working are preferred. A second reason is waterlogging and the general difficulty of water control without organized drainage systems.

Favourable features for cropping are the gentle slopes, good water retention and high cation exchange capacity. Nitrogen levels are generally moderate, 0.1–0.2 percent in savannas falling to 0.05 percent in semi-arid regions. Calcium and magnesium are abundant, but potassium and available phosphorous fairly low. The slow permeability and high CEC means that nutrients in added fertilizers are well retained. Under alkaline conditions the availability of phosphorous and some trace elements is decreased. Correctly managed, many vertisols are capable of supporting annual cropping at moderate to high yields.

Set against these advantages are severe problems and hazards. Tillage difficulties remain with animal-drawn or mechanized cultivation. Root penetration is difficult on the more coarsely-structured types, and cracking can damage roots. On flat groud, the low permeability causes waterlogging. There is a severe erosion hazard owing to the low permeability and high runoff; gullying and sheet erosion can occur on slopes as low as $\frac{1}{2}°$. When completely waterlogged, the topsoil can almost flow away. In semi-arid areas, salinity and exchangeable sodium are further problems.

If machinery is used, the wheels may get stuck in the plastic clay. Repeated use of heavy cultivation machinery can lead to further compaction of the already dense soil, increasing difficulties of root penetration and drainage.

185

Bulk densities over 2.0 have been recorded on compacted soils. A swelling clay makes a bad foundation for roads and other structures, although over basalt, hard rock may be present at no great depth.

Vertisols are topographically suited to irrigation, and can produce a sustained level of production that justifies investment. The main management needs are water control and maintenance of organic matter. Subsurface drainage is impossible, and water must be carried off the land by a system of open drains fed by smaller furrow ditches that cross the contour at an oblique angle; these also serve for erosion control. Whilst formerly set out by ground survey, such a system can now be based on precise photogrammetric contouring at a vertical interval of 1–2 m. Coupled with control of water supply, the soil can be kept at an intermediate moisture level, facilitating cultivation by avoidance of both the hard dry and plastic wet states. Soils maintained under such conditions may lose their polygonal cracking patterns.

The need for maintenance of organic matter is less apparent. Despite their dark colour, organic matter levels of vertisols are only moderate. If allowed to fall there may be a loss of the surface layer of granular structure that aids germination, assists tillage and increases erosion resistance. The use of farmyard manure in addition to fertilizers is particularly beneficial. The inclusion of grass and legumes in the rotation not only adds organic matter through root decay but has the usual effect of grass roots at improving topsoil structure.

Rotations under irrigation, particularly if perennial, should include a substantial proportion of fallows, green manure crops or grasses and legumes. It is noteworthy that the standard rotation on the Gezira Scheme involves nearly 50 percent fallowing. Irrigated rotations can include rice in the wet season, with supplementary irrigation where necessary, and cotton in the dry season. Double cropping of rice gives lower yields than rice in rotation with other crops. Sugar cane is another suitable crop, but shrub and tree perennials are not normally grown. It may be possible to grow rice on very heavy clay types unsuited to other crops.

In soil survey for the irrigation of vertisols the selection of areas will be partly governed by topography. Insofar as there is a choice governed by soil properties, either between different areas or within one large area, some vertisols are considerably more satisfactory than others. Clays with massive structural aggregates are less easily managed, whilst the most favourable soils are those in which the larger peds break down readily into smaller aggregates. The poorly-structured clays are usually darker coloured and occur in lower-lying areas; they are very firm and have a lower permeability. Areas with gilgai micro-relief (see below) should generally be avoided. They tend to have coarse peds which are very hard, have high density and low permeability. In Iraq, Harris (1958) equated gilgai areas with 'bad-structured' soils. Gilgai also present stability problems for roads and are extremely difficult to drain.

Management seeks to maintain or increase the favourable properties of the better natural vertisols and to mitigate the disadvantages arising from seasonal variation in wetness. This is accomplished by maintenance of structure and control of moisture content. Correctly managed, these can be a most productive type of soil. They also have a high hazard potential, and if incautiously handled can result in considerable losses of invested capital.

GILGAI

Some areas of vertisols, possibly not a high proportion of the total, have a micro-relief of low mounds and shallow depressions known as *gilgai*. Introduced by Prescott (1931), this is the only word in international scientific use taken from an Australian aboriginal language, in which it has the meaning of a small water-hole; the singular and plural are the same. When crossing a vertisol plain during the wet season you can suddenly sink up to the knee into hollows filled with soft mud. On air photographs they show as a fine pattern of dots or lines.

The terminology of gilgai, formerly confused, has been clarified by Paton (1974), on which the following is largely based. A gentle rise is a *mound* and a shallow hollow a *depression*; more or less level ground not belonging to either mound or depression is a *shelf*. Mound, shelf and depression may be present in various combinations and proportions (fig. 13(*a*)). In the centre of depressions there is sometimes a *sink-hole*. Both mounds and depressions may be either rounded or linear. In the most common form, round gilgai, both the mounds and shelves are approximately circular and on air photographs form a pattern considerably more regular than random. The wavelength varies from 2 to 100 m and the amplitude from 10 to 100 cm, exceptionally up to 250 cm. Examples with an amplitude as low as 10 cm, hard to detect on the ground, may show a clear pattern on air photographs.

Based on the plan form of the component elements the following types of gilgai are distinguished (fig. 13(*c*)) (based on Hallsworth *et al.*, 1955, as modified by Harris, 1959):

1. *Round gilgai*. Also called normal or melon-hole gilgai. Mounds and/or depressions are approximately circular. Sink-holes may or may not be present. Round gilgai occur on almost level ground.

2. *Mushroom gilgai*. A rare type in which the mounds consist of localized abrupt rises above shelves.

3. *Tank gilgai*. A rare type in which mounds are roughly rectangular. They resemble artificial pools ('tanks') dug to trap rain (which in fact they sometimes turn out to be).

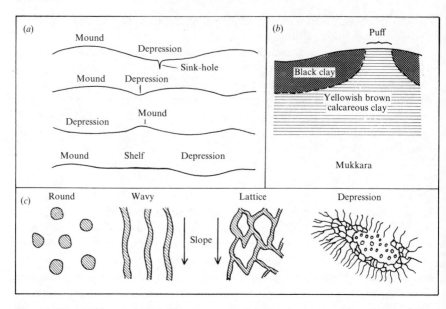

Fig. 13. Nomenclature and classification of gilgai. (*a*) Parts of gilgai; (*b*) mukkara (after Paton, 1974); (*c*) types of gilgai.

4. *Wavy gilgai*. Mounds and depressions occur as bands orientated downslope. They occur on slopes of $\frac{1}{2}°-3°$.

5. *Lattice gilgai*. Mounds and/or depressions form a partially intersecting network. They occur on very gently sloping ground sometimes in positions intermediate between wavy and round gilgai.

6. *Depression gilgai*. These are composite forms, not strictly comparable with the above, which they incorporate. They consist of circular or oval macro-depressions, of a larger order of size than individual gilgai, containing a narrow belt of centripetal wavy gilgai around the sloping margins, passing into lattice and/or normal types in the centres. In Iraq these form the greater part of gilgai areas, and Harris (1958, 1959) subdivided them in lattice types A–D.

In some gilgai the mound has a less dark colour, greyish brown or yellowish brown, smaller peds, a more friable consistence, calcium carbonate concretions are more frequent and the reaction is correspondingly more alkaline; where these features are present the mound is known as a *puff*. Where puffs are present, trenches reveal a three-dimensional soil pattern to which Paton applies the term *mukkara* (= aboriginal 'finger'; singular and plural) (fig. 13(*b*)). The black clay of the depressions is underlain by yellowish brown soil, the depth of the boundary between these diverging away from the depression centre until the yellowish brown material reaches the surfaces at the puff. This is consistent with the interpretation that on the puff the sub-soil has been forced to the surface.

The usual explanation of the formation of gilgai is by soil heaving resulting from expansion on wetting (Hallsworth *et al.*, 1955). Small pieces of the friable surface layer fall into the vertical cracks during the dry season or are washed in at the start of the rains. When the soil is wetted, lateral pressures are developed; these can only be released by expansion upwards, initiating a mound. The formation of one mound gives a locally preferred site for further release of pressure, and hence a regular pattern will tend to form.

Once formed, the sites of mounds are perpetuated and there may be a slow circulation of soil material radially outwards and upwards, finer particles returning to the shelf by surface wash. This explanation is consistent with the high density of shelf and depression soils, caused by compaction under lateral pressure. It could also account for the occasional presence of basalt stones brought to the surface. It is not easy to see how the initial upheaval of the subsoil comes about, nor is it clear why only some vertisols develop gilgai. The origin of the sink-holes is obscure.

The differentiation between round, wavy, and lattice gilgai is clearly related to slope. There is an analogy between patterned ground in polar regions, where stone stripes on slopes give place to stone circles on level ground. The pattern on air photographs of the channels of wavy gilgai gives a clear impression that these serve as drainage channels, whether or not initiated in this way.

Edelman and Brinkman (1962) challenged this steady-state explanation. They noted that gilgai were a localized phenomenon rather than a general characteristic of vertisols, and that they were frequently found in macro-depressions. Their initiation was held to be a once-only process, related to the ripening of alluvial clays (p. 220). Ripening causes a great decrease in volume, resulting in deep, broad and very widely spaced cracks. These become infilled, leading to gilgai formation by the same mechanism of cyclic heaving. The drying cracks of normal vertisols are too closely spaced to initiate heaving. Thus, gilgai should be found only on former lake or swamp beds.

There have been reports that gilgai levelled by bulldozer re-form in a few years. This does not prove contemporary origin, however, as the subsoil differences may remain after levelling.

Some remarkable regolith structures have been recorded near Accra, Ghana (McCallien *et al.*, 1964). There is an horizon succession, from the surface downwards, of normal friable soil, nodular laterite, a quartz stone line, a black cracking clay and weathered schists and gneisses. Road cuttings show that the upper boundary of the black clay is very wavy and contorted, reaching the ground surface in places. There is a surface micro-relief resembling gilgai, although the surface undulations are very much smaller than those of the stone line/black clay boundary. These structures are caused by shrinking, cracking, infilling, and expansion with heaving within the clay, constrained by its overburden.

SOILS ON LIMESTONES

In temperate and Mediterranean climates, limestones are associated with distinctive soil types: rendzinas, (red-)brown calcareous soils and terra rossa. In the tropics the influence of limestone parent material is also substantial, although not always dominant. Soil types developed from it included rendzinas, brown calcimorphic soils, vertisols and latosols.

Rendzinas (= CCTA, FAO) with the same AC profile that is found in temperate climates, also occur in the tropics, on steep slopes. The following profile was described from a rainforest site in New Guinea with 1800 mm rainfall (Haantjens, 1967, p. 57):

Ah   0–13 cm   Very dark greyish brown (10YR 3/2) silty loam, with some limestone grit; subangular blocky structure. Organic matter 6.5 percent, pH 7.8.

AC   13–35 cm   Dark greyish brown (10YR 4/2) silty loam, with limestone grit; crumbly to blocky structure, firm; abrupt boundary. Organic matter 2.9 percent, pH 7.8.

C   35 + cm   Limestone grit.

In dry savanna and semi-arid regions, limestones are likely to develop vertisols or brown calcimorphic soils, accentuating the calcareous tendencies of the profile.

At a rainfall of 1000–1400 mm, at which felsic rocks would give rise to latosols, limestones on gentle slopes may develop into vertisols. Soils of Barbados, developed on coral limestones, show a systematic increase in montmorillonite with decrease in rainfall. On a $\frac{1}{2}°$ slope under 1350 mm rainfall a black (N 2/0) compact clay, very sticky and plastic, passing into rock at 75 cm has been reported (Ahmad and Jones, 1969).

In freely-drained sites with a rainfall above about 1400 mm, or above about 1000 mm on moderate slopes, latosols are usually formed. There may be rock hills, either isolated inselbergs with soil-covered summits and nearly vertical bare rock sides, such as are found in Malaya, or the remarkable relief of cockpit karst developed in Jamaica (Sweeting, 1958). On a slope below a limestone hill in Malaya, with 2500 mm rainfall, a dark reddish brown over dark red (2.5YR 3/6) clay loam, with medium subangular blocky structure and very friable, passes into rock at 180 cm; base saturation is 40 percent, therefore by implication the reaction is acid; further down the same slope, interpreted as a maturity sequence, saturation decreases successively to 15 and 7 percent, and laterite concretions appear (Joseph, 1968).

190

Under a 1700 mm rainfall in Barbados, on a 3° slope, over hard, pure white limestone, Ahmad and Jones (1969) describe the following profile:

Ah	0–15 cm	Dark reddish brown (5yR 3/4) clay; strong medium crumb structure, friable; merging boundary.
Bt1	15–46 cm	Dark reddish brown (5YR 3/4) clay; weak medium blocky structure, friable to firm; merging boundary.
Bt2	46–107 cm	Yellowish red (5YR 5/8) clay; weak coarse blocky structure, friable, sticky and plastic when wet; abrupt wavy boundary.
C	107+ cm	Hard limestone.

Clay content was 80 percent in the topsoil and 96 and 99 percent below, and topsoil organic matter content 4.3 percent. The pH was 7.6 in the topsoil and 8.0 below, and base saturation 100 percent. Kaolinite was the dominant clay mineral. This profile is difficult to classify, having the morphology of a latosol but with an alkaline reaction; it is nearest to a ferrisol.

Latosols on limestones are moderately good agricultural soils, with free drainage but good moisture retention, a well-developed structure, friable consistency, and tending to have a higher organic matter content than soils on siliceous rocks on similar climates. The dominance by kaolinite, and resulting friability, shows that strong leaching overcomes the effects of a calcium-rich parent material.

SELECTED REFERENCES

Vertisols, general. Dudal (1963, 1965), de Vos and Virgo (1969).

Gilgai. Hallsworth *et al.* (1955), Stephen *et al.* (1956), Harris (1958, 1959), Paton (1974).

Limestone soils. Ahmad and Jones (1969).

CHAPTER 10

SOILS OF THE DRY TROPICS

The dry tropics, comprising the semi-arid and arid zones, are the regions in which rain-fed cropping is difficult or impossible, and grazing and irrigated agriculture are important types of land use. The transition from the dry savanna zone takes place at a mean annual rainfall of 500–600 mm and coincides with a fundamental change in soil-forming processes. Above this rainfall limit, the more soluble products of weathering are leached out of the profile. Below it, calcium carbonate and sometimes calcium sulphate and soluble salts accumulate within the profile, giving the zone of pedocals. Associated with this change, the reaction becomes alkaline and base saturation rises to 100 percent in some parts of the profile. The topsoil organic matter content falls to below one percent. These dry lands occupy over one third of the tropics.

The transition from pedalfers to pedocals is associated with a change in soil moisture regime. In both the savanna and semi-arid zones there is an excess of rainfall over potential evapotranspiration for one or more months of the year, during which the profile dries out from above. In the savannas the soil remains moist below this layer of annual drying. In semi-arid regions, provided there is free site drainage, the soil remains permanently dry in depth, the wet season moisture surplus being less than the available water capacity of the profile. There is an absence of biological activity in this permanently dry layer.

Calcimorphic soils, those containing free calcium carbonate, are morphologically distinct from halomorphic soils, the properties of which are dominated by soluble salts. The difference is not primarily climatic in origin but is caused by site drainage. Soluble salts accumulate where there is a net inflow of water, and hence where the water table is within capillary range of the profile. However, in semi-arid climates the soil above the water table experiences leaching for part of the year, so halomorphic soils are more frequently found in the arid than in the semi-arid zone. This chapter covers the zonal soils of the semi-arid and arid regions, comprising calcimorphic soils together with soils of extreme desert regions which, through the absence of pedogenetic processes, are not calcimorphic in the strict sense of the term. The properties and agricultural problems of halomorphic soils are discussed in chapter 11.

PROCESSES

Wherever there is a soil cover chemical weathering is more important than physical, even in extremely arid climates. Because the regolith is only infrequently moist, however, weathering is slower and less intense than in the humid tropics. This results in shallowness, for soils formed *in situ*, and a preponderance of sandy textures. Profiles dominated by both fine and coarse sand are widespread. Relatively fine-textured soils often have a high content of very fine sand and coarse silt (20–100 μm) rather than clay. Textural B horizons are common in semi-arid regions and are probably produced by clay translocation; it has been suggested, but not substantiated, that they might also be caused by greater weathering at the depth which remains moist for the longest period.

In all but extreme deserts, leaching following rains removes soluble salts from freely-drained sites. There may also be weak leaching of bases and silica. In the dry conditions iron passes into ferric forms shortly after it is released by weathering, and so is not leached but remains to colour the soil.

The main process of horizon differentiation is the redistribution of calcium carbonate. This is usually leached from the topsoil, but shows a marked tendency to accumulate at a particular depth to give a Bca horizon. The accumulation may take the form of coatings, filaments, concretions or a layer of massive calcrete. It is tempting to suppose that the depth of accumulation is associated with that of the permanently dry layer, perhaps occurring at the average or maximum depth penetrated with some given return frequency by water in excess of field capacity, but this has not been formally demonstrated. Gypsum may also accumulate to give Bcs or Ccs horizons; these are less common than Bca horizons and occur below them in the profile.

Clay mineral formation is influenced by the conditions of weak leaching. This permits the formation of montmorillonite and other expanding-lattice minerals, provided the necessary elements can be supplied by weathering or inflowing water. Dominance by expanding-lattice minerals is by no means invariable, and clays formed mainly of illite, kaolinite or a combination of these are common; parent material has a considerable influence on clay mineral type.

As a result of slow plant growth but potentially rapid oxidation, equilibrium levels of humus are low. Topsoil organic matter levels are usually less than one percent in semi-arid regions, becoming lower where textures are sandy, and fall almost to zero in the desert zone. Amounts of organic matter are substantially lower than under a corresponding rainfall in the temperate margins of the desert belt. Humus is well incorporated into the mineral soil.

Wind action affects soils by dune formation, deflation and deposition.

Deflation, the selective removal of fine particles, can lead to the residual accumulation of a surface layer of gravel and stones (lag gravel). Wind-deposited materials show a fairly high degree of sorting, frequently with a peak in the silt to fine sand range, 20–50 μm. Very fine particles become too widely distributed in the atmosphere to form local accumulations. Under natural conditions these processes would be confined largely to the true deserts, but as a result of over-grazing wind erosion is common in the semi-arid zone.

Relict features from both wetter and drier Pleistocene climates are wide-spread in dry zone landforms and soils. Alluvial flood-plains, often with a pattern of braided channels, were deposited during wet periods or the transition to drier conditions which followed them. During drier periods vast sand spreads extended well beyond the present desert margins. These include both fixed dunes and more continuous sand sheets, and are likely to be encountered all around the subtropical deserts, including those in West Africa, southern Africa and India (Grove, 1958; Grove and Warren, 1968; Grove, 1969; Goudie *et al.*, 1973).

BROWN CALCIMORPHIC SOILS

CCTA: Brown soils of arid and semi-arid tropical regions
FAO: No specific equivalent; xerosols p.p., krasnozems p.p.

Brown calcimorphic soils are the zonal soils of the semi-arid zone, typically developed under 300–500 mm rainfall but extending up to 600 mm or more where site drainage is imperfect. They are frequently developed on un-consolidated sands, giving very sandy textures, and also on acid igneous and sedimentary rocks. Basic rocks in this zone give rise to vertisols. The vegetation is steppe grassland, with or without trees or shrubs of xeromorphic habit, or thorn scrub.

There is a well-developed ABC or A(B)C profile, usually with a Bca horizon. A dark greyish brown topsoil is clearly visible, with an organic matter content close to one percent under natural conditions but frequently lower if cultivated or heavily grazed. This humic layer merges gradually into a brown, yellowish brown or reddish brown A2 horizon, below which there is a textural or structural horizon of similar colour. If the texture is not sandy there is a weakly to moderately developed blocky or slightly prismatic structure, often with clay skins, and consistency is firm when moist and hard when dry. With sandy loam or loamy sand textures the soil is almost structureless. Calcium carbonate accumulates, usually as concretions, at a depth of between 50 and 200 cm. Calcium sulphate may occur at a greater depth, sometimes within the parent material as a Ccs horizon.

Reaction is weakly acid to neutral in the upper part of the profile, rising to pH 7.0–8.0 in the Bca horizon. Base saturation varies similarly, from

194

between 50 and 90 percent in the A2 horizon to 100 percent in depth. Cation exchange capacity in relation to clay content is moderate, although the total capacity may be low if textures are sandy. The exchange complex is dominated by calcium, sodium being kept at low levels by wet-season leaching. Soil fertility is moderate in fine to medium textured profiles and low where textures are sandy.

Agricultural properties

The correct use of these soils is more a problem of agroclimatology than of pedology. The grazing potential of the natural steppe vegetation is adequate for lowish stocking densities. However, soil fertility is adequate, and with a growing season of about 90 days it is possible in most years to raise a crop of drought-resistant varieties of cereals. Although yields are low the returns per acre considerably exceed even those from improved livestock farming. There is therefore a temptation to practise annual cropping, but rainfall variability leads to recurrent crop failure.

Brown calcimorphic soils are often well suited to irrigation. Sandy textures and consequent low moisture retention are the main problem. Fine sands are preferable to coarse. Calcrete horizons may cause problems (p. 198). There is sufficient leaching from rainfall to reduce the danger of salinization, and a reasonable organic matter level which can sometimes be increased under irrigation. Cotton is the most widespread cash crop. The change in moisture regime is less drastic than is the case when irrigating soils of the arid zone, making it less likely than adverse effects will develop. Furthermore, soil fertility is higher than in the arid zone. On these grounds irrigation of brown calcimorphic soils is pedologically preferable, and likely to be economically more beneficial, than irrigation of arid zone soils.

It is usually held that investment in land improvement should not go to rain-fed cropping in this region, and from some ecological aspects livestock farming is a better form of resource use. Commonly, however, there is already a substantial population dependent on annual cropping, and the reduction in productive capacity that would result from a change to livestock is out of the question. The productivity of these areas can be increased through development and use of improved short-maturing varieties of crops together with normal better farming methods. It is likely that the benefit : cost ratio of this less spectacular form of development will frequently exceed that of irrigation schemes, which in any case are only technically possible over a small fraction of the region.

SOILS OF THE ARID ZONE

With a rainfall of less than 400 mm plant growth becomes sparse or absent and topsoil organic matter falls to 0.5 percent or less. This desert area is of

195

great extent and very low productivity except under irrigation. A substantial proportion of the irrigated areas consist not of sedentary soils but of alluvium or other depositional materials.

The classical sequence of semi-desert and desert soil types was established in the subtropical belt of the USSR, and comprises kastanozems (chestnut soils), sierozems, and grey and red desert soils. Within the tropics, brown calcimorphic soils occupy a position analogous to the kastanozems. The remainder of the sequence is probably applicable to tropical deserts, with the difference that reddish colours become more frequent than grey and organic matter levels for a corresponding rainfall are lower.

Sierozems (sometimes spelt *serozems*; FAO *xerosols*, no CCTA equivalent) typically occur under 200–300 mm rainfall and are the driest soils to have clear profile differentiation. A dark grey topsoil with well-incorporated organic matter passes into a subsoil which may be grey or reddish brown. There is a (B) horizon with a weak blocky structure. Calcium carbonate concretions commence at about 30–50 cm and there may be an accumulation of gypsum at about 1 m. Textures are very dependent on parent material and are frequently sandy. Silt commonly exceeds clay.

Grey and red desert soils (CCTA *sub-desert soils*; FAO *yermosols*) show little profile development other than a thin A horizon low in organic matter, below which is a shallow zone of slightly weathered material with a little calcium carbonate accumulation. These occur under approximately 100–200 mm rainfall.

In the extreme desert zone soil proper is replaced by *desert detritus*. This includes bare rock, skeletal coverings of stones and gravel, desert pavements (regs), blown sand areas including dunes (ergs), and clay plains, the last originating as former lake beds.

A more detailed subdivision of the soil types of arid and semi-arid climates is given by Dan (1973).

The zonal sequence described above is based on profile development as a whole. The FAO classes of dry-climate soils are defined on the basis of the degree of development of the A horizon. In *xerosols* an ochric A horizon is well developed whilst in *yermosols* (Spanish *yermo* = desert) it is weakly developed, the precise difference being defined in terms of organic matter percentage in relation to the sand:clay ratio (FAO–Unesco, 1974, p. 25).

It is misleading to regard dry-climate soils as a succession of zonal belts. The fact that they are shown as such on small-scale maps is partly because so few soil surveys have been done. Large parts of the semi-arid and sub-desert regions are occupied by bare rock, gravel and sand. Elsewhere there is considerable dependence of soil properties on parent material: this applies particularly with respect to texture, which in turn dominates most other soil properties.

An alternative approach to the classification of desert soils of depositional origin was proposed by Western (1972). These soils show little profile development of pedogenetic origin and yet are mature in the sense that their characteristics do not change appreciably with time. Their properties are largely derived from their manner of deposition. Western defined a great soil group of *sedosols* as soils in which depositional features will continue to dominate the nature and utility of the soil under existing climatic conditions unless it is subjected to human influence, i.e. irrigation. Sedosols are divided into families on the basis of mode of deposition, which may be fluvial, lacustrine, littoral, aeolian or anthropic (by irrigation water). As soil series based on pedogenetic features cannot be recognized in deserts they are replaced by 'depositional series' characterized by types of sediment. The main differentiating characteristics employed in classification are texture, degree of depositional stratification and effective soil depth. Salinity, being an impermanent characteristic that can be modified by management, is recognized at phase level. This approach is useful in that at the lower level of classification the properties used are those of significance to irrigation potential, whilst at the family level it provides a genetic basis for rationalizing the soil distribution pattern on the basis of geomorphological processes of origin. It provides an antidote to the unrealism of applying the zonal approach to soils for which climatic influences are weak in comparison with those of parent material.

CALCAREOUS SOILS

CCTA: Soils with calcareous pans.
FAO: Calcic xerosols, calcic yermosols.

Calcareous soils are those in which the problems of land and water use for crop production are primarily caused by a high content of calcium carbonate (FAO, 1973*b*). Their recognition as a class is thus based on one dominant morphological feature; whilst largely confined to semi-arid and arid regions they cut across the zonal arrangement, being found within the zones of brown calcimorphic soils, sierozems, and grey and red desert soils. Quaternary calcareous sediments and limestones are the most common parent materials but they also occur on basalts and other basic rocks.

The essential features of these soils are differentiation of calcium carbonate within the profile and a marked Bca horizon of accumulation. On stony or gravelly soils the carbonate first appears as coatings to stones, followed by inter-stone fillings leading to a massive calcrete horizon that incorporates the stones. In fine-textured soils it commences as diffuse filaments or soft friable concretions, followed by hard concretions, inter-concretionary fillings and massive calcrete. The two accumulation sequences then converge, and may be followed by accumulation in laminar forms above the massive horizon together with thickening of the latter (Gile *et al.*, 1966).

The A horizon of such soils may be calcareous or non-calcareous. There is usually a textural B horizon, either above the Bca or merging into it. Colours in the A and Bt horizons are frequently reddish. The structure is typically blocky in the A horizon and fine blocky to fine prismatic in the Bt and Bca. There may be an horizon of gypsum accumulation below the Bca horizon. The carbonate, where not concretionary and hence in the gravel fraction, occurs mainly as silt and clay. Its active fraction has a high specific surface area and markedly affects moisture characteristics of the soil; hence carbonates must be included in textural analysis of such soils, and not removed by pre-treatment as is the practice with some methods of analysis. The pH is normally 8.0–8.5 in the Bca horizon, and may be above or slightly below neutral in the overlying soil. In a catenary sequence described on a limestone pediment in Iran, under 350 mm rainfall, calcium carbonate contents increased downslope, from 1 percent on the upper pediment to 24 percent on the alluvial plain below (Abedi and Talibudeen, 1974).

Calcrete

Calcrete is a type of duricrust formed of calcium carbonate. Calcretes are largely confined to areas with less than 500 mm rainfall (Goudie, 1973*a*, 1973*b*). They are most often developed over limestones but also occur on other parent materials. The quantity of calcium carbonate present is far greater than could have been derived from the present overlying soil, and so there is the same problem of the origin of the material that arises in the case of laterite (p. 162). Possible mechanisms are downward leaching, precipitation from below by groundwater, and lacustrine sedimentation. In the downward leaching model the carbonate originates from rock weathering, and as the soil profile is lowered by weathering and solution the Bca horizon moves down with it, being redissolved above and precipitated below. This is a steady state explanation, and accords with the observed existence of a continuum from horizons with concretions to thin and thick calcrete layers. There is the objection that weathering and ground lowering in dry regions is very slow. The second explanation is that the carbonate originates by evaporation from calcareous groundwater within the capillary zone immediately above the water table; the water may be replaced in part by inward seepage. This mechanism is analogous to the process of salinization. The third explanation is origin by lacustrine sedimentation and evaporation, possibly with some subsequent redistribution by leaching; this is supported by the fact that many calcretes occur in basin situations and have very gentle slopes (2–4 m/km).

It is certain that many calcretes are relict from different climatic conditions in the Pleistocene. In such cases the calcrete is more correctly regarded as a geological stratum than a pedogenetic horizon. It now forms one of the parent materials from which calcareous soils are developed.

Agricultural problems of calcareous soils

The main agricultural problems of calcareous soils when irrigated are crusting, low availability of phosphorous, problems of potassium and magnesium supply caused by imbalance with calcium, and unavailability of micronutrients. Crusting is particularly likely to occur where textures are silty, and causes difficulties with water infiltration. Phosphorous quickly reverts to insoluble forms owing to the high pH; at least 60 percent of applied phosphorous should be in water-soluble forms. Zinc, iron, manganese and copper tend to become less available with increasing pH; deficiences of iron and zinc are common, resulting in lime-induced chlorosis of leaves. Organic manuring is beneficial to calcareous soils. Alfalfa is a suitable crop, being tolerant of high calcium levels and highly productive under irrigation.

Calcareous soils are common in regions where irrigation is necessary and where there is a severe land shortage; they are widespread in most countries of the Middle East. Moreover, they frequently occur on sites that are topographically suitable for irrigation. Their reclamation and utilization is therefore of considerable importance in such regions. The greatest difficulties arise on first bringing them under irrigated cultivation. If this can be successfully accomplished, then with an adequate drainage system the excess carbonates will gradually be reduced by leaching.

SOILS WITH GYPSUM

CCTA: Soils containing more than 15 percent gypsum
FAO: Gypsic xerosols, gypsic yermosols.

Soils having an horizon for accumulation of gypsum (hydrated calcium sulphate, $CaSO_4 . 2H_2O$) within 100 cm of the surface are occasionally developed in depressions in the arid zone. Usually there is an horizon of calcium carbonate accumulation at a lesser depth, in which case the profile is better classified as a calcareous soil. Although gypsum is non-toxic to plants, root growth is inhibited because of the associated massive structure. Only where gypsum forms over 25 percent of an horizon does it constitute a serious crop limitation (Smith and Robertson, 1962). These soils can be used as a source of gypsum for the reclamation of solonetz.

199

SOIL SURVEY IN ARID REGIONS

Features of resource surveys in arid and semi-arid environments are an orientation towards specific development aims as opposed to general-purpose mapping, and a clear separation between extensive and intensive stages of survey. The main types of survey are pasture resource inventory and the assessment of land and water resources for irrigation. Pasture surveys are necessarily extensive. Irrigation surveys include both extensive coverage to locate and choose between possible irrigable regions, and intensive surveys to determine precisely which land is suitable.

Whether land is potentially irrigable is decided as much on topographic as on pedological grounds. Whilst traditional systems may, by ingenious but labour-intensive means, bring water to quite steeply sloping land, modern technology can achieve the same at higher cost by sprinkler or trickle systems. Much irrigation in less developed countries, however, is by the cheaper basin-flooding method, and development is then largely confined to depositional landforms. As well as being relatively flat, the land must be commanded, that is, below the altitude at which a potential water supply is locally available. The extensive stage of survey is heavily dependent on air photograph interpretation. Most erosional landforms can be ruled out as topographically and often also pedologically unsuitable, and need not be visited in the field.

After a region with a *prima facie* irrigation potential has been located the next stage is one of geomorphological mapping. Quite small differences of elevation and water table depth can be distinguished on air photographs, either directly or as vegetation differences, and mapping can be accomplished by photo-interpretation combined with widely-spaced ground traverses. The maps show the types of depositional landforms present, e.g. flood-plains, terraces, lake beds and sand dunes (either present or relict from different climatic conditions), together with a bounding frame of erosional topography. These landform units provide the basis for detailed soil survey.

The main properties which affect soil suitability for irrigation are texture, depth to a limiting horizon, degree of textural stratification, and salinity. Many other properties, e.g. colour and structure, are either fairly uniform over all soils of an arid area or else vary concomitantly with texture and salinity. With single-flood irrigation, common in arid regions, available water capacity has a dominant effect on crop yields, and this capacity is largely determined by texture and soil depth. It is desirable to use both the US and International particle size limits to separate coarse silt (20–50 μm) from fine silt (2–20 μm) since their properties differ appreciably and there is frequently a high content of material of these sizes. A sandy topsoil is prone to erosion, causes rapid infiltration, and dries out quickly making seed germination difficult. Crops root more deeply than normal under single-flood irrigation. Marked stratification may cause difficulties with water infiltration and rooting. A mild level of salinity can be reduced by leaching but highly saline and saline–alkaline soils are difficult to reclaim.

Salinity has a different pattern of spatial variation from the other differentiating parameters, being concentrated towards depressions. Marked salinity or alkalinity may be used to define soil series, but where salinity levels are moderate, and vary continuously, it may be more convenient to distinguish them at phase level.

SELECTED REFERENCES

Aubert (1962), Buol (1965), FAO (1973*b*), Western (1972), Dan (1973).

CHAPTER 11

HALOMORPHIC SOILS

Halomorphic soils are those with properties controlled by the presence of soluble salts, by the presence of sodium ions in the exchange complex, or both. Unless the degree of halomorphism is slight, these soils cannot be used for agriculture. It is, however, sometimes possible to reclaim them.

They occur on two main types of site: low-lying land in arid climates and marine alluvium. In the former the salinity is caused by evaporation of saline groundwater, in the latter by recurrent seawater flooding. On genetic grounds the former class may be divided into naturally saline soils and those that have become salinized as a result of irrigation.

Halmorphic soils are not very extensive, occupying between one and two percent of the tropics and some five percent of the arid zone, but their agricultural significance is greater than this figure might imply. Much of the arid zone is occupied by lithosols and sands, whilst the halomorphic sites are largely soils which on grounds of slope, profile depth and texture are irrigable. Soil surveys in desert regions commonly show that 10–20 percent of the potentially irrigable areas are affected by some degree of salinity. Coastal halomorphic soils are of less total significance although locally, on densely populated coastal plain and estuary sites, they are in demand for reclamation for padi.

MEASURES OF SALINITY

Salinity refers to the content of readily-soluble salts, namely those more soluble than calcium sulphate. The most common means of measuring salinity is indirectly, as the *electrical conductivity* (EC) of the saturation extract. A soil–water paste is prepared and allowed to stand until chemical equilibrium is reached; the saturation extract is removed by vacuum filtration and its electrical resistance measured. The electrical conductivity in mhos per centimetre is the reciprocal of the resistance in ohms of a conductor, the saturation extract, one centimetre long and with a cross-sectional area of one square centimetre. The conductivity values of soils are usually expressed in millimhos (mmho) and those of irrigation waters in micromohos (μmho), where 1 mho = 1000 mmho = 1000000 μmho. Electrical conductivity is simple to determine but has the disadvantage that it may vary substantially in the course of the year.

It has now become common to determine chemically the content of *total soluble salts* (TSS; sometimes called readily-soluble salts, RSS). This is expressed as a percentage of the dry soil or as parts per million. The relation between these two measures of salinity depends on the types of salts present and their concentration, but for the levels normally found in soils it is approximately given by:

$$TSS \text{ (p.p.m.)} = 0.64 \times EC \text{ (mho/cm)} \times 10^6$$
$$= 640 \times EC \text{ (mmho/cm)}$$

For the lower levels of salinity found in irrigation waters the constant 640 is replaced by 600 (Thorne and Peterson, 1954, p. 379). A further measure for irrigation waters is milliequivalents per litre, which is converted to parts per million by multiplying the values for each ion by its equivalent weight and summing.

Salinity can be detected in the field by the presence of white salt crystals or by taste, a salty taste indicating a TSS level over about 0.5 percent.

The salts commonly present in soils are shown in table 21. Calcium carbonate, a salt in the chemical sense, has a very low solubility, remaining

TABLE 21 *Soluble salts in soils. The carbonates and gypsum, not classified as readily-soluble salts, are included for comparison. After FAO (1973a)*

Salt	Chemical formula (in form found in soils)	Frequency in soils	Solubility	Toxicity to crops	Notes
Calcium carbonate	$CaCO_3$	Very common	Very low	Nil	
Calcium bicarbonate	$Ca(HCO_3)_2$	See notes	Low		Weathering state of calcium carbonate
Sodium carbonate (soda)	Na_2CO_3	Uncommon	High		Only accumulates where gypsum absent
Calcium sulphate (gypsum)	$CaSO_4 . 2H_2O$	Common	Low	Nil	Occurs in hydrated form; used to reclaim solonetz
Magnesuim sulphate	$MgSO_4 . 7H_2O$	Very common	High	High	Occur in association with sodium chloride
Sodium sulphate	Na_2SO_4	Very common	High	Moderate	
Magnesium chloride	$MgCl_2$	Uncommon	Very high	High	Occurs only in soils of very high salinity
Sodium chloride	$NaCl$	Very common	High	High	The most common soluble salt

Magnesium carbonate, potassium carbonate, potassium sulphate, calcium chloride and potassium chloride are rare in soils. Nitrates only accumulate in exceptionally arid deserts.

relatively low when in the form of its weathering product calcium bicarbonate. Hydrated calcium sulphate, gypsum, has a solubility about two orders of magnitude greater than that of calcium carbonate but two orders less that of the readily-soluble salts. Both calcium carbonate and gypsum are non-toxic to crops.

The most common of the soluble salts is sodium chloride, followed by sodium sulphate and magnesium sulphate. The 'fluffy' surface of some saline soils contains calcium and sodium sulphates. The toxicity of salts to plants is approximately inversely proportional to their solubility.

The extent to which the exchange complex is saturated with sodium ions is expressed as the *exchangeable sodium percentage* (ESP), obtained from the content of sodium ions as a percentage of total exchangeable bases in milliequivalents. An alternative measure of sodium saturation is the *sodium absorbtion ratio* (SAR) defined as:

$$SAR = \frac{Na^+}{\sqrt{\left(\dfrac{Ca^{2+} + Mg^{2+}}{2}\right)}}$$

where quantities of exchangeable cations are in milliequivalents. This formula is based on equations of chemical equilibrium and is a more reliable index of the sodium hazard of irrigation waters and its relation to the sodium status of the soil.

PROCESSES

Salinization is the accumulation of soluble salts within the soil profile, whether by natural processes or as a result of irrigation. It is caused by the evaporation of water containing dissolved salts and their precipitation. The water table or the capillary zone above it must reach into the soil profile. It is also necessary that there should be little rainfall, otherwise the salts would be rapidly removed by leaching. Hence salinization is largely confined to depression sites in the arid zone. The horizon of maximum salt accumulation may occur in depth or at the surface, depending on the level of the water table.

Soils subject to seawater flooding in humid climates do not accumulate free salts in large amounts as these are constantly redissolved. In coastal deserts salinization may in part be caused by seepage of seawater beneath the land, but this process is uncommon.

Precipitation of salts in a particular horizon may cause *self-sealing*, giving distorted drainage.

Solonization (alkalization) in the narrow sense of the term is the process whereby the exchange complex acquires an appreciable saturation with sodium ions. This is brought about by prolonged or repeated saturation with water in which sodium predominates over other cations. Since sodium has the lowest replacement ability of the exchangeable bases it is necessary, in accordance with chemical equilibrium laws, that its concentration in the soil solution should substantially exceed that of other cations. It is possible for large amounts of free salts to accumulate within the profile without solonization occurring. The agent of solonization may be groundwater, irrigation water or seawater flooding. It can take place within a time of the order of 3–5 years.

A number of other processes accompany the specific process of solonization as defined above, and are sometimes considered part of the composite process of the same name. As long as free salts are present, the pH remains below 8.5 and the flocculating effect of the salts maintains the structure and permeability of the soil. If the free salts are leached from a solonized soil the pH rises to 8.5–10.0, the clay becomes dispersed and the structure collapses. Clay translocation readily occurs, producing a very impermeable textural B horizon. Under the alkaline conditions humus becomes dispersed through the profile, imparting a dark grey colour. The leaching of previously precipitated salts may occcur naturally, as a result of climatic change, or it may be caused by irrigation water.

Salinization is an easily reversible process, involving the leaching of the soluble salts. Reversal of solonization requires the replacement of sodium by calcium ions and is harder to accomplish, whilst the accompanying clay translocation is irreversible.

Solodization is a composite process thought to accompany the leaching of previously solonized soils. The exchangeable sodium is replaced by hydrogen ions, the soil becomes acid, the clay remains dispersed and there is intense clay translocation. The topsoil becomes sandy and bleached. Wind erosion may follow.

SOIL TYPES

There is relatively little difficulty in classifying halomorphic soils since their morphology is controlled by only two properties, and hence it is only necessary to agree on limiting values for these. There are, however, two common systems of nomenclature.

On the older system, of Russian origin, soils containing free salts, whether or not they are also solonized, are *solonchaks*. Soils without free salts but with appreciable sodium saturation are *solonetz*. Other former names were 'white alkali' and 'black alkali' soils respectively.

The Salinity Laboratory of the US Department of Agriculture established a system with self-explanatory terms. Adopting limiting values for electrical conductivity and exchangeable sodium percentage, chosen on the basis of plant response, they defined the mutually-exclusive soil classes shown in table 22. It is chemically more correct to use the term sodic soil and corresponding derivatives (e.g. saline–sodic soils) in place of alkali soils, but the use of the latter is more common. The term 'alkali' referrs to the sodium ion and is not the same as the use of 'alkaline' to describe high pH values, although alkali soils do have an alkaline reaction.

The FAO classification uses the Russian nomenclature with the American limiting values in modified form. Sodic (alkali) and non-sodic solonchaks are not separated. The CCTA classes of halomorphic soils were devised to accomodate pre-existing mapping units; they are not mutually exclusive, and are unsatisfactory as a general purpose classification. Neither the FAO nor the American system separates halomorphic soils of desert sites from those caused by marine flooding.

The Russian/FAO terms are more suitable for considerations of pure pedology in that they carry an implication of total soil morphology rather than individual parameters. The American terms may be more convenient for purposes of irrigation planning. Both systems of nomenclature have equal status and their concurrent use causes few difficulties.

SOLONCHAKS

CCTA: Saline soils, alkali soils and saline alkali soils p.p.; soils of sebkhas and chotts
FAO: Solonchaks
US Salinity Laboratory: Saline soils, saline–alkali soils

Solonchaks are soils with properties dominated by the presence of free salts. The limiting values on the FAO classification are an horizon within 125 cm of the surface with TSS > 2 percent and/or EC > 4 mmho/cm at some time of the year. They occur exclusively in arid areas, usually in depression sites, and in respect of site are analogous to gleys of humid climates.

Colours are dark grey to greyish brown, modified by white salt crystals

TABLE 22   *US Salinity laboratory classification of halomorphic soils. After Richards (1954)*

EC of saturation extract (mmho/cm)	ESP (%)	Soil type
> 4	< 15	Saline soils
> 4	> 15	Saline–alkali soils
< 4	> 15	Nonsaline–alkali soils

when the soil dries out. If the soil is moist when a profile pit is dug, crystals are precipitated on the sides of the pit. There may be mottling caused by drainage impedance in depth. Solonchaks can have any texture. The clay is flocculated to give a loose granular or blocky structure, and permeability remains adequate. Where salinization is principally in depth the topsoil may have no exceptional properties. Where it is at the surface there may be a white salt crust covering the ground, or alternatively a very loose 'fluffy' surface caused by the growth of long needle-like crystals of sodium sulphate. The pH is 7.0–8.5.

Solonchaks may retain for a time an exchange complex dominated by calcium, or may become solonized. As long as free salts remain these mask the effects of a high exchangeable sodium percentage, and the profile retains the structure and permeability of a solonchak. It is nevertheless important for reclamation purposes to map which soils are solonized, and these may be distinguished as sodic and non-sodic solonchaks or by the US terms saline (or more specifically saline–nonalkali) and saline–alkali soils.

In soil surveys in which large numbers of samples are taken for analysis it is possible to map isopleths showing the distribution of electrical conductivity in the topsoil and at sampled depths. This has been done for an area of 530000 ha in the Sudan (Williams, 1968*b*). Saline soils were concentrated mainly in depressions and especially around the margins of former lakes (relict from more humid climates). Salinity was greatest in the second horizon, 36–100 cm. Topsoils were frequently non-saline, becoming saline only in closed depressions.

SOLONETZ

CCTA: Solonetz and solodized solonetz p.p.
FAO: Solonetz
US Salinity Laboratory: Nonsaline–alkali soils

Solonetz are soils without free salt crystals but with an exchangeable sodium percentage exceeding 15 percent. They frequently occur as small dark-coloured patches amid areas of solonchaks. There is a thin sandy A horizon from which all clay has been eluviated. This overlies a compact, grey, heavy-textured horizon of high density. There is typically a columnar structure, similar to prismatic but with rounded tops to the peds; this is the only soil type in which this structure develops. Where solonization is a consequence of irrigation the structure may not have had time to develop and the B horizon is structureless. The dispersed clay particles become packed close together, creating very small pores and a low permeability. The pH has the exceptionally high values, also not found in other soil types, of 8.5–10.0. Humus is leached from the sandy topsoil and diffused through the B horizon. The lower part of the profile is frequently mottled. For these characteristics to develop it is necessary for a certain amount of clay to be present; sands subjected to the same soil-forming processes do not develop the properties of a solonetz.

For soils of a given texture, permeability is related both to salinity and to exchangeable sodium status. The idealized situation is that saline–nonalkali soils are no less permeable than non-halomorphic soils; saline alkali soils have a partly-reduced permeability, maintained by the flocculating effect of the free salts; whilst permeability becomes very low in nonsaline–alkali soils. From permeability measurements on 157 soil samples of the Gezira, Sudan, Williams (1968c) found that the actual situation did not agree with this model; saline–nonalkali soils had a lower average permeability than non-halomorphic soils, whilst soils high in exchangeable sodium were very impermeable whether saline or not.

SOLODS

CCTA: Solonetz and solodized solonetz p.p.
FAO: Solodic planosols

Solods have an A horizon of bleached sand with a irregular lower boundary, which tongues into a compact B horizon with columnar structure. They are acid and not sodium-saturated. These soils are very infrequent. It has been held that a number of soils in northern and central Australia with marked duplex textural profiles originated by solonization followed by leaching and solodization.

In Malaŵi and Rhodesia there is a soil type known locally as 'mopane soils' from the fact that the vegetation is dominated by *Colophospermum mopane*. This tree is xerophytic and shallow-rooting. Mopane soils occur at low altitudes, under a rainfall of 400–650 mm with very high temperatures, and are extremely poor agriculturally. They were formerly thought to be solonetz. In a study of the Sabi Valley, Ellis (1950) found that neither with respect to salts nor to sodium ions or pH did they differ from more fertile adjacent soils carrying *Acacia*. A sandy topsoil, 3–15 cm thick, overlay an impervious subsoil with 40 percent clay and a pH of 8.3; exchangeable sodium was only 0.7 m.e./100 g, an ESP of 4 percent. These appear to be nearer solods than solonetz. It is likely that *mopane* occurs on more than one kind of soil (Thompson, 1960). It is also found on shallow sandy soils, in the Lower Shire Valley of Malaŵi and in Wankie Game Reserve in Rhodesia.

AGRICULTURAL PROBLEMS OF HALOMORPHIC SOILS

Salinity affects crop growth primarily through osmotic effects and second-arily through the toxic effects of certain ions. Water uptake by plant roots is brought about by the difference in osmotic pressure between the solution within root hairs and the surrounding soil solution, and a high salt concentra-tion in the latter inhibits this mechanism. The chloride ion is toxic to some plants. Boron may accumulate in irrigated soils; although an essential micro-nutrient, it becomes very toxic at concentrations of more than 4 p.p.m. The sodium ion is not toxic, but where present in large amounts it may disturb the nutrient balance.

A total soluble salt content of 2 percent, or an electrical conductivity of 15 mmho/cm is the approximate limit at which the yields of even salt-tolerant crops become seriously affected. At much lower levels of salinity the growth of most crops is stunted, the leaves become blue-green and yields are depressed. Levels for the salt and boron tolerance of different crops were established by the US Salinity Laboratory, taking as a criterion of intolerance the level at which yields fell by 50 percent. Tables of salt tolerance are published (Richards, 1954, p. 67; Allison, 1964, p. 160; Rauschkolb, 1971 p. 24). Highly salt-tolerant tropical and subtropical crops, yielding satis-factorily at EC levels of up to 15 mmho/cm, are date palm, barley, cotton and some species of fodder grasses. Barley and cotton in particular are widely grown where there is a salinity hazard. Moderately salt-tolerant crops include maize, sorghum, sunflower, most vegetable crops and alfalfa. Some cultivars of rice can be grown in mildly saline conditions, but yields are depressed. Citrus fruits have a low salt tolerance.

The main injurious effects of solonetz soils are caused by poor physical structure. The dense, impermeable B horizon inhibits root growth, gives poor aeration, and disturbs control of irrigation water. Where irrigated they flow when wet and then dry out to give a smooth, structureless surface like a polished floor. Caustic alkalinity is also a hazard, and some plant roots and organic matter are dissolved at pH values exceeding 9.0. Cotton, barley, alfalfa and some grasses are tolerant of moderately high ESP levels. In the Gezira, Sudan, cotton was found to give good yields on sodic vertisols with ESP levels up to 25 percent and satisfactory crops up to 35 percent, but this is in part associated with the otherwise favourable properties of vertisols (Robinson, 1971). Fruit crops are very sensitive to sodium satura-tion. As the main problems are caused by low permeability, ESP values are a very imperfect guide to soil suitability.

Salinity and sodium status are impermanent characteristics which can be altered by reclamation procedures. Therefore when classifying land for irrigation purposes, these should be kept separate from other criteria of suitability such as texture and permeability.

IRRIGATION AND SALINIZATION

The irrigation of arid areas without measures to prevent salt accumulation has in the past led to the salinization of extensive areas. Not all irrigation water is used by plants, and part of it percolates downwards to groundwater. The addition of water in excess of the supply under natural conditions causes a rise in the water table. When the capillary zone above this reaches the level of plant roots, evaporation and transpiration occur; if the groundwater is saline, as is common in arid areas, salts are precipitated. The effects on crops are not immediate, and satisfactory growth continues for some years whilst the level of salinity increases. Yields are reduced and then approximately circular bare patches appear in which germination has failed, surrounded by belts of stunted growth. The salinized patches, originating in slightly lower lying areas, then spread to render large areas uncultivable. Subsequently dark-coloured alkali patches, known as slick spots, may develop within the salinized areas.

The most extensive example of salinization occurs on the alluvial plains of the Indus and upper Ganges. In the 1890s, prior to the construction of modern irrigation works, the water table lay at a substantial depth, about 20 m in the Punjab. The construction of large-scale canal irrigation works and the irrigation of vast areas of former desert led to the transfer onto the land of a large amount of water which had formerly flowed via the rivers to the sea. The water table began to rise at 30–50 cm a year. At about the time of the Second World War it reached to within three metres of the surface over substantial areas, and salinized patches appeared and spread. These have a spectacular appearance from the air, their grey rounded forms cutting across the rectangular field patterns of planned irrigation settlements. On the ground the low bunds marking the former field boundaries may remain within the salinized areas. By 1960 some 15 percent of the Pakistan Punjab had been rendered uncultivable and land was being lost at a rate of 20000–40000 ha a year. Salinization is worst toward the centres of the interfluves where the groundwater is more saline, seepage from the rivers reducing the salinity closer to the channels.

Salinization can only occur in arid regions. Where more humid regions are irrigated any salts are removed by leaching from natural rainfall, but waterlogging can be a problem.

The salinization hazard is now well recognized and with good management need not occur again on a major scale. The measures necessary for its prevention are drainage, the application of sufficient water to leach salts, and avoidance of the use of saline water. A system of deep drains is necessary to prevent a rise in the water table and so permit applied water to drain downwards through the profile. To produce leaching of salts it is necessary to apply water in excess of the irrigation requirement, the amount which will be lost by evaporation from the soil surface and transpiration by crops. The percentage excess is termed the *leaching requirement* (LR) defined as:

$$LR = \frac{EC_{iw}}{EC_{dw}} \times 100$$

where EC_{iw} is the electrical conductivity of the irrigation water and EC_{dw} that of the drainage water, taken as the value of conductivity that can be tolerated with the root zone by the crops being grown (Richards, 1954). Thus for crops where $EC_{dw} = 8$ mmho/cm and irrigation water with a conductivity of 2 mmho/cm, the leaching requirement is 25 percent, and the water applied is 125 percent of the estimated crop use plus bare soil evaporation. Greater amounts must be applied to initially saline soils.

Irrigation waters are divided into four classes, of low, medium, high and very high salinity, with class limits at EC values of 250, 750 and 2250 μmho/cm. Low and medium salinity waters are preferable, and 2250 μmho/cm, approximately equivalent to 1500 p.p.m. total soluble salts, is the upper limit for standard practice. Besides the salinity hazard, potential irrigation supplies are appraised in respect of sodium, boron and bicarbonate hazards. The sodium hazard is assessed by the sodium absorption ratio; permissible values vary with salinity and other circumstrances but should not normally exceed 13. To avoid the danger of boron accumulation in the soil the content in irrigation water should not exceed 2 p.p.m. The bicarbonate hazard arises from the possibility of precipitation of calcium carbonate in the soil accompanied by the release of sodium. It is measured as the residual sodium carbonate content in milliequivalents per litre. Water with a residual sodium carbonate content exceeding 2.5 m.e./l is unsuitable for irrigation, and values of 1.25–2.5 are marginal. Recommended limits for various water quality criteria for irrigation waters are given by Rauschkolb (1971, p. 30).

There are areas in which saline groundwater is the only (or the cheapest) source of irrigation supplies, and experiments in the use of saline waters for irrigation have been conducted (Kelley, 1963; Boyko, 1966, 1968). In some cases it is possible to mix saline groundwater with fresh canal water. Another possibility is to alternate between saline and non-saline water; saline irrigation is practised until salinity builds up to a level at which yields fall; the next two or three irrigations are then fresh water to leach out the salts. The cost of desalinating seawater is of the order of £0.50 per cubic metre, equivalent to £10000 per hectare for an application of 2 m depth (1975 values). At present this is economically out of the question in normal circumstances, but there is clearly a vast reserve if desalination costs can be substantially reduced (e.g. using solar power) and land becomes a more scarce resource.

Reclamation

Non-sodic solonchaks can be reclaimed by leaching out the soluble salts. As the soil is permeable this is a relatively straightforward procedure. The first stage is to install deep drains. Water considerably in excess of the irrigation requirement is then applied, usually by basin and flood irrigation. Owing to the high water requirement, the reclamation procedure is taken successively round different areas. After salts have been reduced to acceptable levels the land can be returned to cultivation under normal irrigation procedures, ensuring the salinization does not recur by the practices noted above, then monitoring soil changes by a programme of annual or longer term sampling and analysis.

Saline coastal soils can be reclaimed by the procedures established in the Netherlands, of diking, draining out of seawater and flooding with fresh water, followed by the growth of ameliorative crops. Pumping may be necessary to keep the water table low. Land productivity in the tropics does not normally justify the reclamation of land below sea level.

Solonchaks are more difficult and expensive to reclaim owing to their low permeability. The need is to bring about replacement of sodium by calcium ions in the exchange complex, followed by removal of the released sodium. The main difficulty is to establish drainage through the impermeable B horizon and so get water to enter the soil. Soluble calcium salts are applied, most commonly gypsum; this releases calcium ions in solution, which replace sodium. Calcium carbonate has too low a solubility to be of use in the early stages of reclamation. If the soil already contains calcium carbonate, sulphur may be used; this reacts to form calcium sulphate and carbonic acid, the former then acting in the same manner as gypsum and the latter increasing the acidity. Where impermeability is serious, there is a technique of first flocculating the soil by flooding it with saline water of a lower sodium absorption ratio than that of the saturation extract of the soil being reclaimed; this is followed by a series of leachings with successively less saline water, ensuring that on each application the SAR of the soil comes into equilibrium with that of the applied water, otherwise loss of permeability may result (Reeve and Bower, 1960).

SELECTED REFERENCES

Richards (1954), Thorne and Peterson (1954), Allison (1964), FAO (1973*a*).

CHAPTER 12

GLEYS AND ALLUVIAL SOILS

There is an overlap between gleys, alluvial soils, 'padi soils' and vertisols. Not all alluvial soils are gleyed; and whilst gleys often incorporate some materials of alluvial origin, many do not show the features characteristic of alluvial soils, namely depositional bedding and immature profile development. Padi is most widely cultivated on alluvial soils, but sometimes on other gleys and, with terracing, on originally non-gleyed soil types. Some vertisols occur in low-lying sites with a high water table, but by no means all dark-coloured clays of the tropics possess the distinctive properties of vertisols.

In soil classification, it is generally accepted that vertisolic properties take precedence over gleying. Padi soils, whilst a valid and necessary class for the study of fertility, are not a soil mapping unit. The main problem of overlap therefore arises between gleyed alluvial soils, non-gleyed alluvial soils and non-alluvial gleys. Here a practical consideration dictates that alluvial properties shall form the higher category of classification: gleyed and non-gleyed alluvial soils occur in associations, often indistinguishable from each other on air photographs. Hence the FAO classification (after initially including 'fluvic gleysols' in the unamended 1968 version) has defined the *fluvisol* (alluvial soil) class in such a way as to cover both gleyed and non-gleyed profiles, whilst the *gleysol* class excludes profiles formed on recent alluvial deposits.

In the following discussion, non-alluvial gleys will be considered first, followed by organic soils, alluvial soils, acid sulphate soils, and lastly distinctive features of soils under padi.

GLEYS

CCTA: Mineral hydromorphic soils
FAO: Gleysols

Comparatively little attention has been given to gleyed soils of the tropics. Most texts discuss alluvial soils, but give no attention to gleys as such; and, except for the special case of padi soils, there has been comparatively little research on them.

This neglect could be justified if gleys were of small extent or low agricultural potential, but neither of these is the case. Both in the rainforest and

213

savanna zones, gleys occupy something of the order of 10 percent of the total area. There are physical difficulties to their cultivation owing to high clay content, together with a severe flood hazard; but in terms of plant nutrients and organic matter, gleys are frequently superior to the freely-drained soils with which they are associated.

Gleys occupy poorly-drained sites in the savanna and rainforest zones. Two kinds of site may be distinguished: valley floors and broader swamp zones. In valley floors, gleys are the lowest member of the catena, and the material from which they are formed is partly colluvial in origin. In extensive low-lying areas, gleys are more likely to have some alluvial characteristics.

Dark grey or mottled clays are a widely distributed feature of the gently undulating plains typical of the savanna zone. They are frequently recognized by a vernacular term, e.g. *mbuga* (Swahili), *dambo* (Ci-Cewa), although this is more a translation of 'meadow', grazing land, than of the soil type as such. In the savannas the valley floor is often continuously concave in cross-section, whereas in the rainforest zone a level flood plain is found even in first-order streams. In the savannas, the clays are often bordered by zones of mottled sandy soils or by laterite outcrops.

In terms of classification, gleys grade into soil types in which gleying is present, usually in the lower horizons only, but does not dominate the profile. Such soils may be called *gleyed*, e.g. gleyed ferruginous soil. In the FAO system the requirement for gleysols is that hydromorphic properties, such as bluish grey colours or prominent mottling, should dominate the morphology within 50 cm of the surface; this corresponds to the depth at which crops would become severely affected by water-logging. The prefix 'gleyic', e.g. gleyic acrisols, is applied to profiles in which hydromorphism is subsidiary to other properties. Rather surprisingly, a 50 cm limit is again specified, and from the definitions alone it appears impossible to have gleyic soil types that are not gleysols; only one introductory sentence to the definition of 'hydromorphic properties' makes the distinction clear (FAO–Unesco, 1974, p. 29).

In gleys of the savannas, the topsoil usually consists of a very dark grey to black heavy clay. In some profiles this continues to 2 m or more, whilst in others it gives place to a pale grey horizon with prominent mottling. Close to the surface, grass roots confer a medium blocky or even crumb structure. In depth the structure is very coarse prismatic, with peds typically $20 \times 20 \times 40$ cm, covered with thick clay skins. The soil is hard when dry, plastic and very stiff when moist, and sticky when wet. Whitish flecks or concretions of calcium carbonate may occur in depth. A few small black rounded iron concretions are sometimes found.

If the upper-catena soils are very sandy, the gley may be a pale grey bleached sand or loamy sand. A brown clay with an angular blocky structure was noted by Milne (1947, p. 208). In general, however, the morphology of gleys varies less with parent material than does that of the corresponding upper-catena soils. There are often substantial similarities between gleys of the tropics and those of temperate latitudes.

In the rainforest zone, gleys remain waterlogged for all but short periods. The topsoil is typically a mixed organic and mineral (muck) horizon; this overlies a mottled clay or sandy clay, below which is material subject to permanently anaerobic weathering.

Gleys compare favourably with corresponding upper-catena soils in respect of nearly all analytical properties. They have substantially higher organic matter levels, a higher cation exchange capacity and base saturation, and usually also higher phosphate and potash levels. Savanna zone gleys are weakly acid in the upper horizons, becoming approximately neutral in depth. These properties, whilst partly the result of the heavy texture and slower organic matter decomposition, originate also by the transfer in solution of ions from higher parts of the catena.

Agricultural properties

In the savanna zone of Africa, the potential of gleys is greatly under-utilized. They are normally under permanent grazing, with communal rights. Overgrazing is normal, trampling and erosion common, and positive measures of pasture improvement rare. Good pasture management requires, as a minimum, that these grasslands should be seasonally rested, by taking the cattle onto hill pastures during the wet season and possibly onto crop residues after harvest. For more sophisticated management, fences and rotational paddock grazing systems are necessary. Correctly managed, the fertility of gley soils is such as to permit stocking densities as high as two Livestock Units per hectare.

Many years ago Milne wrote: 'The question whether means cannot be devised for realizing its potential fertility, in spite of annual waterlogging and forbidding physical properties, demands serious consideration . . . A tentative suggestion may be made that the heaviest *mbuga* soils are well fitted to grow a large bulk of grassy material such as uba sugar-cane or Napier grass. This could be used both as a cattle fodder after making into silage in pits . . . and as a raw material for compost making . . . If the *mbuga* lands can by some such scheme be fully utilized and the produce devoted to the sandy soils . . . the purpose will gradually be achieved of *bringing fertility uphill*, counter-current to the natural tendencies' (Milne, 1947, pp. 263–4). Systems of mixed farming, using the fertility of gleys for meat and milk production, and physically restoring some nutrients to the upper catena soils by compost or, perhaps more realistically, farm-yard manure, have yet to be widely developed. The major obstacle is not, probably, that of agricultural education, but of making enabling institutional arrangements with respect to land rights and tenure.

An alternative use for gleys is for padi cultivation. Growth is satisfactory, with bunding but without irrigation, in the wetter parts of the savannas, and there is a good urban market for rice in most African countries. With labour abundant and a rising population there is pressure to expand rice production. However, whilst considerably more productive than even intensive livestock management in terms of calorific yield, and hence human 'carrying capacity', padi cultivation neither satisfies the equal or greater dietary need for animal protein nor integrates with farming of the upper-catena soils. It is better policy to confine padi cultivation to more extensive areas of gleys, particularly those where irrigation is possible, and to utilize valley-floor gleys for intensive livestock production.

In Asia, gleys in humid climates are almost invariably under padi. Where they occur in narrow valley floors, these are converted into broad steps each with a level terrace and water-retaining bund, giving a most picturesque effect as of contours. Here the problem is not under-utilization of the gleys but the reverse: so prized is the padi land that what is referred to as 'high' land is much less valued and suffers relative neglect.

In the rainforest zone it is possible to use gleys for tree crops, for example oil palm, through lowering the water table by deep drainage ditches. This is the practice on some estates, where rice production is of no interest and the high value of the crop justifies the expense of ditch digging and maintenance. In a narrowly economic analysis, drainage and the cultivation of perennials is currently more profitable than padi. On more general grounds of resource use, however, particularly where smallholder production is involved, it is better to allocate gleys to padi and restrict cash crops to the upper catena soils.

ORGANIC SOILS

CCTA: Organic hydromorphic soils, organic non-hydromorphic soils
FAO: Histosols

Organic soils are less extensive in the tropics than in the temperate zone, and have a lower agricultural potential. They are divided into *peats*, consisting largely or entirely of organic matter, and *mucks*, consisting of mixed organic and mineral matter. This distinction is not recognized in the FAO classification, in which the limits for-histosols are set at an horizon at least 20 cm thick containing over 20 percent organic matter if the clay content is zero, rising proportionately to over 30 percent where the mineral fraction has over 60 percent clay.

On the basis of site, montane, river and coastal organic soils may be distinguished. Montane peats are only reached in the tropics at high altitudes, 3000 m or more, and are of very small extent. On river flood-plains the soils are non-organic gleys if flooding is only seasonal, only developing into peats with permanent or nearly-permanent submergence; papyrus (*Cyperus papyrus*) and phragmites (*Phragmites communis*) are common vegetation types. On coastal plains, organic soils, mainly mucks, develop under mangrove swamps. Organic soils cover less than one percent of the tropics, becoming locally important in inland swamps such as the Sudd and Lake Bangweulu, and coastal plains such as the west coast of Malaya and the coast of Surinam–Guyana.

There are severe problems with the agricultural use of tropical organic soils. Flood control is expensive because of the large seasonal and random variations on river discharge. If drained, the soils undergo rapid oxidation leading to subsidence. In brackish water coastal sites, acid sulphate soils may develop (p. 225) and trace element deficiencies can occur.

Locally, small areas on flood-plains are used for intensive vegetable cultivation. In Malaya, pineapple growing was formerly on latosols, and led to severe soil erosion; the industry has now been transferred almost entirely to peat soils on the Pontian coastal plain of south-west Johor, the water table being kept by drains at about 1 m depth. Organic soils are not well suited to padi. Low-quality firewood from mangrove swamps can be turned into charcoal, whilst from inland swamps, reeds for thatching are a useful product.

ALLUVIAL SOILS

(Profile 6, p. 433)

CCTA: Juvenile soils on riverine, lacustrine and fluvio-marine alluvium
FAO: Fluvisols

Soils derived from alluvium occupy a distinctive and important place in tropical agriculture. The alluvial plains of Asia have for many centuries supported high population densities; the population of the Indo-Gangetic Plains approaches that of the whole of Africa, whilst the Irrawaddy, Mekong and many other rivers of South-East Asia also carry large populations. In Africa and South America the productivity of alluvial soils is more often potential than actual. The fact that the land on which they occur is flat gives them a special potential for irrigation, and alluvial soils are widely used for land development schemes.

Alluvial soils occupy some two percent of Africa, over three percent of South America, and an unknown but certainly greater proportion of tropical Asia. They cover a quarter of the area of India, and a high proportion of the agriculturally productive land of both Pakistan and Bangladesh. Because of their flatness and high water table they are particularly, but by no means exclusively, used for padi cultivation.

Together with this distinct potential alluvial soils present special problems of description, survey, evaluation and management. They are an azonal soil class, possess relatively few properties in common, and these properties vary greatly over short distances. Hence description by means of modal profiles is difficult, and special sampling methods are necessary in survey. Management is particularly concerned with the effects, and control, of water in the profile.

Definition and types

If narrowly defined, alluvial soils are those showing indications of concurrent or very recent sedimentation. They have (A)C profiles, the properties being determined primarily by sedimentation, relatively little modified by pedogenetic processes other than humus formation.

There are, however, a large number of soils derived from alluvium which are no longer subject to flooding, and in which appreciable horizon development has taken place; they have A(B)C or even ABC profiles. Thus, on the 'older alluvium' of the Ganges Plains in Uttar Pradesh, a textural as well as a structural B horizon is not uncommon. The term *alluvial soils* is used here in this broader sense, genetically defined as soils derived from alluvium, including calcimorphic and gleyed profiles, but excluding alluvially-derived vertisols and halomorphic soils.

The FAO definition of *fluvisols*, although no longer (as in the original 1968 version) necessarily requiring current deposition, is more restrictive than the above genetic definition, requiring one or more characteristics of immaturity (an irregular organic matter profile, fine stratification) unless current deposition occurs. Some soils derived from older alluvium would fall into one of the 'mature' soil types, e.g. cambisols, luvisols or xerosols.

Alluvial soils *sensu lato* may be subdivided on the basis of genesis, maturity, site or properties. Genetically they may be of fluvial, lacustrine or marine origin. With respect to maturity, those which are and are not subject to current flooding may be distinguished; division of the latter group on the basis of extent of modification of the profile presents difficult problems in the definition of class limits.

The main types of site are alluvial flood-plains, river terraces, alluvial fans, former lakes, coastal plains and deltas. This classification is not wholly coterminous with that of genesis, for whilst the first three sites are wholly fluvial in origin, the remainder contain both fluvial and lacustrine or marine sediments.

In terms of properties, a major distinction is between *calcareous* and *non-calcareous alluvial soils*, the former having a pH above 7.0 and free calcium carbonate in the profile. On the FAO system the former are *calcaric* fluvisols and the latter *eutric* or, with base saturation below 50 percent, *dystric fluvisols*. Other distinctions of use on local surveys are between *gleyed alluvial soils*, with gleying reaching to the surface, and profiles ungleyed or gleyed only in depth; and between sandy, silty and clayey profiles. Finally there are soils of alluvial origin with special properties, including acid sulphate soils.

Properties and processes

The most significant property associated with alluvial soils is that the land on which they occur is flat. It is this which gives them a special potential for padi cultivation and for irrigation. A second common, but not universal, property is a high water table, hence the gleying within the profile. Textures vary from coarse sand to heavy clay; a high degree of particle-size sorting in any one horizon is characteristic, and depositional bedding common. Hardpans may occur, particularly of calcrete. In India there is a widely recognized distinction between 'older alluvium', *bhangar*, in which calcrete (*kankar*) is abundant, and newer alluvium, *khadar*, in which it is infrequent. The reaction is less acid than non-alluvial soils of corresponding climates, and the pH is frequently 6.0–8.0. For example in Bangladesh, with a monsoon rainforest type of climate, typical pH values are 6.0–6.5 on the older alluvium and 7.0–8.5 on the newer alluvium of the Ganges; on the currently inundated areas of Brahmaputra alluvium, however, it may fall to 5.5 (Islam, 1966). Organic matter levels are slightly higher than for non-alluvial soils of comparable texture and climate.

No conditions may be called typical, but a widespread situation (e.g. in Uttar Pradesh) is that of a recent alluvial soil under a 500–1000 mm rainfall, with the water table rising seasonally up to 1–2 m depth. Such a profile is likely to be yellowish brown to dark greyish brown. The texture is dominated by fine sand or silt, with occasional coarse sandy layers. The heavier-textured horizons have a weak medium blocky structure and a firm consistency, becoming hard when dry, whilst sandy horizons are structureless. The pH is weakly acid to weakly alkaline at the surface, perhaps rising to 8.0 in depth, with base saturation varying correspondingly between 90 and 100 percent. Calcium carbonate concretions and/or mottling occur in depth. The organic matter level varies from 2.0 to as low as 0.5 percent, depending on cultivation history and topsoil texture.

Contrasting textural and drainage types in the same climatic zone include coarse sandy, structureless profiles, typically brown in colour and more acid; and black heavy clays, with a strong very coarse blocky structure, sometimes with vertisolic tendencies.

Processes can be divided into those of sedimentation, initial soil formation, processes normal to the climatic zone, and processes peculiar to alluvial soils.

Sedimentation takes place where the speed of flowing silt-laden water is reduced. The particle sizes deposited are dependent on the velocity of flow. A basic distinction on flood-plains is between coarse sandy sediments of levees and braided channels, fine sands and silts of the main part of the plain, and clays of backswamps and other depressions. The patterns are further complicated by intermittent changes of course of the main channels and, in the longer term, lowering of the active flood-plain to leave terraces. In deltas, levees form branching patterns and swamps are relatively more extensive. On coastal plains (and the shore of some large lakes) a common pattern is a series of progradational sand bars separated by swales with gleyed soils; such a pattern occurs for example on the east coast of Malaya (Nossin, 1962), the coasts of Ghana (Boughey, 1957*b*), the Guianas (Brinkman and Pons, 1968), and south-east facing reaches of the shore of Lake Nyasa. Mangrove swamps have a different depositional mechanism, that of vegetation retarding the flow of seawater or estuarine river water, giving clays with a high content of partly-decomposed vegetation.

Under traditional flood irrigation the land acquires an annual layer of silt, which in some areas long under such methods has reached a thickness of one metre or more. Where large dams have been constructed much of this sediment unfortunately fills the reservoir instead.

In some soils of semi-arid and arid regions a process of *micro-lamination and self-sealing* occurs. Platy silt-sized fragments, particularly micas, become arranged during deposition in a horizontal overlapping manner. This gives marked anisotropy, making the soil almost impervious to vertical water movement. Such layers can be broken by tillage but may re-form each time the soil is flooded, either naturally or by irrigation (Radwanski, 1968).

There are two main processes termed *initial soil formation*: ripening and homogenization. *Ripening* affects clays. Clays deposited under water are loosely packed, contain much water and are in reduced forms; they have a low load-bearing capacity, and are termed *mud-clays*. Ripening processes include the draining and evaporation of excess water, development of drying cracks, and loss of part of their organic matter by oxidation; this leads to consolidation and subsidence of the clay, with an increase in its strength. Accompanying these changes is the oxidation of some of the iron compounds to ferric forms. The ripening process is irreversible: once consolidated, the clay if flooded will not take up water to the former amount (Pons and Zonneveld, 1965).

Homogenization is the elimination of features of depositional layering, mainly textural but also features of chemical composition. In the early stages the main agents are biological: plant roots, worms, termites, micro-organisms and, in mangrove soils, crabs. Over a longer period, clay translocation comes to play a part.

In Bangladesh, many soils of active flood-plains remain submerged from June to September. This gives rise to a distinctive group of processes, which may quite probably occur in other deltaic areas. Peds and pores are conspicuously coated with grey, reduced, fine silt and clay, which unlike normal clay skins is not, or is only weakly, orientated. The material may be derived by translocation from the topsoil, but is deposited when the soil is submerged. These gleyed clay skins have been termed *gleyans* (Brammer, 1971). Some soils, calcareous when deposited, subsequently become decalcified but without any corresponding lime deposition in depth. Decalcification probably takes place under submergence (Brammer, 1968). Annual submergence leads to a variety of other effects, including the change of iron, manganese, nitrate and sulphate ions to reduced forms, with consequent fertility problems, and changes in structure, permeability, and other physical properties (Islam, 1966).

Normal pedogenetic processes commence concurrently with intermittent sedimentation, and come to have a greater effect if it ceases. The extent of leaching of salts, carbonates and bases depends on the rainfall and depth of the water table. With a water table at a depth of about 100 cm, calcimorphic properties can occur under a rainfall of 600–750 mm, which under free drainage would give ferruginous soils. In dry climates, re-deposition of calcium carbonate in depth is common, whilst salinization occurs in depressions.

Distribution patterns

The spatial variation of properties in alluvial soils may be rationalized by reference to climate, sediment source and depositional pattern, each factor operating on a different scale.

The climatic factor, primarily rainfall, differentiates between broad areas with different chemical properties. Thus in the Indian sub-continent there are broad distinctions between the arid-zone flood-plains of the Punjab, Sind and Rajasthan, intermediate conditions of Uttar Pradesh and the deltas of the eastern Deccan, and the wet monsoonal alluvium of the Ganges–Brahmaputra delta. In the arid zone, reaction is normally alkaline, and calcrete and saline soils are common. Alluvial soils of climates with a savanna regime, including those monsoonal in origin, are characterized by neutral to weakly alkaline reactions, sometimes with calcium carbonate concretions in depth. High rainfall zones have numerous gleyed alluvial soils, but reaction may become moderately acid where older alluvium has been leached.

Locally, alluvial soils may vary according to the rock type in the source-basins for their sediments. This affects texture and some other properties, including clay minerals. A clear example occurs in the lower Shire valley of Malawi. On the right bank of the river the alluvium is derived not from the Shire itself but from streams (or sheet-flooding) from the consolidated rocks to the west. The alluvial soils show a relation to the adjacent rock outcrops. Thus, opposite to Karroo sandstones are sandy, infertile soils, whilst adjacent to basalts the alluvium is clayey and montmorillonitic, forming an alluvial type of vertisol (Muir and Stephen, 1956; Young, 1959/60).

On a local scale, the main differentiation of alluvial soil types is between coarse, medium and fine-textured profiles, and between those with a high and a low water table. These properties are determined by the processes of sedimentation.

Soil survey in alluvial areas

In regions where alluvial soils form only one mapping unit among many, i.e. relatively narrow flood-plains amid high land, the normal survey procedure is to show the whole alluvial area as one mapping unit, a soil complex. Some indication may be given of the types of profile that occur, but on account of the complexity of the distribution pattern, no attempt is made to map these. This procedure is justified, in that it supplies sufficient information for planning the allocation of land use types. A refinement is to include random cross-valley traverses to give an estimate of the relative extent of different types (e.g. sandy profiles and clays). This leaves open the option of subsequent more detailed survey if justified by planning requirements.

The extensive areas in which alluvial soils are the only or the predominant soil type offer special problems of survey. In such cases it is the non-alluvial land that is separated for special purposes, e.g. grazing, and often excluded from detailed survey.

There are two approaches to the survey of such alluvial plains: grid survey and survey based on photo interpretation. In grid survey, auger bores are taken at regularly spaced intervals, with profile descriptions (often simplified to exclude invariable properties) and samples at standard depths. Mapping units can be established based on a few properties, e.g. texture, salinity and gleying, with boundaries between classes set at limiting values chosen on the basis of agricultural significance. Such an approach is appropriate for plains with an apparent high degree of landform and soil uniformity, and has been used in the Sudan. It is also the only practicable way if the land is flooded prior to the construction of proposed drainage works.

Where possible, however, a method based on mapping of landform units by air photo interpretation is preferable. It has been widely demonstrated that pedologically significant landform units on alluvial plains can be identified and delineated on air photographs, making use of tone, photographic texture, pattern (e.g. of channel remnants) and small height differences. This method has been successfully applied in the deltas of Senegal (Tricart, 1956), Niger, the coastal plain of the Guianas (Brinkman and Pons, 1968) and over a huge area of the Indus Plains (Hunting Surveys

Corporation (Canada), 1958; Hunting Technical Services, 1966; Holmes and Western, 1969). On the Indus Plains, mapping units recognized include (fig. 14):

1. Active flood-plains.
2. Meander plains. Former active flood-plains of abandoned river channels, recognizable by abundant meander scars.
3. Levees. Fairly continuous zones along the margins of 1 and 2.
4. Cover flood-plains. Extensive areas occurring beyond levee belts, in which deposition was by relatively stagnant escaping floodwater, and appearing uniform on air photos.
5. Backswamps. Clayey depressions occupying the lowest parts of the cover flood-plain, and having a darker tone on photographs.
6. Infilled channels.

Each of these units possesses a distinctive range of soil textures. By determining these it is possible to map *textural associations*, each consisting of a specified range of texture profiles (e.g. sand; 10–50 cm sand over clay; 50–200 cm sand over clay). An example of the mapped pattern of textural associations is given in fig. 14.

Mapping based on air photograph interpretation of depositional units is preferable to grid surveys wherever it can be applied. Its advantage is to replace systematic sampling of a single highly heterogeneous area by stratified sampling of more homogeneous mapping units. Only when the limit of resolution by photo-interpretation has been reached are grid surveys justified. A corollary of this is that training for the survey of alluvial areas should include a study of sedimentation processes.

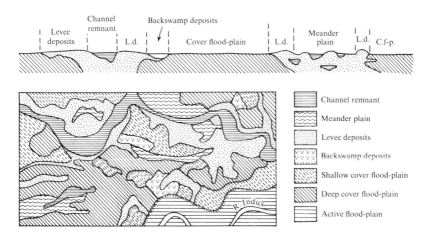

Fig. 14. Depositional cross-section and map of textural associations on the Indus alluvium. Based on Holmes and Western (1969).

Agricultural evaluation

There are two common land-use situations on alluvial plains. Either the land is largely to entirely unused, and under jungle, swamp or desert vegetation; or the whole plain is continuously cultivated and densely settled. The former situation is more common in South America and Africa, the latter in Asia. One reason for this contrast is the need for control of water, through flood protection, drainage and irrigation. Until control is effected by some organized community, individual cultivators will be at the mercy of flooding and other hazards.

Given a water control system, the two great agricultural advantages of alluvial soils are the ease of water distribution and relative freedom from erosion problems, both deriving from the property of flat land. Long-distance transport of irrigation water can be by canal, and field distribution by the cheapest method, that of basin flooding.

The natural fertility of the profile is usually not high, except in special circumstances such as alluvium derived from basic rocks. The high fertility of temperate zone alluvium is not found in the tropics; reasons for this include the often sandy texture, the fact that the sediment is derived from rocks already highly weathered and contains a high proportion of quartz and muscovite, and the more rapid destruction of organic matter (Edelman and van der Voorde, 1963). The low available water capacity of sandy profiles is sometimes counteracted by continuous availability of water in depth. Alluvial clays are more fertile, but difficult to cultivate. Other common problems are compactness, the formation of pans, natural salinity and, under irrigation, salinization.

Notwithstanding these limitations, alluvial soils have a high development potential by reason of comparative freedom from erosion, ease of irrigation water distribution, and potential for double cropping. Crops particularly associated with alluvial plains are padi and jute, both requiring standing water. Sugar cane is a field perennial which, whilst not confined to alluvium, can offer high returns per acre provided drainage is adequate. Citrus, other fruits, vegetables, and perennial tree crops (including rubber, cocoa and oil palm) can also be grown if the water table is kept at 1.5–2.0 m depth by a network of drainage ditches, supplemented by pumping where necessary. Coconuts can tolerate the moderate salinity found in coastal sands. In India and Pakistan, double cropping of annual crops is the usual practice where perennial irrigation is available, with complex rotations of wheat (in winter), rice, maize and other grains, cotton, vegetables, fodder crops and sugar-cane.

Intensive use carries with it the need for heavy fertilization, a fact insufficiently appreciated in peasant smallholder irrigation schemes. On grounds of efficient resource use, there would be a case for not making water available to farmers without evidence of adequate fertilizer application. Farm-yard manure can usefully contribute to this; the need for fodder can be met by alfalfa, which can have very high yields under irrigation. In the case of very sandy profiles, the soil may be regarded almost as a medium for hydro-ponic crop growth, with crop requirements added chemically. To treat most alluvial soils in this manner, however, is a misuse of their natural biological potential, and intensive cropping should be accompanied by green manuring, farm-yard manure or other methods of organic matter maintenance.

ACID SULPHATE SOILS

CCTA: Not distinguished separately
FAO: Thionic fluvisols

Acid sulphate soils are soils of coastal plains that become exceptionally acid (pH 2.0–3.5) on drainage. They are sometimes called 'cat-clays', a term descriptive of their prominent yellow mottling. They are mainly found in brackish water environments, usually formerly under mangrove. Examples on freshwater swamps have occasionally been reported.

The profile is usually a clay or silty clay with a substantial organic matter content, sometimes high enough to be called a muck. When submerged, colours are dark grey to bluish black if largely mineral gleys, becoming dark brown with increasing organic matter. Semi-decayed roots are common. The waterlogged soils have a pH of 5.6, and may smell of hydrogen sulphide.

If such soils are reclaimed by drainage, prominent pale yellow (2.5Y 8/4) mottles develop, and the pH falls below 3.5. Vertical cracking may give a strong coarse prismatic structure to the mottled horizon, whilst the soils in depth remain structureless and impermeable. A surface layer of peat may overly the mottled 'cat-clay' horizon.

The cause of the acidification is the oxidation of pyrites, FeS_2, with the formation of sulphuric acid. Sulphates in seawater are reduced by bacteria under anaerobic conditions to form hydrogen sulphide, H_2S; the presence of a substantial amount of organic matter is necessary for bacterial growth. The hydrogen sulphide reacts with iron compounds in the soil to form ferrous iron sulphide, FeS_2. Oxidation on contact with air causes the forma-tion of ferric sulphate, which produces the yellow mottling and staining, and sulphuric acid. Further reaction of the latter with clay minerals forms aluminium sulphate, which may cause aluminium toxicity.

Acid sulphate soils have developed where apparently promising coastal plain and estuarine gleys have been reclaimed. Once acidification has occurred it is difficult to reverse. Neutralization of the acidity by liming

225

is a technical solution, but the large quantities of lime required (100 t/ha or more) make it uneconomic. Leaching out of sulphides takes many years. Where the acidity of drained soils is moderate, pH 3.5–5.0, they are used for rice but give poor yields. True acid sulphate soils with pH values below 3.5 are frequently abandoned.

Soils likely to 'go acid' should be identified prior to reclamation; this can be done by smell, analysis of sulphur content, or by vegetation and soil profile indicators. Large amounts of semi-decayed roots in the profile are a bad sign. In Sierra Leone, *Rhizophera racemosa* mangrove swamps always go acid, whilst other *Rhizophera* species and most *Avicennia* soils do not (Jordan, 1964). Once identified, such soils should not be drained. Reclamation without drainage for padi is possible, and practised in the Mekong delta, but yields are low and do not justify substantial investment.

PADI SOILS

A world area close to 100 million hectares, equivalent to a square 1000 by 1000 kilometres, is devoted to wet rice or padi cultivation, 85 percent of which is in Asia. Because of the periods of submergence required by the rice plant, the soils on which it is grown acquire common characteristics which justify treating them as a unit, albeit of a taxonomically different nature to other soil types.

By far the most extensive soils used for padi are alluvial soils, together with gleyed profiles and organic soils of alluvial and coastal plains. Through terracing, various types of latosols are brought under padi, e.g. in the Szechwan Basin of China; and through irrigation, soils of the semi-arid and arid zones. Other soil types used are vertisols and andosols.

Padi soils are characterized by a thick bluish grey reduced horizon, over-lain by a thin brownish oxidized horizon; the former is 30–80 cm thick and the latter varies from a few millimetres to a few centimetres (fig. 15 (*a*) and (*b*)). Sometimes a very thin, discontinuous layer of ferric iron can be detected between these two horizons. In profiles which were originally freely-drained, the reduced horizon passes in depth into oxidized subsoil; at the junction between the latter horizons there is a layer 5–20 cm thick containing precipitated iron and manganese, as concretions or a pan. In soils originally poorly drained, the reduced horizon continues to depth. Thus, freely-drained profiles become changed by artificial flooding into surface water gleys, whilst poorly-drained profiles remain as groundwater gleys.

The main processes producing this distinctive profile are reduction, oxidation, solution, leaching, and precipitation of iron compounds. In all soils there is a demand for oxygen from micro-organisms and for root uptake. Once that available from soil air is used up, organisms take oxygen

226

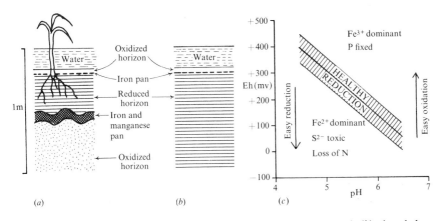

Fig. 15. (*a*) and (*b*): horizons in padi soils, (*a*) where freely drained in depth, (*b*) where below the water table. Based on Greene (1960) and Grant (1964). (*c*) Eh/pH relations in padi soils. After Greene (1963).

from other compounds, reducing them. There is some downward movement of oxygen from floodwater, but the demand from the soil exceeds this rate of flow. Hence the soil is differentiated into a thin surface layer with oxygen present and the main, reduced, part of the profile with no free oxygen.

The anaerobic conditions cause the reduction of iron and manganese compounds to ferrous and manganous forms, in which conditions they are moderately soluble. These compounds are taken into solution, carried downwards by leaching, and oxidized and precipated on contact with oxygen if present in depth. It has been estimated that it takes 50–100 years for an iron and manganese pan to develop (Dudal, 1966).

Other processes common in padi soils are siltation, the effects of terracing, and the formation of a plough-sole. Repeated flooding by muddy water leads to the accumulation of a surface layer of clay and fine silt; in soils long under cultivation, such layers may be a metre or more thick, completely changing the original soil texture. Where sloping land has been levelled by terracing, the profile is initially modified by cut-and-fill operations; subsequently, washing of topsoil from the higher terraces and re-deposition on the lower causes further modification, tending towards truncated and buried profiles respectively: A plough-sole is a compacted horizon formed at the lower limit of cultivation as a result of repeated pulverization and treading.

Redox potential and fertility

The oxidation–reduction, or redox, potential is a measure of the relative oxidizing or reducing power of a solution. It is measured by inserting a platinum electrode and measuring the potential difference against another electrode of known potential. The condition in which the equilibrium

$$H_2 \rightleftharpoons 2H^+ + 2 \text{ electrons}$$

exists is arbitrarily fixed at a redox potential, Eh, of 0.0 volts. An increase in positive redox potential indicates a progression towards more strongly oxidizing conditions, and a decrease (or increase in negative potential) a progression towards reducing conditions. Redox potential is not a routine measurement in soil analysis as there is difficulty in obtaining replicable results; in freely-drained soils it generally lies between +400 and +700 mv (millivolts).

For a given reaction there exists a redox potential at which ions of the oxidized and reduced forms are present in solution in equal concentrations. The equilibrium between ferrous and ferric ions for the half-reaction $Fe^{2+} \rightleftharpoons Fe^{3+} + e^-$, occurs under acid conditions (pH < 4.0) at an Eh of +770 mv. Since oxidation–reduction reactions involve OH^- and H^+ ions they are affected by pH. It has been shown that the solution concentration of active ferric ions varies with pH, decreasing by approximately 10^3 for unit increase in pH (Jeffery, 1960, 1961; Greene, 1963, pp. 3–4), and the equilibrium equation can be written in the form:

$$Eh = 1.033 - 0.18\text{pH} - 0.06\log[Fe^{2+}]$$

Thus, reduction occurs more readily as the soil becomes more acid (fig. 15(*c*)).

In freely-drained soils, the Eh is between +400 and +700 mv, and thus iron is dominantly in ferric forms. On flooding the soil for padi there is a decrease in redox potential, over about three weeks, to about +100 mv, accompanying reduction of ferric oxides by organic matter. This leads to a condition in which both ferrous and ferric ions are present in solution in the reduced layer.

The redox–acidity condition of a flooded soil affects nutrient availability and toxic elements. At a pH of 7.0, nitrates are reduced at an Eh of about +225 mv, ferric compounds at +120 mv, and sulphates are reduced to sulphides below −150 mv. Hence in the absence of free oxygen, nitrates are not found, mineralization proceeding only as far as the ammonium ion. Hydrogen sulphide, which is toxic to plants, is only produced under extreme reducing conditions, and these will not be reached as long as iron is present; thus hydrated ferric oxides act as a buffer, preventing the development of an excessively low redox potential.

If the soils are too highly oxidized, dominance by the ferric ion leads to phosphate fixation; this situation can be corrected by the addition of organic matter as green manure. Toxic reduction, with the production of sulphides and excessive loss of nitrogen by denitrification, is only likely to occur in acid, sandy soils in which no free iron is present; it can only be corrected by addition of soil containing ferric oxide or by making the soil less acid. Between these two states is a condition termed 'healthy reduction' (Greene, 1963) in which both ferric and ferrous iron are present in solution, and nutrient availability is adequate (fig. 15(*c*)).

In padi fields, nitrogen is fixed from the atmosphere, probably by *Azotobacter* in association with blue-green algae. This is one of the processes which confers on padi soils the unusual property of being able to sustain cropping indefinitely. Nevertheless, nitrogen is the most common deficiency in padi. The main cause of nitrogen loss is by denitrification, leaching being a secondary cause. The ammonium ion is nitrified in the oxidized layer, and passes downward by leaching; in the reduced layer it is denitrified by anaerobic micro-organizms and lost as nitrogen gas. It is desirable to maintain the oxidized layer as thick as possible. Draining and re-flooding the soil during padi cultivation increases nitrogen availability. There is usually a good response to fertilization at a level of 20 kg of nitrogen per hectare. Nitrates if present are rapidly lost by leaching, whereas the ammonium ion is held as an adsorbed cation and is more stable; hence, fertilizer for padi should contain nitrogen in the form of ammonium, e.g. ammonium sulphate, ammonium phosphate or urea (Patrick and Mahapatra, 1968; de Datta and Magnaye, 1969).

Flooding increases phosphate availability, possibly by reducing ferric phosphate to the more soluble ferrous form. There is only sometimes a response to phosphate fertilization.

Degraded padi soils

Degraded padi soils (*akiochi* in Japan) is a name given to profiles in which the reduced layer is pale grey and the oxidized layer absent, the topsoil being grey to whitish. At 50–100 cm depth there is an accumulation of ferric oxide concretions. When such soils are flooded, hydrogen sulphide is produced. The padi roots lack the coating of ferric iron which is present in normal soils and which protects them from hydrogen sulphide damage. The texture of degraded padi soils is sandy and the reaction acid. Padi yields are low.

These conditions are caused by the absence of mobile iron from the profile, which consists mainly of quartz; the normal buffering action of iron does not occur, and redox potential falls to negative values. Degraded soils are more likely to develop from alluvial deposits derived from acid igneous rocks or sandstones, which contained little iron at the time of

229

deposition. Such iron as may be present has been completely leached. Rozanov and Rozanova (1968) suggest that absolute accumulation of silica, derived from groundwater, may occur, but this is questionable. Such soils were originally called 'rice podzols' (Thorp, 1936), and the resemblance is more than superficial.

Evaluation

Padi soils are a specialized topic, not often of direct concern in soil survey, although of considerable importance in evaluating the productive potential of tropical countries. Moreover, the major problems of closely-settled padi lands concern population increase, land tenure and other social matters. Nevertheless, padi can have a part to play in land development, particularly where alluvial soils and gleys of valley floors occur in juxtaposition to high land suitable for other crops. In such environments the potential of padi cultivation to produce the carbohydrate needs of a family from about one hectare can form a valuable component in a mixed farming system.

The strongest influences on padi yields are the correct depth and timing of water application, followed by those of fertilizer and seed variety. Soil properties are often of lesser significance. It is, however, possible to identify soil properties that, given these other conditions, permit sustained high yields, and other properties which are unfavourable. Properties conducive to high yields are:

(1) A heavy texture, preferably with clay + silt > 70 percent. Sandy soils are likely to be deficient in nitrogen and other nutrients, have low moisture retention, and may develop into degraded profiles.

(2) Moderate acidity, optimally a pH of 5.5–6.5. There are many different varieties of padi, tolerant of a wide range of pH, about 4.0 to 8.0, but yields decline at the extremes. On acid soils, pH increases following flooding. Acid terraced latosols require heavy fertilization.

(3) A moderate but not high level of organic matter. Alluvial soils and inorganic gleys are preferable. Cultivation of peats and mucks, although practised out of necessity in deltas, lead to a variety of problems, including danger of the development of acid sulphate soils if the land is drained.

(4) Absence of salinity. Some padi varieties can tolerate mild salinity, but water control measures should seek to remove hazards of seawater flooding or saline groundwater.

(5) If mechanized cultivation is intended, adequate support for wheels or tracks is necessary. Firm alluvial soils and well-consolidated clays are satisfactory, while peats, mucks and unripened muds are not. Where recent muds are reclaimed they should be drained for a sufficient time for the ripening process to produce a firm clay.

SELECTED REFERENCES

Gleys. Milne (1947).

Alluvial soils. Edelman and van der Voorde (1963), Brinkman and Pons (1968), Holmes and Western (1969), Slager *et al.* (1970).

Acid sulphate soils. Van Beers (1962), Moorman (1963), Bloomfield *et al.* (1968), International Institute for Land Reclamation and Improvement (1973).

Padi soils. Greene (1960, 1963), Dudal (1966, 1968*b*), Grant (1964), Grist (1965), Patrick and Mahapatra (1968), de Datta and Magnaye (1969).

PART THREE

SOIL CLASSIFICATION AND EVOLUTION

CHAPTER 13

CLASSIFICATION

Scientists who are otherwise reasonable and unemotional are liable to behave quite differently when discussing this topic.

Mulcahy and Humphries (1967)

There is no universally accepted system of soil classification and nomenclature, comparable to that which exists, for example, for rock types or biological species. Differences exist between the approaches used in the Communist and Western worlds, and between French and English speaking sectors of the latter. Even within the limited realm of the English speaking countries, three systems are in common use.

This confusing situation has arisen through national soil survey organizations establishing their own classifications, and the failure to date of any one attempt at international standardization to gain universal acceptance. The countries of former French Africa have long worked on a common system, but a different one was developed in the Belgian Congo. Apart from the Ghana system, partly used in other West African countries, British Commonwealth territories used *ad hoc* local classifications. India and Ceylon formerly used modifications of the US 1938 nomenclature, whilst Australia developed a completely independent system.

One national and two international systems have some measure of international recognition. In 1960 the US Soil Survey produced a scheme entitled *Soil classification. A comprehensive system. 7th approximation*, usually known as 'the 7th approximation'. This attempted to avoid the confusion that had arisen through the same terms being defined in different ways by sweeping away earlier systems and starting with a clean sheet and a new nomenclature. The new principles of the system, together with the authority of the US Soil Survey, attracted much international interest and some measure of adoption, particularly among less developed countries not in a position to assess its merits. The first international scheme originated when the Commission for Technical Co-operation in Africa (CCTA) produced in 1964 the *Soil map of Africa 1:5000000*. The explanatory monograph to this, the work of the Belgian J. L. D'Hoore, contains a classification system that is an attempt to combine those of French (ORSTOM) and Belgian (INEAC) work in Africa. The only fully international scheme is that produced by the FAO in 1974 for use in their *Soil map of the world*; it is also principally the work of a Belgian, R. Dudal.

In working on tropical soils one is likely to encounter these three systems, as well as the ORSTOM and other national schemes. Moreover, classifications are themselves, when they are based on recognition of the genetic relations between soil types, contributions to knowledge. This chapter is an introduction to the more common systems, giving particular attention to the problem of the classification of the latosols.

DEFINITIONS

Some terms used in the discussion of individual systems may first be defined. *Artificial systems* are based on the use of a small number of selected properties, or a single property, in differentiating classes. *Natural systems* are based on all properties of the soil, regarded not as independent variables but as an entity (Kubiena, 1958). *Morphological systems* are based on the properties of the soil profile itself, irrespective of their origin. They may be based only on properties obtainable from field survey or may include those for which laboratory analysis is necessary. *Genetic systems* are based on the presumed origin of the soil; they should strictly be defined in terms of processes but in practice, owing to the difficulty of observing these, they are based upon soil-forming factors. Most soil classifications have an *hierarchical* structure, comparable in this respect with botanical classification, rather than a *multi-variate* structure. The successive levels of grouping are the *categories* of the classification. Hierarchical classifications may be *descending*, in which the soil individuals are successively subdivided according to prescribed differentiating criteria; or *ascending*, in which the individuals are grouped on the basis of similarities, the lowest-category groups being successively aggregated at higher levels.

INTERNATIONAL CLASSIFICATIONS

The CCTA Soil map of Africa classification (1964)

Main account: D'Hoore (1964), with representative profiles.
Summaries: D'Hoore (1968b); Ahn (1970), pp. 213–19.

This scheme originated from a joint project by the Inter-African Pedological Service and the Commission for Technical Co-operation in Africa (CCTA) for a soil map of the continent on a scale of 1:5000000. It is largely an attempt to reconcile the ORSTOM and INEAC systems (see below), and shows no apparent features taken from English-speaking countries. A map in seven sheets was successfully completed, although many records were lost as a result of the Congo disturbances of 1961. The classification has since been used in a number of African countries, and by the Land Resources Division of Britain.

A summary of the legend is given as table 23. There are 16 main groups, four of which are subdivided to give 25 sub-groups and 63 soil types. The soil types are shown on the map as simple units, of one type, or as binary or ternary associations in which the symbols are listed successively and the colour shading is that of the most extensive type.

The classification is a natural one and in large part genetic (except for the anomalous use of colour to subdivide ferrallitic soils). The subdivision of groups into soil types is frequently on the basis of parent material. 'Non-differentiated' units are included to cover areas for which there is no information. Each soil type is defined by a descriptive paragraph setting out its morphological and analytical features, sometimes with quantitative values, together with the environment under which it occurs. Thus the definition of *brown soils of arid and semi-arid tropical regions* is as follows (slightly abbreviated):

Brown soils, darkened by organic matter in the greater part of the profile, under steppe vegetation, without an A2 horizon but having a textural, structural or colour B horizon. The reserve of weatherable minerals is often considerable. They contain appreciable quantities of 2:1 lattice clay minerals. The cation exchange capacity of the mineral fraction is medium to high, and is more than 50 percent saturated in horizons B and C. They often contain free carbonates. The organic matter content is low (< 1 percent) but very well distributed throughout the profile. These soils are formed under hot and dry climates where annual rainfall rarely exceeds 500 mm. They are often formed on aeolian deposits.

Where precise values are given they are intended as approximate but not rigid limits. Thus, unlike the situation in artificial classifications, a profile with all other attributes of a particular soil type but exceeding one limiting value is not necessarily excluded. If the properties of a profile conflict with the environmental factors under which it occurs then the former take precedence. The philosophy of a natural system is that such a conflict should not arise, and if it does then the system is based on imperfect understanding and should be modified. This approach breaks down where relict features are prominent.

The subdivision of raw mineral and weakly developed soils is proportionately more complex than in other classifications, accounting for one third of the total number of soil types; this can be justified on the ground that these soils cover more than one third of the continent. The two most extensive latosol groups, ferruginous and ferrallitic soils, are taken from the ORSTOM system, and are clearly separated from each other; ferruginous soils have a B horizon base saturation of over 40 percent, and ferrallitic soils under 40 percent.

The major misunderstanding of the system is the placing of ferrisols as an intermediate group between ferruginous and ferrallitic soils. Ferrisol is a term taken from the INEAC classification, in which a broad group of

TABLE 23 *Legend of the Soil map of Africa classification (D'Hoore, 1964)*

	Map symbol
RAW MINERAL SOILS	A
Rock and rock debris	
Rocks rich in ferromagnesian minerals	Aa
Ferruginous crusts	Ab
Calcareous crusts	Ab'
Not differentiated	Ac
Desert and detritus	
Sands (ergs)	An
Clay plains	Ao
Desert pavements, residual	Ap
Desert pavements, transported	Ap'
Not differentiated	Ar
WEAKLY DEVELOPED SOILS	B
Lithosols (skeletal soils) and lithic soils	
On lava	Ba
On rocks rich in ferromagnesian minerals	Bb
On ferruginous crusts	Bc
On calcareous crusts	Bc'
Not differentiated	Bd
Sub-desert soils	
Not differentiated	Bf
Weakly developed soils on loose sediments not recently deposited	
Not differentiated	Bh
Juvenile soils on recent deposits	
On volcanic ash	Bn
On riverine and lacustrine alluvium	Bo
On fluvio-marine alluvium (mangrove swamps)	Bp
On wind-borne sands	Bq
Soils of oases	Br
CALCIMORPHIC SOILS	C
Rendzinas, brown calcareous soils	Ca
Soils with calcareous pans	Cb
Soils containing more than 15 percent gypsum	Cc
VERTISOLS AND SIMILAR SOILS	D
Derived from rocks rich in ferromagnesian minerals	Da
Derived from calcareous rocks	Db
Of topographic depressions, not differentiated	Dj
PODZOLIC SOILS	E
Not differentiated	Ea
HIGHVELD PSEUDO-PODZOLIC SOILS	F
Not differentiated	Fa
BROWN AND REDDISH BROWN SOILS OF ARID AND SEMI-ARID REGIONS	G
Brown soils of arid and semi-arid tropical regions	
On loose sediments	Ga
Not differentiated	Gb
Brown and reddish brown soils of arid and semi-arid Mediterranean regions	
Not differentiated	Gn

TABLE 23 *Continued*

	Map symbol
EUTROPHIC BROWN SOILS OF TROPICAL REGIONS	H
On volcanic ash	Ha
On rocks rich in ferromagnesian minerals	Hb
On alluvial deposits	Hc
Not differentiated	Hd
RED AND BROWN MEDITERRANEAN SOILS	I
Red Mediterranean soils, not differentiated	Ia
Brown Mediterranean soils, not differentiated	Ib
FERRUGINOUS TROPICAL SOILS (fersiallitic soils)	J
On sandy parent materials	Ja
On rocks rich in ferromagnesian minerals	Jb
On crystalline acid rocks	Jc
Not differentiated	Jd
FERRISOLS	K
Humic	Ka
On rocks rich in ferromagnesian minerals	Kb
Not differentiated	Kc
FERRALLITIC SOILS (*sensu stricto*)	L
Dominant colour: yellowish brown	
On loose sandy sediments	La
On more or less clayey sediments	Lb
Not differentiated	Lc
Dominant colour: red	
On loose sandy sediments	Ll
On rocks rich in ferromagnesian minerals	Lm
Not differentiated	Ln
Humic ferrallitic soils	
Not differentiated	Ls
Ferrallitic soils with dark horizon	
Not differentiated	Lt
Yellow and red ferrallitic soils on various parent materials	
Not differentiated	Lx
HALOMORPHIC SOILS	M
Solonetz and solodized solonetz	Ma
Saline soils, alkali soils, and saline alkali soils	Mb
Soils of sebkhas and chotts	Mc
Soils of lunettes	Md
Not differentiated	Me
HYDROMORPHIC SOILS	N
Mineral hydromorphic soils	Na
Organic hydromorphic soils	Nb
ORGANIC SOILS, NON-HYDROMORPHIC	O
Organic soils of mountains	Oa
ASSOCIATIONS AND COMPLEXES	R
Complex of volcanic islands	Ra

kaolisols is divided into less-weathered ferrisols and highly-weathered ferralsols, distinguished by the presence and absence respectively of clay skins. The ferrisols have no equivalent on the French system, and are in fact developed mainly on basic rocks. The placing of them as transitional between ferruginous and ferrallitic soils is genetically a misunderstanding and results in overlapping definitions (Young, 1968*b*). There is an over-emphasis on climatic zonation and insufficient recognition of the distinctive influence of basic rocks. A further disadvantage is use of the silica:sesquioxide and silica:alumina ratios as differentiating criteria, since these are not routine analytical determinations.

This is an outstanding synthesis, and served a particularly valuable purpose in bringing the ORSTOM and INEAC approaches more fully to the attention of Anglophone countries. Its aim was to serve as a basis for international exchange of pedological and agronomic information, and might have become still more widely adopted had work on it not been seriously interrupted by political events. Although considerably dependent upon analytical properties, there are nevertheless sufficient morphological criteria for provisional class identifications to be made on the basis of field data. It remains a viable system for converting local surveys into widely recognized mapping units, and a rich source of information on the relations between soil properties and genetic factors. In respect of the identification of natural soil units it is better than the FAO system, and it is only for the political reason of the greater international authority of the FAO that the latter is to be preferred as a medium for international communication.

The FAO classification (1974)

Main account: FAO–Unesco (1974), superceding Dudal (1968*a*). For representative profiles, see the explanatory volumes of the FAO–Unesco *Soil map of the world 1 : 5 000 000*.

The FAO–Unesco *Soil map of the world* was produced by a small group at the FAO in Rome, headed for the greater part of the period by R. Dudal. The legend for this map was the outcome of a long series of regional and international discussions and field correlation meetings, started in 1961 and passing through five draft stages prior to 1968, when the fifth draft was presented to the Ninth International Congress of Soil Science (Dudal, 1968*a*). Supplements were issued during 1969 and 1970. The definitive version appeared in 1974 as *Volume I, Legend* of the *Soil map of the world 1 : 5 000 000* (FAO–Unesco, 1974). It is now the basis for the 18 sheets and ten explanatory volumes of this map, and is used on FAO soil surveys. It is the most authoritative system for international use.

The FAO itself disclaims that this is a classification system, holding only that it is a map legend. It is stated that 'the legend of the Soil Map of the World is not meant to replace any of the national classification schemes but to serve as a common denominator. Improving understanding between different schools of thought could profitably lead to the adoption of an internationally accepted system of soil classification and nomenclature', whilst among the objectives of the map are 'to promote the establishment of a generally accepted soil classification and nomenclature' (FAO–Unesco, 1974, p. 2). It is further stated that 'no international consensus could be obtained on how the various soil units, which are the basic elements of the system, should be grouped; the list of soil units . . . is therefore a monocategorical classification [*sic*] of soils, and not a taxonomic system subdivided into categories at different levels of generalization' (*ibid.*, p. 10).

These statements may be taken as disclaimers offered by an international organization in the interest of political propriety. All national soil classifications, as also the CCTA system, were produced primarily for use in soil maps. Moreover, the FAO text subsequently ignores its caveat and refers to 'the soil classification used for the Soil Map of the World' (*ibid.*, p. 11). There is slightly more force to the assertion that the 'groups' of soil units are for convenience of presentation and do not have equivalent levels of generalization. However, in practice all but two of the groups (luvisols and acrisols) can be identified with natural soil classes; the system of map symbols and colouring employed strengthens the impression that the groups are in fact classes at a higher level of classification. In practice, therefore, the FAO legend is a classification system with two categories.

The basic element is the *soil unit*, of which 106 are recognized (table 24). Each soil unit is designated by an adjective and a noun, in most cases ending in -ic and -sols respectively, e.g. orthic ferralsols. The units are combined 'on the basis of generally accepted principles of soil formation' into 26 groups, identified by the noun element, e.g. ferralsols. In addition there are 12 *phases* which may be applied to any soil unit. Three *textural classes* are recognized (coarse, medium and fine textured), and three *slope classes* (level to gently undulating, rolling to hilly, and steeply dissected to mountainous). The mapping symbol is of the form illustrated by Fo6–2b, in which Fo is the dominant soil unit (F = ferralsols, Fo = orthic ferralsols), 6 is the number of the *soil association*, the soil units included within which are given in tabular form in the full legend, 2 is the texture class and b the slope class.

Although many of the soil groups are natural soil types, this is structurally an artificial classification. It makes use of the principle of diagnostic horizons borrowed, together with most of the horizon names and definitions, from the US 7th approximation. The diagnostic horizons are defined by precise limiting values and play an important although by no means exclusive part in the identification of soil units (table 25).

J	FLUVISOLS	S	SOLONETZ	L	LUVISOLS
Je	Eutric fluvisols	So	Orthic solonetz	Lo	Orthic luvisols
Jc	Calcaric fluvisols	Sm	Mollic solonetz	Lc	Chromic luvisols
Jd	Dystric fluvisols	Sg	Gleyic solonetz	Lk	Calcic luvisols
Jt	Thionic fluvisols			Lv	Vertic luvisols
				Lf	Ferric luvisols
		Y	YERMOSOLS	La	Albic luvisols
G	GLEYSOLS			Lp	Plinthic luvisols
		Yh	Haplic yermosols	Lg	Gleyic luvisols
Ge	Eutric gleysols	Yk	Calcic yermosols		
Gc	Calcaric gleysols	Yy	Gypsic yermosols		
Gd	Dystric gleysols	Yl	Luvic yermosols	D	PODZOLUVISOLS
Gm	Mollic gleysols	Yt	Takyric yermosols		
Gh	Humic gleysols			De	Eutric podzoluvisols
Gp	Plinthic gleysols			Dd	Dystric podzoluvisols
Gx	Gelic gleysols	X	XEROSOLS	Dg	Gleyic podzoluvisols
		Xh	Haplic xerosols	P	PODZOLS
R	REGOSOLS	Xk	Calcic xerosols		
		Xy	Gypsic xerosols	Po	Orthic podzols
Re	Eutric regosols	Xl	Luvic xerosols	Pl	Leptic podzols
Rc	Calcaric regosols			Pf	Ferric podzols
Rd	Dystric regosols			Ph	Humic podzols
Rx	Gelic regosols	K	KASTANOZEMS	Pp	Placic podzols
				Pg	Gleyic podzols
		Kh	Haplic kastanozems		
I	LITHOSOLS	Kk	Calcic kastanozems		
		Kl	Luvic kastanozems	W	PLANOSOLS
Q	ARENOSOLS			We	Eutric planosols
		C	CHERNOZEMS	Wd	Dystric planosols
Qc	Cambic arenosols			Wm	Mollic planosols
Ql	Luvic arenosols	Ch	Haplic chernozems	Wh	Humic planosols
Qf	Ferralic arenosols	Ck	Calcic chernozems	Ws	Solodic planosols
Qa	Albic arenosols	Cl	Luvic chernozems	Wx	Gelic planosols
		Cg	Glossic chernozems		
				A	ACRISOLS
E	RENDZINAS				
		H	PHAEOZEMS	Ao	Orthic acrisols
				Af	Ferric acrisols
U	RANKERS	Hh	Haplic phaeozems	Ah	Humic acrisols
		Hc	Calcaric phaeozems	Ap	Plinthic acrisols
		Hl	Luvic phaeozems	Ag	Gleyic acrisols
T	ANDOSOLS	Hg	Gleyic phaeozems		
				N	NITOSOLS
To	Ochric andosols	M	GREYZEMS		
Tm	Mollic andosols			Ne	Eutric nitosols
Th	Humic andosols		Orthic greyzems	Nd	Dystric nitosols
Tv	Vitric andosols	Mg	Gleyic greyzems	Nh	Humic nitosols

242

TABLE 24 *Continued*

					F	FERRALSOLS
V	VERTISOLS	B	CAMBISOLS		Fo	Orthic ferralsols
					Fx	Xanthic ferralsols
Vp	Pellic vertisols		Eutric cambisols		Fr	Rhodic ferralsols
Vc	Chromic vertisols	Bd	Dystric cambisols		Fh	Humic ferralsols
		Bh	Humic cambisols		Fa	Acric ferralsols
		Bg	Gleyic cambisols		Fp	Plinthic ferralsols
		Bx	Gelic cambisols			
Z	SOLONCHAKS	Bk	Calcic cambisols		O	HISTOSOLS
Zo	Orthic solonchaks	Bc	Chromic cambisols			
Zm	Mollic solonchaks	Bv	Vertic cambisols		Oe	Eutric histosols
Zt	Takyric solonchaks				Od	Dystric histosols
Zg	Gleyic solonchaks	Bf	Ferralic cambisols		Ox	Gelic histosols

TABLE 25 *Diagnostic horizons and properties used in the FAO classification. The descriptions given are indications of the general meaning; full definitions are given in FAO–Unesco (1974, pp. 23–7). Only selected diagnostic properties are listed.*

Diagnostic horizon	Meaning
Histic H horizon	Peaty; organic matter > 20 percent.
Mollic A horizon	Organic matter > 1 percent, base saturation > 50 percent.
Umbric A horizon	Organic matter > 1 percent, base saturation < 50 percent. Mollic and umbric are the normal A horizons of humid areas.
Ochric A horizon	Pale in colour, organic matter < 1 percent; the normal A horizon of dry areas.
Argillic B horizon	Textural B horizon; clay skins present.
Natric B horizon	Alkali; exchangeable sodium percentage > 15 percent, columnar structure.
Cambic B horizon	Resulting from weathering *in situ* rather than illuviation; otherwise called a (B) or 'structural B' horizon; includes most B horizons that do not have the requirements of argillic, natric, spodic or oxic horizons.
Spodic B horizon	The sesquioxide-rich horizon of podzols.
Oxic B horizon	Highly weathered, with few or no weatherable minerals, no clay skins, clay minerals dominantly kaolinite and free sesquioxides, apparent cation exchange capacity of the clay fraction < 16 m.e./100 g.
Calcic horizon	Enriched in secondary calcium carbonate.
Gypsic horizon	Enriched in calcium sulphate.
Sulfuric horizon	Rich in sulphides; pH < 3.5.
Albic E horizon	Bleached material, usually sandy.

Diagnostic property	Meaning
Aridic moisture regime	Profile not moist throughout for more than two consecutive months nor moist in some part for more than three.
Ferric properties	With red mottles or concretions; apparent cation exchange capacity of the clay fraction < 24 m.e./100 g.
High salinity	Electrical conductivity of the saturation extract > 15 mmhos/cm.

The names of the soil groups are mainly borrowed from widely-recognized traditional nomenclature; thus rendzina, ranker, andosol, vertisol, solonchak, solonetz, chernozem, podzol and lithosol are used in their customary senses, albeit with artificial limiting values. Terms that have been used with widely different meanings have been dropped, notably brown earth/brown forest soil and any reference to 'podzolic' and 'lateritic'. The adjectival terms have Latin and Greek roots, and have the same meanings when applied to any higher-category group.

Brief descriptions of the meanings of the main classes, with special reference to those common in tropics, are as follows:

Fluvisols. Alluvial soils; developed from recent alluvial deposits, with depositional rather than pedogenetic horizons. Gleyed profiles are included.

Gleysols. Gleys; hydromorphic properties dominate others. Gleyed profiles on recent alluvium are classed with fluvisols. Nine other groups include *gleyic* units, in which hydromorphic properties are subsidiary (the printed definitions fail to make a clear distinction between gleysols and gleyic units, but the intention is clear).

Regosols. Weakly developed soils on unconsolidated materials, usually sands; lacking diagnostic horizons except possibly an ochric A horizon.

Lithosols. Soils with continuous hard rock at < 10 cm depth. Profiles with rock commencing at between 10 and 50 cm belong to the *lithic phase*, and very stony profiles to the *stony phase*.

Arenosols. Very sandy soils which have an identifiable B horizon; clay < 15 percent. *Ferrallic arenosols* are the sandy equivalent of ferralsols; *albic arensols* are lowland tropical podzols or 'bleached sands'.

Rendzinas. Shallow calcareous soils on limestones.

Rankers. Weakly developed shallow soils on consolidated rock.

Andosols. Soils developed from recent volcanic materials.

Vertisols. Dark cracking clays.

Solonchaks. Saline soils; having a high salinity (table 25).

Solonetz. Alkaline soils; having a natric B horizon. If the presence of free salts prevents the formation of a columnar structure, as required for a natric horizon, the profile remains a solonchak.

Yermosols. Desert soils; having an aridic moisture regime (table 25) and a very weak ochric A horizon (organic matter <0.5 percent).

Xerosols. Semi-desert soils; having an aridic moisture regime and a weak ochric A horizon (organic matter 0.5–1.0 percent).

Kastanozems. 'Chestnut soils' of the temperate steppe zone.

Chernozems. Black earths of the temperate steppe zone.

Phaeozems. 'Prairie soils'; as chernozems, but less dark.

Greyzems. Grey forest soils of cool temperate latitudes.

Cambisols. Primarily intended as the equivalent of those brown earths which do not have an argillic horizon, but have a cambic B ('structural B')

244

horizon. The defining criteria in fact permit occasional identification in the tropics.

Luvisols. Having an argillic B horizon with base saturation > 50 percent. *Ferric luvisols* show ferric properties (table 25) and correspond approximately to the ferruginous soils of the CCTA and ORSTOM classifications.

Podzoluvisols. Intermediate between podzols and luvisols; having an argillic B horizon with an irregular upper boundary.

Podzols. The traditional meaning, having a spodic B horizon. Lowland tropical podzols (bleached sands), however, are albic arensols.

Planosols. Having an albic E horizon with hydromorphic properties, and a slowly permeable B horizon.

Acrisols. Having an argillic B horizon with base saturation < 50 percent. These include tropical soils insufficiently weathered to be ferralsols but more strongly leached than luvisols. *Orthic acrisols* (normal), *ferric acrisols* (with ferric properties, table 25) and *plinthic acrisols* (with plinthite – in practice hardened plinthite, i.e. ironstone) are common in the savanna zone.

Nitosols. These are soils derived from basic rocks in the humid tropics. The root means 'shiny' and refers to the strongly developed clay skins which are in practice the diagnostic feature. This criterion has unfortunately been dropped from the definition, which in its final form refers only to a deep argillic B horizon with merging boundaries. They are subdivided into *eutric nitosols* (base saturation > 50 percent), *dystric nitosols* (base saturation < 50 percent) and *humic nitosols*.

Ferralsols. The INEAC term ferralsol is adopted for the highly weathered soils of the humid tropics, corresponding to the ferrallitic soils of the CCTA system. They are defined as possessing an oxic horizon, the definition of which is less demanding than in the US 7th approximation, essentially requiring a cation exchange capacity of the clay fraction of less than 16 m.e./100 g, i.e. virtual absence of 2:1 lattice clay minerals, and very few or no weatherable minerals. They are subdivided into *orthic ferralsols* (normal), *xanthic ferralsols* (yellow), *rhodic ferralsols* (red to dusky red), *humic ferralsols* (high in organic matter), *acric ferralsols* (extremely low cation exchange capacity, <1.5 m.e./100 g of clay) and *plinthic ferralsols* (with plinthite).

Histosols. Peats and mucks; having a histic A horizon.

The following working procedure enables most tropical soils to be placed into a soil group:

(i) Note whether any of the given 'special features' are present, in which case the profile is likely to belong to the class indicated:

Derived from alluvium	*Fluvisols*
Gleyed at <50 cm depth, gleying dominant	*Gleysols*
Rock at <10 cm depth	*Lithosols*
Derived from recent volcanic materials	*Andosols*

245

Dark cracking clays	*Vertisols*
Free salts present	*Solonchaks*
Strongly alkaline but free salts absent	*Solonetz*
With strongly bleached eluvial horizon:	
No spodic horizon	*Albic arenosols*
With spodic horizon	*Podzols*
Peaty	*Histosols*
Very sandy	See (ii)

(ii) Very sandy profiles (clay <15 percent) are likely to be *albic arenosols* if they are deep bleached sands; *cambic, ferralic* or *luvic* arenosols if a B horizon is identifiable; or *regosols* if there is little or no profile development.

(iii) If none of the special features listed in (i) are present and the soil occurs in a dry climate (wet season <3 months), it is likely to belong to one of the following:

Clearly visible humic A horizon	*Xerosols*
Very weakly developed humic A horizon	*Yermosols*
No profile development (unconsolidated materials)	*Regosols*

(iv) If none of the special features listed in (i) are present and the soil occurs in a humid climate, it is likely to belong to one of the following:

Strongly developed clay skins; developed on basic rocks	*Nitosols*
Argillic B horizon with weakly to moderately developed clay skins:	
Base saturation of B horizon > 50 percent	*Luvisols*
Base saturation of B horizon <50 percent	*Acrisols*
No clay skins, very few or no weatherable minerals	*Ferralsols*

Of critical importance among the humid tropical soils is whether the B horizon is argillic or oxic, the definitions of both of which run to over six properties. Fortunately for field survey purposes they are distinguished by the presence or absence respectively of clay skins. An argillic B horizon should be heavier-textured than the topsoil whereas an oxic B need not be. Confirmation of an oxic B horizon rests on laboratory analysis of the cation exchange capacity of the clay fraction, but this can again be estimated in the field; the required absence of 2:1 lattice clay minerals can be assessed from moist consistence, oxic horizons being notably friable, non-stiff and non-sticky in relation to the texture.

Plinthite is defined as 'an iron-rich, humus-poor mixture of clay with quartz and other diluents . . . which changes irreversibly to an ironstone hardpan . . . on exposure to repeated wetting and drying . . . When irreversibly hardened the material is no longer plinthite but is called ironstone.' As such material almost invariably *is* hard, this last qualification effectively disposes of the misconceived definition of plinthite as formulated in the US 7th approximation and ostensibly adopted. Soils with an ironstone horizon within 100 cm of the surface belong to the *petroferric phase* if it is massive or cemented nodular laterite (p. 155), or the *petric phase* if non-cemented nodular laterite.

The FAO system has some of the defects of all artificial classifications, that not all natural units are not identified as classes, and that very similar soils which happen to fall on either side of some limiting value may be placed far apart. The 1968 formulation contained internal inconsistencies, many of which have since been amended. Whilst containing features adopted from the 7th approximation, the FAO system reduces its main defects, notably by considerable simplification.

Greatest among its merits is that of possessing the authority of the leading international organization in soil science. It is a compromise document, incorporating features and nomenclature from various national systems. In particular it succeeds in achieving a moderately successful rapport between artificial and natural classifications, being outwardly an artificial system but with parameters chosen in an attempt to define natural classes.

The Soil map of the world

At the time of writing, eleven of the eighteen sheets of this map have appeared. It is an impressive document; with the exception of Africa, this is the first time that soils information in anything like this detail has been collated for continents. The areas mapped are *map units*, which consist of 'soil units or associations of soil units occurring within the limits of a mappable physiographic entity' (FAO–Unesco, 1974, p. 4). The method of compilation was essentially that of taking national soil maps produced by governments, or in some cases individuals, and interpreting the legends in terms of the FAO system; a consequence of this procedure is that certain soil units are more likely to appear in some countries than others. The wide variations of accuracy of survey show up clearly as differences in complexity of the soil patterns. As would be expected at the 1:5000000 scale, most map units are associations, of one dominant and up to three associated soil units, sometimes with up to three inclusions; the divide between associated soils and inclusions is set at 20 percent of the area of the map unit. It is estimated that when complete for the whole world there will be approximately 5000 different map units.

There are some surprises, for example the allocation of the middle Ganges Valley to eutric cambisols (which elsewhere appear in the Andean foothills at latitude 40° S), but the broad outlines of the pattern conform with expectations. Vast areas of the Amazon and Congo Basins are occupied by orthic and xanthic ferralsols. In the drier parts of the savanna zone, ferric and chromic luvisols predominate, the former in West Africa, the latter in the Indian Deccan. Vertisols duly occupy such areas as the Deccan Traps

247

and the Ethiopian Highlands. Deserts are made up largely of yermosols, regosols and the 'miscellaneous land units' of desert detritus and dunes. Xerosols are unexpectedly rare, apparently because most of the semi-arid zone is covered with sand, giving arenosols. Belts of fluvisols show at the 1:5000000 scale along major rivers only.

NATIONAL SYSTEMS

The French (ORSTOM) classification

Main accounts: Aubert (1965a); Commission de Pédologie (1967).
Summaries: Aubert (1965b, 1968); in English, Aubert (1964).

The ORSTOM (Office de la Recherche scientifique et technique d'Outre-mer) or French classification is used throughout the countries of former French Africa. It is a natural system, based not on single parameters or diagnostic horizons but on the evolution of the profile as a whole.

The classification is hierarchical, with ten classes subdivided into sub-classes, groups and sub-groups (table 26). Features used in division at the class level are degree of evolution as indicated by horizon differentiation,

TABLE 26 *The French system of soil classification*

A. Classes

I. Sols minéraux bruts
II. Sols peux évolués
III. Sols calcomagnésimorphes
IV. Vertisols et paravertisols
V. Sols isohumiques
VI. Sols à mull
VII. Podzols et sols podzoliques
VIII. Sols à sesquioxydes et à matière organique rapidement mineralisée
IX. Sols halomorphes
X. Sols hydromorphes

B. Partial subdivision of Class VIII

Class VIII. Sols à sesquioxydes et à matière organique rapidement mineralisée
  Sub-class 1. Sols rouges et bruns méditerranéens
    Group a. Sols rouges méditerranéens non lessivés
    Group b. Sols rouges méditerranéens lessivés
    Group c Sols bruns méditerranéens
  Sub-class 2. Sols ferrugineux tropicaux
    Group a. Sols faiblement ferrallitiques
    Group b. Sols ferrallitiques typiques
    Group c. Sols ferrallitiques lessivés
    Group d. Sols ferrallitiques humifères

type of weathering as shown by the composition of the clay fraction, humus type, and two features which fundamentally affect processes, hydromorphy and halomorphy.

The main interest in respect of tropical soils is the subdivision of Class VIII, soils rich in sesquioxides and metallic hydroxides (table 26B). Subclass 1 covers Mediterranean soils. The two other sub-classes, *sols ferrugineux* and *sols ferrallitiques*, are divided into six groups and 24 sub-groups. *Sols ferrugineux* are rich in free iron but lack free alumina, and have a B horizon base saturation of over 40 percent. Sols ferrallitiques contain free alumina, have a silica: alumina ratio of less than 2.0 and a low base saturation. As applied to *sols ferrugineux*, the term *lessivés* indicates that there is a textural and iron-enriched B horizon. *Sols ferrallitiques lessivés* are defined as being more acid at the surface than in depth; the modal sub-group is leached only in bases whilst other sub-groups have clay and iron lessivage. Other criteria used at the sub-group level are the presence of concretions, indurated laterite and gleying. The *sols ferrallitiques typiques* are subdivided into red, yellow and yellow-over-red sub-groups.

The *sols faiblement ferrallitiques* include the sub-groups *modal* and *ferrisolique*, the latter defined as containing illite or weatherable minerals and having a strongly-developed structure. The *modal* sub-group is genetically intermediate between *sols ferrugineux* and *sols ferrallitiques* having the characteristics of the latter but to a less marked degree (cf. weakly ferrallitic soils, p. 146). The *ferrisolique* sub-group, however, in fact owes its properties not to less intense weathering but to the influence of base-rich parent materials, although this is not explicitly recognized in the classification. Confusion between the ORSTOM *ferrisolique* sub-group and the *ferrisols* of the INEAC classification was responsible for the ambiguous treatment of ferrisols in the CCTA system.

In detail the French system can be faulted for an over-emphasis on climatic zonation as compared with lithogenic influences. In principle it has the property inherent in natural systems, that class boundaries are imprecise and allow for subjective interpretation, a feature which prevents its wide acceptance in the English-speaking world. Set against this drawback for practical mapping purposes is the advantage that, through the use of genetic natural units at all levels, it is possible to make general statements about the soil classes in any category. It is not a classification of its own sake, but an embodiment of views about the nature of soil evolution, and has been of particular value as a contribution to understanding of the development of latosols.

The United States 7th approximation (1960, with subsequent amendments)

Main account: Soil Survey Staff (1960, supplemented 1967), including representative profiles.

Classification

Summaries: Aubert and Tavernier (1972), the best short account; Smith, G. D. (1963, 1965, 1968).

The full title of this system is *Soil classification. A comprehensive system. 7th approximation.* Whilst produced for use in the United States it was received with considerable interest throughout the world and for a time was adopted by a number of organizations as an international system.

The history of the first six approximations is known only to a few; they date from 1951 onwards, and the appellation refers to the fact that each was circulated to a large number of soil scientists for comment and testing. The '7th approximation', as it is usually known, completely abandoned all previous soil classifications. The view was taken that traditional names, such as podzol and chernozem, has been used with such varying meanings that to retain them would cause confusion. A new terminology was therefore devised, the only definitions of which are those contained in the new system. For example, whilst many meanings have been attached to 'podzol' there is only one definition of its approximate equivalent, a *spodosol*. The definitions in the 1967 supplement supercede those of the 1960 volume.*

The classification is hierarchical, with six categories: orders, sub-orders, great groups, sub-groups, families and series. There are 10 orders, 47 sub-orders and 225 great groups. It is highly artificial, with limiting values for classes exactly specified. The definitions of classes are lengthy and achieve precision by the use of legal phraseology. For example, *ultisols* 'have an argillic horizon but have no fragipan and have base saturation (by sum of cations) of less than 35 percent at 1.25 m below the upper boundary of the argillic horizon, or 1.8 m below the soil surface, or above a lithic or paralithic contact, whichever is shallower, or...', continuing in this fashion for a further 160 words. A feature is the use of diagnostic horizons, defined in lengthy but precise terms. For each parameter employed there are precise quantitative limits given, together with specification of the analytical procedures to be employed.

Most tropical soils fall into the orders of aridisols, entisols, inceptisols, alfisols, ultisols and oxisols. *Aridisols* are moist (above wilting point) for less than three consecutive months; they contain soils of both arid and semiarid climates, including halomorphic types. *Entisols* lack horizons of pedogenetic origin; this class is used to cover not only alluvial soils but also sands with little profile development and rocky soils of mountainous areas. *Inceptisols* lack an argillic horizon; whilst primarily belonging to the temperate zone they include a tropical suborder *tropepts*. *Alfisols* and *ultisols* have an argillic horizon, the base saturation being above 35 percent in alfisols and

---

* A further, possibly definitive, version is in preparation but had not appeared at the time of going to press.

below in ultisols (with some exceptions). The most highly weathered tropical soils are *oxisols*; these must possess an oxic horizon, the requirements for which are exacting. Alfisols, ultisols and oxisols are divided into sub-orders mainly on the basis of moisture regime, as follows:

	Moist > 90 days, dry > 90 days	*Dry <90 days*
Suborder root:	Ust-	Ud-
Alfisols:	Ustalfs	Udalfs
Ultisols:	Ustults	Udults
Oxisols:	Ustox	Orthox

In broad terms, freely-drained soils of the rainforest zone are likely to be orthox if they lack an argillic horizon, or udults if they possess one. Soils of the savannas can be ustox if very highly weathered, but are more commonly ustults if strongly leached or ustalfs if less leached. If there is no argillic horizon but the requirements for an oxic horizon are not met the soil is classed as a tropept. The former 'red-yellow podzolic soils' of humid subtropical eastern continental margins are udalfs. Oxisols are approximately equivalent to (although more narrowly definied than) ferralsols in the FAO system. Alfisols and ultisols correspond approximately to the FAO luvisols and acrisols respectively.

Disadvantages of the system, apart from its nomenclature, include the extreme complexity, excessive reliance on laboratory analysis, and dependence on parameters of the annual soil moisture and temperature regimes for which data are rarely available. A further difficulty in field survey is that owing to the rigidity of the class boundaries, natural landscape units commonly contain more than one higher-category class (Webster, 1968*a*, 1968*b*). It is unfortunate that this ill-conceived classification should have the authority of so influential a body as the US Soil Survey, and its use outside the United States is unnecessary and undesirable.

The US 1938 classification

Main accounts: Thorp and Baldwin (1938); Baldwin *et al.* (1938).

The terminology of this system was widely used in the English-speaking world until recently. In particular it was responsible for the use of the epithet 'lateritic' to mean rich in sesquioxides and 'podzolic' to mean with clay and/ or iron eluviation. *Red-yellow podzolic* was applied to soils developed in the south-eastern United States, which have a textural B horizon and sometimes translocation of iron. The classification has not been adopted internationally, and is now superceded within the United States by the 7th approximation.

Classification

The INEAC classification

Main account: Sys *et al.* (1961).

Summaries: Sys (1960, 1967); Tavernier and Sys (1965), including representative profiles.

This system was devised for use in the former Belgian Congo by INEAC (Institut Nationale de l'Étude Agronomique au Congo). Its principal interest lies in the subdivision of the order of *Kaolisols*, soils dominated by 1:1 lattice clays and hydroxides. The sub-orders are:

> *Hydrokaolisols* (gleyed)
> *Humic kaolisols* (high-altitude)
> *Hygrokaolisols* (of rainforest climates, always moist, base saturation normally < 25 percent)
> *Hygro-xerokaolisols* (of moist savannas, base saturation < 50 percent)
> *Xerokaolisols* (of dry savannas, base saturation > 50 percent)

Thus four of the five sub-orders are based on climatic zonality. Within each sub-order, however, there are sub-groups of ferrisols and ferralsols (e.g. hygroferrisols, hygroferralsols). The *ferrisols* have an horizon with a well-developed blocky structure and clay skins, and represent a less advanced stage of weathering. The *ferralsols* are weakly structured or structureless, have few or no weatherable minerals in the profile, a low silt:clay ratio, and a clay fraction with a cation exchange capacity of less than 16 m.e./100 g; they result from weathering proceeding to an advanced stage. The emphasis is on successive stages of weathering, although it is recognized that one of the causes for 'retardation' of weathering may be basic parent material. This system is superior to the French in respect of the more humid tropical climates, and incorporates ideas on age and stage of weathering that are of general interest.

The Ghana classification

Summaries: Brammer (1962); Ahn (1970, pp. 202–8).

This system was devised by C. F. Charter and used mainly in Ghana and Nigeria. The classes of freely-drained soils in the humid zone are as follows:

Latosols:	Savanna ochrosols
	Forest ochrosols
	Forest oxysols
Basisols:	Forest rubrisols
	Savanna rubrisols
	Forest brunosols
	Savanna brunosols

252

Latosols (as used in this system) have parent material of intermediate to felsic composition, dominantly kaolinitic clay minerals and a weak grade of structure. *Basisols* are the less extensive but more fertile soils developed from basic parent materials; there is some montmorillonite present and a well-developed blocky structure. The latosols are divided into ochrosols and oxysols. *Ochrosols* have a topsoil pH of 5.5–7.0, some becoming more acid in depth; they are slightly to moderately leached. *Oxysols* are strongly leached, with a pH of < 5.5 throughout the profile. *Savanna ochrosols* are most frequent in the savanna zone, with a rainfall range of 1000–1400 mm, and *forest ochrosols* in the semi-deciduous forest zone, up to 1600 mm. *Forest oxysols* occur mainly in the evergreen forest zone, with a rainfall over 1800 mm, but soils with these properties also occur under 1500–1800 mm rainfall on very siliceous rocks. The basisols are normally reddish, *rubrisols*, in the forest zone, and brown, *brunosols*, in the savannas, the two other types being infrequent. On rocks of intermediate composition, e.g. hornblende–biotite grandodiorites, forest rubrisol–ochrosol intergrades occur. This system was valuable in being one of the few to give prominence to the influence of parent material. It is natural classification, based on the recognition of two main axes influencing soil genesis, those of rainfall and rock composition. In this respect it is a better basis for understanding the differentiation of this group of soils than the ORSTOM system, which over-emphasizes the zonal factor, and that of INEAC, which emphasizes stage of weathering rather than the direct influence of parent material.

The Australian classification

Main Account: Northcote (1960, revised 1971).

The earlier of the two Australian classifications was of a generalized descriptive type (Stephens, 1962). The later, employed in the 1 : 2000000 sheets of the *Atlas of Australian soils*, is methodologically of interest in being an artificial system based almost entirely on morphological properties identifiable in the field (the one exception is the use of reaction). It has a descending hierarchical structure with five categories, subdivision in each category being based on single properties. The division of the highest category is based on the texture profile, defined as the variation of texture with depth. The classes are uniform textural profile (U), with similar texture throughout; gradational (G), having a gradual increase of clay with depth; duplex (D), having an abrupt textural contrast between the A and B horizons; and organic (O). In the next lower category, uniform profiles are subdivided on the basis of texture (coarse, medium, fine, and fine cracking clays), gradational profiles are separated into calcareous and non-calcareous, and duplex profiles are subdivided on the basis of colour of the B horizon. The lowest

Classification

category is termed a *principal profile form* (PPF), of which 855 have been identified. The properties possessed by any principal profile form may be ascertained by reference to the criteria used in successive subdivision. Thus a typical ferruginous soil (in the sense described on p. 135) might be PPF Dr4.11, this coding indicating a duplex texture, a red B horizon with well-developed peds, no A2 horizon and an acid reaction.

This is an extreme form of an artificial system and illustrates the advantages and limitations of such an approach. The key is easy to use, and profile can be unambiguously and rapidly placed into its class from field survey data only. The classes, however, are very numerous, and do not form natural units even in the higher categories. The system is not suitable for international use, but its key is useful as a model for the construction of national systems if this approach is favoured.

USSR systems

English-language summaries: Basinski, 1959; Ivanova and Rozov, 1960; Tiurin, 1965; Gerasimov, 1968; Rozov and Ivanova, 1968.

The several USSR classification systems are little used outside the Soviet Union. A world soil map has been published (USSR Academy of Sciences, 1964) but the information on the tropics is based heavily on inference from climate. The approach is genetic, with a multivariate structure based on the influences of climate, drainage and parent material. The main aspects that have been of international influence, including with respect to the FAO system, are those concerning soils of the subtropical semi-arid and temperate steppe zones.

Systems in use in the Indian subcontinent

India has for long divided its soils into three major groups: red soils, black soils and alluvial soils (Raychaudhuri, 1962, 1963; FAO, 1965b; Raychaudhuri and Govinda Rajan, 1971). The red soils are latosols *sensu lato*, whilst the black are vertisols of the Deccan lavas. This has been aptly compared to a classification of the buildings of Oxford into those built of stone, brick and concrete – it is valid so far as it goes. The red soils are divided into red loams (probably equivalent to ferruginous soils), red sandy soils (sandy ferruginous and/or weathered ferrallitic soils), both of which occur in the medium-rainfall zone of the Deccan, and laterite soils (leached ferrallitic soils) of the high-rainfall Western Ghats. The black soils comprise shallow, medium and deep types. A map on 1:7000000 is given in Raychaudhuri and Govinda Rajan (1971). Indian soil science is considerable in volume but has been largely directed towards experimental work on soil fertility, and soil survey is comparatively little advanced in relation to the size of the country.

254

Large areas of Pakistan and Bangladesh have been covered by soil survey, and both countries have made important contributions to knowledge of alluvial soils (Hunting Surveys Corporation (Canada) Ltd, 1958; FAO, 1965*b*, pp. 11–15; Islam, 1966). Much of this work is unfortunately not yet widely available.

Sri Lanka has a long-established soil classification system, probably related initially to the US 1938 system, and the entire country is covered by reconnaissance survey (Panabokke and Moorman, 1961; De Alwis and Panabokke, 1972–3). In the Wet Zone the predominant soils are 'red and yellow podzolic soils', some with laterite, and in the Dry Zone, 'reddish brown earths'. These correspond to leached ferrallitic and ferruginous soils respectively. The upper rainfall limit for soils with ferruginous properties occurs at mean annual values of about 1750 mm, an unusually high value the reason for which has not been explained. Other types of latosols present are 'red and yellow latosols' on unconsolidated sediments (sandy ferruginous soils, perhaps with arenosols), 'reddish-brown latosolic soils' derived from basic rocks (ferrisols) and 'red yellow podzolic soils with dark B horizon' with a high organic matter content (humic ferrallitic soils). Vertisols, alluvial soils, gleys and halomorphic soils are all present. This small country is a microcosm of soils of the humid tropics.

COMPARISON

The following authoritative correlation tables have been published:

Based on	Equivalents in	Reference
FAO	Eight national systems	Dudal (1968*a*, pp. 11–18)
		FAO–Unesco (1974, pp. 14–20)
FAO	ORSTOM, 7th approximation	Aubert and Tavernier (1972)
ORSTOM	7th approximation	Aubert and Tavernier (1972)
US 1938	7th approximation	Douglass *et al.* (1969)

A general discussion of the relation between the CCTA and other systems is given by D'Hoore (1968*b*), and between the ORSTOM system and the 7th approximation by Duchaufour (1963).

DISCUSSION

Many soil types are widely-recognized natural units. These include lithosols, gleys, andosols, vertisols, solonchaks, solonetz and organic soils. Each has distinctive genetic factors, morphological features or both, although classifications differ with respect to marginal limits. For these soil types the differences between systems are not of classification but of nomenclature.

255

Classification

The main classification and correlation problems concern soils not possessing any such distinctive properties. These include, however, the most extensive and agriculturally productive soils of the tropics. Differences between systems originate in the understanding of what are the natural units of soil formation, and the relative emphasis placed on environmental factors, soil-forming processes and profile morphology; these differences are expressed in the soil properties that are considered to be of importance in the higher categories. For example, within the latosols (in the broad sense employed here, see p. 132) there are three groups of properties which may be regarded as fundamental: firstly, the composition of the clay fraction and hence the cation exchange capacity, properties indicative of the degree of weathering; secondly, the reaction together with the base saturation, indicative of the intensity of leaching; and thirdly, the presence or absence of a textural B horizon with clay skins, usually assumed to indicate clay translocation. These groups do not necessarily vary together. Thus, not all soils with a low cation exchange capacity of the clay fraction have a low base saturation, nor do they necessarily lack a textural B horizon.

As the various classification systems emphasize different groups of properties there is substantial overlap between classes. It is therefore impossible to give exact equivalents, and correlation can only be achieved by reference to the properties of soil series or other low-category classes.

Artificial systems have their place, being useful in situations where it is desired that every soil profile shall be unambiguously allocated to a class on the basis of a limited number of properties. This is necessary first, in computerized classification, and secondly, where classification is left in the hands of non-professional grades of staff. But such systems are inherently unsatisfactory in achieving the main purpose of a classification: that of ordering knowledge so that the properties of the soils in each class can be easily remembered and their relationships understood.

The limitations of artificial systems were aptly castigated by C. F. Charter, in a government paper written at a time when they were beginning to make their appearance. 'Their guiding principles appear ... emotionally feeble, lacking in inspiration and most unlikely to form an incentive for enthusiastic research. They are unpleasantly reminiscent of the filing systems of Government Departments' (Charter, 1957b).

To serve as a medium for the organization, transfer and advancement of knowledge a soil classification must have a genetic basis. This principle was establsihed by the Russian work that formed the foundation of modern soil science and has always been accepted by French-speaking peoples. The recent trend towards 'filing cabinet' systems, for which the 7th approximation was responsible, is retrogressive. It is noteworthy that soil textbooks, concerned with the organization of knowledge as it stands at the time, frequently employ natural soil classes with a genetic basis, reserving descriptions of one or more formal classification systems for a separate section.

Natural soil classifications require skill to use, since they involve an understanding of soil genesis. They have blurred edges to the classes; that is, there will be profiles which cannot be definitely allocated to a class. Artificial systems solve such cases by a decision based on some arbitrary limiting value, whereas in fact the profiles concerned are genuinely intermediate.

Besides the ordering of knowledge in general, soil classifications serve two specific purposes, soil mapping and agricultural planning. In mapping, the use of systems with rigid limiting values makes it difficult to avoid the proliferation of soil associations, mapping units containing two or more classes. Soil maps are necessarily based on natural landscape units, which correspond to soil classes based on genetic principles. Agricultural planning is concerned with predicting the behaviour of the soil under given conditions. To do this requires a knowledge not only of the existing static condition of the soil, its morphology, but of the processes acting upon it, in order to estimate how these will respond to changes brought about by agricultural practices. It is inherent in natural systems that each class is characterized by a particular set of processes, since these are part of its definition.

It is for these reasons that a natural classification is used in this book. The nomenclature employed is given in table 18 (p. 128), the rationale of the six main groups outlined on pp. 127–131, and the division into soil types discussed in the respective chapters on these groups. This is not intended as a formal classification but as a means of identifying the main genetically-based soil types of the tropics. Like all natural systems it is not immutable but subject to change in the light of advances in knowledge.

Many attempts have been made by individuals to establish new classifications, but only those recognized by national soil survey organizations have achieved common recognition within a country. Similarly, to achieve international recognition requires the authority of the FAO and the International Society of Soil Science. The FAO system has the former although not yet the latter. Moreover, as well as being politically a compromise it attains a reasonable balance between the use of natural units and of artificial properties as limiting values (although erring in the direction of the latter). The CCTA system has served as a common means of communication for some ten years and is superior to the FAO scheme in respect of its recognition of natural units, but it neither comprehends extra-tropical soils nor has the necessary authority. With the completion of the FAO–Unesco *Soil map of the world* it is likely that the FAO classification will achieve general acceptance and gradually supplant other systems as a means of international communication.

Countries with no strongly-established national classifications could suitably adopt the FAO system as a basis for higher-category grouping. For the most part, however, national soil survey organizations will continue to operate their own systems. What is desirable is a recognition by all countries that insofar as it is wished to transmit information internationally, the equivalents in the FAO classification should be given. This applies not only to soil maps but with equal or greater force to reporting the results of agronomic experimental work. There is an analogy with the position in biology, in which plant and animal names in national languages are widely and usefully used but are always accompanied by the Latin equivalents, for the standardization of which an international organization exists. Owing to overlapping definitions the analogy with soils is not exact, but such phrasing as, for example, 'red loams (approximately equivalent to FAO *ferric luvisols*)' can be used. It will be a matter of regret if this belated first attempt at international standardization fails to gain acceptance.

SELECTED REFERENCES

Omitting both the large literature on general principles of soil classification and the sources of particular systems given above, the following may be noted as containing discussions and comparisons of different classification systems: Pédologie (1965), FAO (1968), D'Hoore (1968*b*).

CHAPTER 14

PROBLEMS OF SOIL EVOLUTION

Quand on ne disputait pas, l'ennui était excessif.
Candide

The topics discussed in this chapter concern the origin and development of the features found in tropical soils, including their relations with the present environment and the influence of features inherited from the past. It is within the broad group of latosols that questions of this nature are the least resolved. Possibly this is because in many other soil types, the morphology is dominated by one process distinctive to that soil, for example salinization, in the formation of solonchaks or sedimentation in the case of alluvial soils. In contrast, the same kinds of process are operating in all latosols, and the differentiation of properties between the various latosol types results from differences in absolute and relative rates and intensities of processes.

The questions discussed are mainly of scientific interest rather than practical significance. There are, however, indirect aspects of an applied nature. First, an understanding of soil evolution assists in rationalizing observed distribution patterns, and hence in soil survey. Secondly, a knowledge of the processes which have brought the soil to its present condition, and which are currently acting upon it, is an aid in predicting how it will respond to changes brought about by management.

AGE AND STAGE OF WEATHERING

The *age* of a soil is the time, in years, since the materials of which it is formed were part of the unaltered parent material. The *stage of weathering* is the degree to which the soil has been altered by weathering from its condition in the parent material. These two concepts, often confused, are distinct.

Soil age strictly applies only to a part of the soil, and the topsoil will normally have a greater age than the deeper horizons. Consider the case of a soil developed from consolidated rock on erosional topography (fig. 16(a) – see p. 264); it is assumed that 'regolith' and 'rock' are separated by a planar weathering front. Two processes are acting: rock weathering, which converts rock to regolith, and erosion, assumed to be by surface wash, which lowers the ground surface. At time 1, a quartz grain at *P* has just been weathered from parent rock to soil; the horizon within which it lies, that immediately overlying the weathering front, has an age of only a few years. At time 2, *P* has passed into the middle of the regolith profile and new material at *Q* has just been weathered. At time 3, *P* is in the topsoil, *Q* in the middle of the profile and *R* has just been weathered. The age of the topsoil at time 3 is the interval, in years, between times 1 and 3.

Concurrently with the conversion of 'rock' to 'regolith', weathering processes are altering the composition of the regolith. Point *P* consists of slightly weathered rock at time 1, highly weathered rock at time 2 and soil at time 3. Thus at time 3, as well as having different ages, points *P*, *Q* and *R* are at different stages of weathering.

This model holds good if the soil depth is not constant but increases with time. The transfer of material downslope by soil creep causes modifications. If it is assumed that creep causes downslope movement parallel to the ground surface, in a manner analogous to laminar flow, then the net effect will be that material at an equivalent depth has a greater age lower on a slope than near its crest. Soil creep is in fact largely confined to the upper part of the regolith, usually the top metre or less, as indicated by shadow rock structures and quartz veins *in situ* below this depth. Vertical mixing, i.e. transfers of material perpendicular to the ground surface, is brought about by termite action and clay translocation. It is largely confined to the top 1—2 m and tends to homogenize the age of this upper layer.

For the case of transported, sedimentary materials soil age is differently determined. The soil age is the time that has elapsed since deposition. In the case of recurrent sedimentation, as on active flood-plains, the soil will be youngest in the topsoil and age will increase with depth. There is then no necessary relation between age and stage of weathering, since the alluvial materials may be highly weathered at the time of deposition.

The distinction between original parent material (parent rock) and proximate parent material has already been noted (p. 16); the latter is the weathered regolith material on which pedogenesis *sensu stricto*, as soil profile differentiation, acts. Soil age refers to the time elapsed since the soil was in the condition of original parent material, i.e. since the commencement of rock weathering. It is also possible to refer to the age of the proximate parent material; care is again necessary to distinguish between age and stage of weathering.

The distinction between original and proximate parent materials is also useful in the case of pediments or other low-lying sites, if it is thought that the soil is developed from colluvial deposits derived originally from rock higher on the slope. The colluvium is then the proximate parent material. In the case of alluvial soils the concept is stretched, but can sometimes usefully be applied, inasmuch as alluvial deposits derived from basins of different rock types can be identified.

Soil age is straightforward as a concept but can rarely be determined. Thus use of carbon–14 dating is difficult owing to the high probability of contamination from the topsoil or from upslope. Occasionally an horizon can be dated by buried archaeological remains. For the most part, however, soil age is inferred from its stage of weathering.

Rock material subjected to weathering passes through a progressive series of changes in particle size distribution, mineralogy, chemical composition and other properties; the exact nature of these changes varies with rock type and climate. The position in such a series reached by a soil or a part of the regolith is its stage of weathering.

One well-established series of changes is the succession of weathering zones in granite (Ruxton and Berry, 1957; Berry and Ruxton, 1959). These are, in order from the parent rock upwards: Zone IV, weathered rock (> 90 percent solid rock), Zone III, core-stones with fine material (50–90 percent rock), Zone II, debris with core-stones (< 50 percent rock) and Zone I, debris without core-stones. In the perspective of pedology this series ends at a relatively early stage of weathering.

Van Wambeke (1962) proposed three criteria for classifying tropical soils and (proximate) parent materials according to what he called age but which is more correctly stage of weathering. These criteria are soil structure, the silt:clay ratio and the percentage of weatherable minerals. His evidence is based on comparison of soils on old erosion surfaces with those on dissected topography.

'Young' parent materials, provided they are clayey, have angular blocky peds with clay skins, whilst old materials lack well-developed planes of weakness; this is additional to the fine crumb structure possessed by most latosols. The silt:clay ratio is above 0.15 in younger materials and below this in older. The proportion of weatherable minerals (other than quartz, but including muscovite) decreases with soil age; for the light fraction (specific gravity < 2.69) the division between older and younger materials is at 3 percent of the total soil or 15 percent of the 50–250 μm fraction. 'Old' parent materials are massive or weakly structured, have a silt:clay ratio below 0.15 and less than 3 per cent weatherable minerals. Young materials have *at least one* of the following characteristics: (i) a well-developed blocky structure with clay skins; (ii) a silt:clay ratio above 0.15; (iii) more than 3 percent weatherable minerals. Subsequently van Wambeke (1967) added a fourth criterion, that soils on old, stable landscapes usually contain an horizon in which the cation exchange capacity of the clay fraction is less than 12 m.e./100 g.

The first criterion, on soil structure, is given by van Wambeke as a fact of observation. A possible rationale for it is that the presence of clay skins indicates current release of clay by weathering and its translocation. The second index rests on the assumption that the silt fraction consists mainly of weatherable minerals and the clay fraction of secondary material. The reason for the third index is apparent. If the fourth index is added, its explanation would be that in materials at an early stage of weathering there is always sufficient montmorillonite present to raise the exchange capacity of the clay above 12 m.e./100 g.

These are not criteria of age, as stated by van Wambeke, but of stage of weathering. There is some confusion as to whether it is the soil or the parent material that is referred to; specifically, it is not clear if the criteria should be applied to the B horizon or to material at about two metres depth. There are exceptions to each criterion; sandy soils do not develop a blocky structure, some shales release residual silt-sized quartz particles to give high silt:clay ratios, whilst some recently deposited alluvial materials are themselves low in weatherable minerals. These possibilities are recognized by the fact that only one of the criteria for 'younger', i.e. less highly-weathered, material need be fulfilled. There is an element of circular argument in the establishment of the criteria, since there is little independent evidence of the age of erosion surfaces and they are partly identified as such by the presence of a deep and highly-weathered regolith.

Ruxton (1968) proposed the silica:alumina ratio as an index of the degree of chemical weathering of rocks. The ratio was found to correlate well with silica loss and with total element loss. Whilst put forward with respect to the early stages of weathering, this index may also be applicable to the later stages found in soils. One of the criteria for indentifying ferrallitic soils on the French and CCTA classifications is a silica:alumina ratio of less than 2.0.

Notwithstanding reservations concerning the danger of circular reasoning, there is a substantial amount of observational evidence that is at least consistent with the use of these five criteria, applied with regard to local special circumstances, as indices of stage of weathering. They form one basis for classification of the latosols, and are given prominence in the INEAC system. There is general agreement on all classification systems that the ferrallitic soil or its equivalent (ferralsol, oxysol, etc.) possesses the features indicating an advanced stage of weathering.

The distinction between age and stage of weathering can be applied to the two types of ferrallitic soil. Both types have reached an advanced stage of weathering. In leached ferrallitic soils the material is not necessarily old, but is highly weathered because weathering has been intense and rapid. In weathered ferrallitic soils the advanced stage of weathering results from the high absolute age of the material, and the long period of time that it has been subject to weathering processes of moderate intensity.

MATURITY AND EQUILIBRIUM

Two further concepts, also sometimes confused with each other, are soil maturity and equilibrium. A *mature soil* is one in which appreciable horizon differentiation of pedogenetic origin has taken place. A soil in *equilibrium* is one the properties of which remain unchanging with time.

Maturity is a relative concept, the chief use of which is to distinguish *immature soils*, in which horizon differentiation that could develop in the environment concerned has not yet taken place; these include recent alluvium and other materials in which such horizons as may exist are of sedimentary origin. The fact that a soil is called mature signifies only that some pedogenetic horizons are distinguishable; a mature soil may undergo further horizon differentiation, both in intensity and type. In practice soils are regarded as mature when a humic A horizon is clearly developed. Desert detritus and rock cliffs present a special case; since there are no processes capable of causing profile differentiation, immature and mature soils are identical and the concept is inapplicable.

The condition of equilibrium, or steady state, implies that processes are no longer causing change in the soil considered with reference to the ground surface although the entire profile may be slowly moving downward. The concept was first applied to soils by Nikiforoff (1949).

Fig. 16(a) and (b) illustrate the concept with reference to soil depth for the simplified case in which there is assumed to be a sharp boundary between 'rock' and 'regolith'. In Fig. 16(a) rock weathering and erosion are acting at the same rate and the soil is in equilibrium with respect to the property of depth. In Fig. 16(b), between times 1 and 2, rock weathering lowers the weathering front by a greater amount than erosion lowers the ground surface; the soil depth is therefore greater at time 2 and the soil is not in equilibrium. Fig. 16(c) and (d) show a textural B horizon produced by clay traslocation. In (c) the degree of development of the eluvial and illuvial horizons and the position of their boundaries with respect to the ground surface remain unchanged with time, and the soil is in equilibrium. In (d) the horizon becomes more strongly developed with time. The same concept can be applied to chemical properties, (e.g. base saturation or reaction), properties related or organic matter, or to the complete soil profile.

Equilibrium with respect to more persistent properties, such as profile depth and texture, is almost impossible to measure directly owing to the slowness of change, and must remain a matter of inference. Properties which change more rapidly, notably organic matter, can be monitored over time, and the equilibrium concept has important applications with respect to agricultural use.

A soil which is in equilibrium is necessarily mature; one which is not in equilibrium may or may not have reached the condition of maturity. An

263

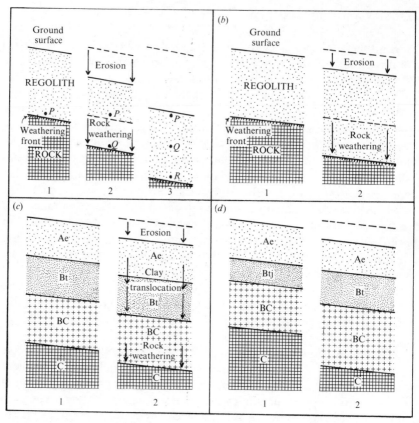

Fig. 16. Soil age and soil equilibrium. For explanation see text.

immature soil cannot be in equilibrium, except for special cases where there is no pedogenetic development, such as desert detritus. A mature soil may or may not be in a condition of equilibrium.

A case which is commonly misinterpreted is that of lithosols on steep slopes. These have shallow profiles owing to rapid natural erosion, and are sometimes held to be immature, on the grounds that they lack the horizon differentiation found on soils of gentle slopes having the same climate and parent material. This interpretation is incorrect, since it implies that erosion is not a soil-forming process. Such soils are both mature and in equilibrium. They have AC profiles, which represent the maximum horizon differentiation attainable under their total environment, including slope angle. The very fact that the depth of lithosols varies over a slope between relatively narrow limits, about 10–50 cm, suggests that a state of equilibrium between rate of rock weathering, surface wash and soil creep has been reached.

Any soil can be interpreted either in equilibrium terms, by comparing its properties with the present factors and processes, or in terms of features inherited from different conditions in the past. There is no general agreement, especially with respect to the latosols, as to the relative importance of these two types of explanation.

THE ORIGIN OF HIGHLY-WEATHERED REGOLITH MATERIAL

The gently undulating plateaux that are widely found on Basement Complex shield areas frequently carry a thick and highly-weathered regolith. This feature is extensive in Africa, and also found in central Australia (Mabbutt, 1965) and in the Mato Grosso and Guyana Shields of South America. It occurs under climates ranging from semi-arid to the savanna–rainforest transition zone. The regolith thickness is typically 10–50 m, exceptionally up to 100 m, and may vary substantially over short distances; these variations in thickness may show little relation to surface topography (Thomas, 1966a). The existence of two separate levels, the ground surface and the weathering front, has been geomorphologically called 'double planation surfaces' (Büdel, 1957).

The regolith material is crushable by hand, but except within a few metres of the surface it retains the feel of highly-weathered rock rather than soil. It is weathered *in situ*, as indicated by shadow rock structures and undisturbed quartz veins. Chemically it is characterized by a loss of more readily weatherable elements, including the bases and silica, and a consequent relative gain in iron, aluminium and minor resistant materials such as titanium. The silica:sesquioxide ratio is substantially below that of the underlying weathered rock. This highly altered regolith forms the proximate parent material of weathered ferrallitic soils.

The main changes in chemical composition take place in a layer quite close to the weathering front, where fresh rock passes into what would morphologically be described as 'partly weathered rock'. This layer of change may be separated by many metres from the level at which highly weathered rock passes into soil. Fig. 17 shows the variation with depth of some properties of a six metre deep regolith layer on dolerite under a savanna climate in Rhodesia. The main changes in silica, iron and alumina content take place at two levels within the weathered rock, the lower one immediately above the weathering front at 6 m depth. The composition changes again at 4.5 m, the level at which clay is first produced. The change to what is morphologically soil occurs between 3 and 4 m, where there is a further increase in clay, a loss of bases and consequent change in pH, but no appreciable change in silica, iron and alumina. These three main constituents, and hence the silica:sesquioxide ratio, remain almost constant from the surface to 4 m, through the whole of the soil and well into the weathered rock.

Hence, where there is a deep regolith the main stages of rock weathering

265

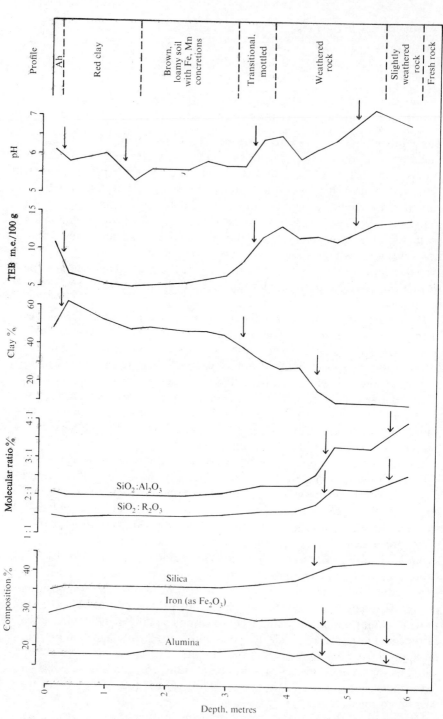

Fig. 17. Variation with depth of soil and regolith properties on dolerite, Rhodesia. Based on data in Ellis (1952).

266

occur at a substantially greater depth than soil formation *sensu stricto*, in the sense of profile differentiation within the upper two metres. By inference, weathering and soil formation are separated in time. This feature is termed *pre-weathering*.

There are two explanations of the origin of pre-weathered material: first, that it results from inequality between rates of rock weathering and surface erosion, such that the regolith becomes steadily deeper with time; secondly, that it was produced by weathering during a previous cycle of erosion.

The first explanation is that for a prolonged period the rate at which the weathering front advanced into rock exceeded the rate at which the ground surface has been lowered by natural erosion (fig. 18(*a*)). Deeply-weathered regolith covers occur on erosion surfaces at a relatively advanced stage of geomorphological evolution, on which the maximum valley side slopes do not exceed 2°–3°. These gentle slopes cause very slow denudation but do not affect the rate of weathering. Hence the regolith has been in a state of disequilibrium for a long period, still continuing, and has progressively increased in thickness. The upper layer of the regolith, from which the soil profile is formed, is therefore both old and highly-weathered. Where slopes

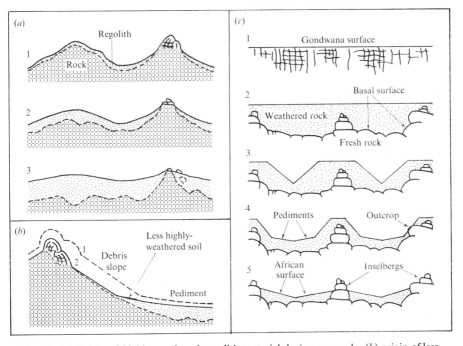

Fig. 18. (*a*) Origin of highly-weathered regolith material during one cycle; (*b*) origin of less highly-weathered soil around base of inselberg; (*c*) the two-cycle weathering theory ((*c*) after Ollier, 1959).

267

are steeper denudation is more rapid, rates of rock weathering and ground lowering are in equilibrium, and the deep regolith does not occur. At the margins of inselbergs (fig. 18(*b*)) the adjacent pediment has only recently been exposed by retreat of the inselberg; its upper part is also steeper and undergoes erosion by surface wash (pediment regrading). The regolith in this peripheral zone is therefore renewed, and is shallow and less weathered.

The second explanation is called the *two-cycle weathering theory*. It had its origins in geomorphological investigations of the origin of inselbergs and was first applied to soils by Ollier (1959; also Radwanski and Ollier, 1959). The theory states that present-day soils are formed from regolith material weathered during a previous cycle of erosion. Ollier accepts King's chronology of erosion surfaces (p. 28), according to which most of the gently undulating plateaux belong to the African surface. He assumes that the preceding, Gondwana, surface remained undissected for a long period, during which the regolith was weathered very deeply (fig. 18(*c*), times 1 and 2). Variations in depth of penetration of the weathering front are related to differences in lithology and, in particular, joint spacing. Rejuvenation led to dissection of this regolith, followed by the formation of the African surface at a lower level. Higher parts of the unweathered rock were stripped of their cover and exposed to form inselbergs. The greater part of the surface, however, was formed within the previously weathered material (fig. 18(*c*), time 5).

The consequences for soils are demonstrated for the case of a Uganda soil catena found in an inselberg-and-pediment landscape, the Buwekula catena. Over the greater part of the catena there is a thick regolith and the material within the soil profile is low in weatherable minerals. Close to the inselberg margins, however, the regolith is shallower and its upper layer much less weathered. Thus in the fine sand fraction, close to the inselberg foot feldspars form 50–75 percent of the light fraction, magnetite about 50 percent of the heavy fraction, and among the heavy mineral suite the less resistant minerals epidote and biotite are common. In the lower members of the catena the corresponding figures are 0–5 percent feldspars and 5–10 percent magenetite, whilst the main heavy minerals are zircon and tourmaline, both extremely resistant to weathering. The less weathered basal zone of the regolith above the weathering front therefore reaches the surface adjacent to inselbergs, in a manner analogous to a geological outcrop, having been revealed by the stripping of its previous highly-weathered upper zone.

In comparing these two explanations, the former may be called the one-cycle theory. Both the one-cycle and two-cycle theories are consistent with the observed pedological features; the pedological evidence presented in support of the two-cycle theory can equally well be explained on a one-cycle basis. The other type of evidence adduced in favour of a two-cycle origin is geomorphological: that it accords with two-cycle explanations of the observed variations in depth of weathering and of the origin of inselbergs. Without discussing these questions in detail, it may be noted that if wide variations in depth of weathering can arise in an early cycle of erosion there is no reason why the same mechanisms should not operate in the present cycle. There are certainly difficulties in explaining the origin of inselbergs, but the question of whether they are formed by subterranean weathering or subaerial erosion remains a matter of dispute (e.g. King, 1948, 1966; Thomas, 1965, 1966*b*).

The two-cycle theory requires a particular set of circumstances: that there should have been two erosion surfaces, developed at a height separation less than that of the regolith thickness weathered during the former. These two surfaces are often separated by scarps 300–600 m high, which exceeds the greatest known weathering depths of the present. The difference in erosion resistance between weathered regolith and rock is many orders of magnitude, and for erosion to have come to a halt within the non-resistant regolith would require powerful operation of base-level control. By contrast, the one-cycle theory requires only that the plateaux shall have been in existence sufficiently long for weathering beneath it to have penetrated deeply. It is not necessary to call upon pre-Tertiary time to achieve the observed depths of weathering. If it is allowed that the weathering front advances by as little as 0.1 mm per year, then the two million years of Pleistocene time is sufficient for attainment of the maximum observed depth of weathering penetration, some 200 m. Available evidence about rates of alteration (e.g. Leneuf and Aubert, 1960) is compatible with rock material reaching an advanced stage of ferrallitization in substantially less than one million years.

It is not in dispute that, in areas where there is a deep regolith, chemical and mineralogical changes to the rock preceded soil formation by a long period of time; that is, that the proximate parent material is pre-weathered. The question is if pre-weathering needs to be related to a previous cycle of erosion. Studies of detailed variations of weathering indices, laterally and with depth within the regolith, may well provide further evidence. At present the two-cycle theory remains unproven, and on the grounds of Occam's razor the one-cycle explanation, involving fewer hypotheses, is to be preferred.

RELICT FEATURES IN SOILS

There is evidence of the presence in some soils of *relict features*, those inherited from past conditions of climate or landforms that differed substantially from those of the present; the concept is a relative one, resting on the interpretation of what is a 'substantial' difference. Some writers would go

269

further and say that relict features are very widely distributed, and definitely absent only in such cases as lithosols on steep slopes and recent alluvial soils.

Some soil properties are relatively persistent whilst others respond rapidly to changes in the environment. Organic matter content and properties associated with it are the least persistent, being liable to substantial change in 1–5 years. The relative proportions of the exchangeable bases, their total amount and hence base saturation, and the soil reaction can be changed slightly less rapidly than organic matter, but within the order of time of a decade. Clay mineral type and consequently cation exchange capacity are considerably more persistent, and changes in them are to some degree irreversible; thus desilication, with the formation of kaolinite and gibbsite, takes place more readily than resilication. Still longer periods are required for substantial changes in texture. The most persistent features are the presence of resistant primary minerals, quartz stones and laterite. The past history of a soil is more likely to be revealed by the more persistent properties, and hence mineralogy is the most widely used type of evidence in studies with this aim.

The two features most frequently held to be relict are laterite horizons and stone lines. Laterite is an extremely persistent feature, and in some cases has undoubtedly been inherited from past conditions of different climate and topography. Even thin laterite layers appear able to retain an identity as horizons whilst the ground undergoes considerable lowering. Fölster's theory of the development of stone lines by a series of alternating humid ('stable') and drier ('unstable') climatic phases has been noted (p. 169).

A more general case of explaining soil formation in terms of past climatic changes is the theory of periodic phenomena, or K-cycles, developed in Australia (Butler, 1959, 1967; Walker, 1962). Mabbutt and Scott (1966) applied this approach to two soil catenas under a humid savanna climate in Papua. On the basis of texture, stone content, clay minerals, and content of carbonates and sodium they distinguish eight 'layers', each with distinct properties. Using as evidence the relations between these layers within individual profiles and laterally within the catena they suggest a complex sequence of changes in climate and process; it involves alternations of weathering and regolith stripping on the steeper upper parts of the slopes, of weathering, aggradation and dissection of the footslopes, and of accumulation and leaching of bases within the soils. Conclusions of the same type were reached by Watson (1964/5) from a study of a granite catena in Rhodesia. The evolutionary sequence that he suggests extends back into Tertiary time and involves three phases of erosion or planation, two of deposition, one of weathering and one of laterite formation (table 27). 'Soil formation' is attributed to the seventh phase. Similarly Fölster *et al.* (1971) explained the morphology of ferrallitic soils in southern Sudan by

TABLE 27 *Phases in the evolution of a catena on granite in Rhodesia, according to Watson (1964/5)*

Main Phase	Age
7 Erosion of M(mineral) and G(gley) horizons with some replenishment by termites carrying material to the surface; soil formation over the whole catena; ferricrete formation in profile 5	Quaternary
6 Aggradation of locally derived material later differentiated into M and G horizons	Quaternary
5 Erosion of Kalahari Sand; truncation of W horizon during this phase or phase 2	Quaternary
4 Formation of ferricrete at base of Kalahari Sand	Upper Pleistocene
3 Deposition of Kalahari Sand	Middle Pleistocene
2 Epeirogeny: formation of weathering crust (W horizon); truncation of W horizon at this time or during phase 5	Tertiary to Pleistocene
1 Pediplanation and formation of the African surface; accumulation of stones at or near the surface	Tertiary

recourse to a complex sequence involving climatic changes with successive phases of deepening by weathering and truncation by erosion.

It is not known how widely explanations of this type are applicable. It is quite possible that the same soil profile or catena might be approached by one person seeking a steady-state explanation and by another looking for relict features and for both to produce plausible explanations of observed features. We do not have actual examples of this, but an interesting comparison is that between Watson's study and an analysis of a catena in Zambia by Webster (1965). This catena is also gently sloping and under a savanna climate but is developed on gneiss. Webster's explanations are largely in terms of present-day processes. A partial explanation of the difference in findings is that Webster was concerned largely with pedological features within the upper two metres, which in Watson's analysis are treated as one layer among many, the 'mineral horizon'.

The K-cycle approach is used largely by scientists working in, or with experience of, Australia; those who study soils in West Africa are also prone to find explanations in terms of relict features and phases of development. It is hard to assess to what extent it is the observed soil features which compel such interpretations and to what extent it is the attitude which is contagious. Complex regolith cross-sections such as those shown by Fölster (1969, figs. 12, 14) certainly appear to call for a multi-phase explanation; their like is not present in East and Central Africa.

271

CHAPTER 15

THE SOIL CATENA

Few concepts have proved more useful in tropical soil science than Geoffrey Milne's formulation, in 1935, of the catena. Whilst applicable to soils in any climate, it seems that it is in the tropics that topographically controlled variation in soil properties is most strongly developed and most regularly repeated over the landscape. Many of the leading contributions to tropical soil genesis have been based on studies of catenas, whilst in the applied field it is much used in soil survey. The catena is also employed as a unit for studying tropical vegetation.

A *catena* is a succession of soils down a slope.* It usually extends from interfluve crest or hill summit to valley floor. The ground surface must slope downward continuously, except for micro-relief, from the crest to the base of the catena. The essential feature which gives genetic unity to a catena is that water and soil material can move laterally downslope. It is incorrect to apply the term, as has occasionally been done, to altitudinal sequences of erosion surfaces or river terraces, which lack this feature.

There has been frequent discussion as to whether the term should be restricted to topographic sequences developed on one parent material only (see e.g. Watson, 1965). Such a restriction would greatly limit the applicability of the concept, quite apart from the difficulty of determining the parent material in some cases. Slopes developed on two or more rock types should therefore be included. Where required, a distinction may be made between *simple catenas*, developed on a single rock type, and *compound catenas* developed slopes formed in two or more rock types.

PROCESSES OF CATENARY DIFFERENTIATION

The soil catena is primarily a function of the factor of relief, together with the indirect effects of relief upon hydrology. In the case of compound catenas the parent material factor may also be an important cause of intra-catenary differentiation of soil properties.

The soils in a catena are acted upon by the soil-forming processes of level

---

* The original definition was 'a regular repetition of a certain sequence of soil profiles in association with a certain topography' (Milne, 1935).

sites. Differentiation of properties as between the different parts of the catena is brought about by additional processes consequent on the slope. These latter processes are erosion and deposition, processes arising from the position of the water table, and the lateral eluviation of materials in solution.

The processes of natural erosion on slopes are soil creep, surface wash, solution and rapid mass movements (landslides). In tropical climates ranging from semi-arid to moist savanna, surface wash is the most important. Under rainforest climates the relative importance of creep and wash is not known, although surface wash is certainly substantial. Solution is important in all climates, its rate probably increasing with rainfall. Rapid mass movements occur on very steep slopes in all climates, and continue to recur on slopes of moderate steepness in the rainforest zone (Young, 1972a).

The rate of erosion by surface wash increases with steepness of slope and distance of overland flow, i.e. from the slope crest. The rate of increase with steepness lies somewhere between linearly with slope angle and as the square of the angle, varying with local conditions. With respect to distance the increase is approximately in proportion to the square root of distance from the crest. It is, however, not only the transporting power of surface wash that is significant but the net removal of soil material. On the upper parts of slopes denudational processes remove material from the soil surface and thus cause renovation of the soil material by rock weathering beneath (the former concept of a 'belt of no erosion' along interfluve crests has been disproved). Lower down a slope, soil material is being brought in from upslope as well as removed downwards, whilst still further down there may be intermittent deposition. This difference in net erosion rates between the upper and lower parts of a slope leads to a tendency for the regolith to become shallower and less highly weathered towards the crest. On convex slopes this tendency is partly counteracted by the effects of the downslope increase in angle. On pediments, which are generally concave in profile form, the effect is strongly developed, causing less highly-weathered soil to occur close to the hillslope–pediment junction (p. 267).

Colluvium is material transported by surface wash and deposited on the lower part of a slope. In soil descriptions the term is often used rather freely to describe any parent material on the lower, concave part of a slope. It is true that the regolith frequently thickens downslope but this can be brought about by differential exposure to renewal of weathering, as described above, rather than by deposition. If a slope is continuously re-treating deposition cannot occur, other than shallow deposits caused by short-term variations in rainfall intensity; the erosion necessary to create the slope on which deposition occurs would not take place unless the transporting processes were able to remove incoming sediment. Concavities by no means necessarily cause deposition, their profile form being a response to the downslope increase in transporting power of surface wash. Colluvial deposits of significant thickness can only accumulate if there has been a change in environmental conditions, either climatic change or accelerated erosion caused by cultivation. There circumstances may well be common. Nevertheless, parent material should not be described as colluvium unless there is definite evidence of deposition.

The second control of slope form upon soil-forming processes is its influence on the position and seasonal fluctuations of the water table. The most common circumstance is a valley floor with the water table at or close to the surface throughout the year, and a belt on the concavity in which it rises close to the surface in the rains and falls in the dry season, this zone of fluctuation falling deeper in the soil profile upslope. In the upper, and usually the greater, part of the catena the soil profile remains above the permanent water table throughout the year. In humid climates this effect causes gleying of the lower part of the catena, whilst in arid climates it may also be associated with natural salinization. A freely-drained upper part and a gleyed lower part is the most important division in a catena.

The third cause of catenary differentiation is the movement of substances laterally down the slope in solution. The particular significance of this process in the tropics was recognized in early studies of the catena (e.g. Greene, 1945, 1947; Milne, 1947). 'This [lateral] leaching and redeposition of material constitutes a physical link between the members of a catena which is closely analogous to the physical link between the A and B horizons of a soil profile' (Greene, 1947).

The downslope movement of water within the soil is termed *throughflow*, and that of dissolved substances *lateral eluviation*. All the substances involved in vertical leaching, including silica, iron and the exchangeable bases, are also carried laterally. Whether lateral translocation of clay particles occurs in appreciable amounts has not been proven. One of the principal effects of lateral eluviation is on clay mineral formation; the increased concentrations of silica and bases in the soil solution toward the base of slopes permits the synthesis of montmorillonite in areas climatically unfavourable for its formation. There is sometimes a considerable concentration of plant nutrients, derived from the upper and middle parts of the catena, in the gleyed soils of valley floors, a fact of much agricultural significance (p. 215).

A further consequence of lateral eluviation is the formation of laterite by precipitation of iron compounds at seepage sites. A common feature in the savanna zone is a pseudo-outcrop of laterite at the valley floor margin (*D* in fig. 9, p. 159). Watson (1964/5) identified a laterite layer at 3 m depth in the regolith in a position just above the dry season water table.

It is possible to distinguish two groups of influences which bring about catenary development. *Static causes of differentiation* are effects on processes brought about by site differences *per se*, irrespective of the position of the site within the catena. Examples are the effects of slope angle and of the depth of the water table. *Dynamic causes of differentiation* are effects on processes that arise from the position of the site with respect to the slope. These concern the processes of downslope transport, both of solid materials by soil creep and surface wash, and of substances in solution. It is possible to conceive that on an extremely permeable parent material there might be no lateral transport, but static causes of differentiation would nevertheless cause the development of a catena. Conversely, dynamic causes may be expected to be of particular significance in humid climates and on impermeable rocks (Young, 1972*b*).

Catenary evolution may be explained in either steady-state or historical terms. In the steady-state explanation it is assumed that form differentiation has been caused by the same processes as are presently active. In the historical approach it is assumed that the soil carries a record of past changes, e.g. of climate, giving features no longer being produced under present conditions but which have resisted obliteration.

CATENARY FORM

The basic unit for studying soil catenas is the catenary diagram, a cross-section of the soil horizons constructed from a series of profile descriptions at intervals down a slope (fig. 19(*a*). In soil survey these are auger

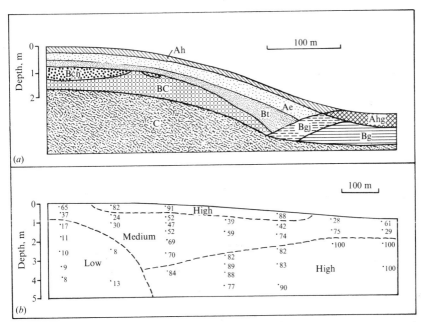

Fig. 19. (*a*) Catenary horizon diagram; (*b*) catenary isopleth diagram, showing data for percentage base saturation on a granite catena in Rhodesia, after Watson (1964/5).

observations whilst in pedogenetic studies at least a proportion of them are pits. An alternative method is to plot values of selected properties and to draw isopleths (fig. 19(*b*)).

The analysis of catenary form is usually carried out in terms of the identification of horizons, considered in their whole lateral extension, as the basic descriptive unit. A single horizon, if traced downslope, in the simplest case may remain unchanged (fig. 20, diagram (*a*)). It may remain unchanged in properties but thicken (*b*) or thin (*c*), or its boundaries may become shallower (*d*) or deeper (*e*). It may terminate downslope, or a new horizon commence (*f*). It may be replaced by another horizon, either commencing from the base (*g*) or the top (*h*). Finally an horizon may undergo gradual change in properties whilst retaining continuity and identity (*i*), a situation comparable to facies change in geology.

If such changes are analyzed for all horizons, three situations are common. These are *uniform sectors* of the catena, in which there is no downslope change in the succession, depths and properties of horizons; *sectors of gradual change*, in which one or more horizons undergo change, e.g. become progressively thicker, deeper in the profile, sandier or less acid; and *sectors of rapid change* in which, over a short distance, some horizons terminate or are substantially modified whilst others appear (Young, 1972*b*).

The primary division of most catenas is into a freely-drained upper part and an imperfectly- to poorly-drained lower part, the former usually occupying the greater part of the length of the slope. The freely-drained part may sometimes be subdivided into a crest section, a uniform sector in which the soil remains identical to that on the interfluve crest, and an slope section, on which some type of modification to the crest soil has occurred. The lower

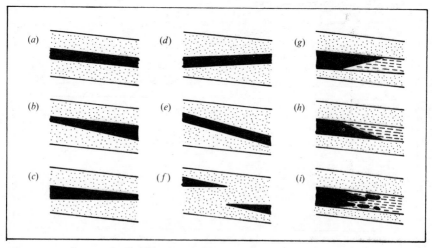

Fig. 20. Possible catenary changes to a soil horizon.

part is usually subdivisible into imperfectly- and poorly-drained sections.

Since no two soils and no two slopes are identical, it is still more true that all catenas are different. Nevertheless certain *catenary types* can be identified, comparable, and in part associated with, the types of slope form recognized in climatic geomorphology. Catenary types are not restricted to a single parent material. Catenas on different rock types, which therefore differ in respect of their freely-drained, upper catena soils, may exhibit similar patterns of intra-catenary change. Some of the most widespread and distinctive catenary types in the tropics are as follows:

(1) Savanna catena on Basement Complex rocks, with no inselberg (fig. 21(a)). The ground surface is smoothly convex–concave and the maximum slope gentle. The catena is characterized by sectors of gradual change. Reddish, clayey profiles on the crest section become yellower on the slope section; a lens of sandy material, bleached or mottled, occurs at the valley-floor margin, giving place to black clay in the valley centre. This is the catena most often recorded in the classic descriptions of Milne (1947; cf. also Webster, 1965; Young and Brown 1962; Young, 1968c). It is present in savanna climates of monsoon origin in Asia, e.g. in the 'Dry Zone' of Sri Lanka (Panabokke, 1959). It is developed on gentlly-undulating plateaux under a savanna climate, on rocks varying in composition from felsic to intermediate. On felsic rocks the crest soil is yellowish red but the down-

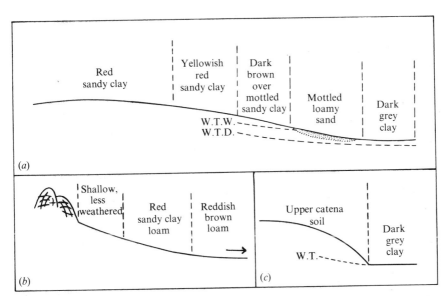

Fig. 21. Some types of soil catena: (*a*) savanna climate, Basement Complex rocks, gently undulating plain; (*b*) inselberg and pediment savanna catena, lower members of the catena similar to (*a*); (*c*) rainforest climate. W.T.W. and W.T.D. = wet season and dry season water table positions.

277

slope change to paler and less red profiles still occurs. It is commonly supposed that this colour change is caused by increasing hydration of iron minerals, although Webster (1965) failed to detect any such difference (p. 87). The sandy belts of valley-floor margins broaden into lobate expansions around valley heads. One possible explanation is that this is caused by differential deposition by surface wash, fine material being carried on into the valley centre. Certainly surface wash is rapid in this section. Another explanation is that the sands are residual, resulting from differential lateral eluviation of clay. This problem requires investigation. The black clay of valley floors, known in East Africa as *mbuga* or *dambo* clay, is a gley, with the water table permanently at or close to the surface; it does not normally have the properties of a vertisol, although this is a common misconception.

(2) Inselberg and pediment savanna catena. This catenary type is commonly developed on granite (fig. 21(*b*)). The valley floor and valley-floor margin members are similar to type (1) above. The main freely-drained soil, occupying the greater part of the length, occurs in the centre of the catena. At its crest is an inselberg or kopje. Adjacent to the latter is a relatively narrow belt of shallower soils containing a much higher proportion of weatherable minerals than lower on the slope. The possible causes of this feature are discussed above (p. 265). (Milne, 1936; Radwanski and Ollier, 1959.)

(3) Rainforest catena with convex slope and flat valley floor (fig. 21(*c*)). This is developed on various rock types under a rainforest climate wherever the valley form known geomorphologically as a *Sohlenkerbtal* is found (Louis, 1964). In this type of slope, a convexity intersects a level flood-plain, with little or no concavity. The convexity and valley floor are both uniform sectors. The valley-floor margin is a sector of rapid change, sometimes passing from freely-drained to a strongly gleyed profile in 2–3 m. The abrupt valley-floor margin is maintained by lateral erosion when the valley is flooded after storms. (Young, 1968*c*.)

(4) Lithosol catenas on steep slopes. This is an agriculturally unimportant although extremely widespread type. The slope is sufficiently steep, and basal river erosion sufficiently active, to produce a catena consisting largely or entirely of lithosols. Often, even on rectilinear slopes, the soil thickens towards the base.

It should become possible to identify other catenary types (cf. table 28). There is a measure of similarity between the catenas described on granitic gneiss by Nye (1954/55) and on granite by Watson (1964/5); both are on gentle and fairly rectilinear slopes and on both (unlike types (1) and (2) above) an increase in sand on the lower slope persists to the valley centre. The catena described by Morison *et al.* (1948) consists of two uniform sectors, a plateau underlain by laterite and a broad flood-plain, separated by a gently-sloping section in which the laterite horizon thins and softens.

TABLE 28 *Some studies of soil catenas in the tropics, listed in order of decreasing mean annual rainfall*

Climatic type	Mean annual rainfall (mm)	Parent material	Country	Reference
Rainforest	2300	Limestone	Malaya	Joseph, 1968
Rainforest	1400–2200	Intermediate to basic igneous	Ivory Coast	Delvigne, 1965
Semi-deciduous forest	1500	Charnockite	Sri Lanka	Panabokke, 1959
Moist savanna	1400	Sandstone	Brazil	Askew et al., 1970
Moist savanna	1300	Biotite gneiss	Zambia	Webster, 1965
Moist savanna	1200	Granitic gneiss	Nigeria	Nye, 1954/55
Moist savanna with two wet seasons	1200	Granite	Uganda	Radwanski and Ollier, 1959
Moist to dry savanna		Mainly Basement Complex	East Africa	Milne, 1947
Moist to dry savanna	800–1200	Basement Complex (?)	Sudan	Morison et al., 1948
Dry savanna		Granite	Tanganyika	Milne, 1936
Dry savanna	850	Granite	Rhodesia	Watson, 1964/5
Semi-arid	250–350	Tertiary sedimentaries	Kenya	Makin et al., 1969
Semi-arid to arid	200–300	Stabilized sand dunes	Sudan	Williams, 1968d

279

Another characteristic pattern occurs below laterite breakaways (Moss, 1965). The dune catena described by Williams (1968c) is *sui generis*, self-sealing by sodium clays causing surface runoff on an initially highly permeable parent material.

Morison (1949) put forward the hypothesis that intra-catenary differentiation of soil properties reaches its greatest expression in the savanna zone, under 1400–1750 mm rainfall. Toward both wetter and drier climates the complexity of the soil pattern and the number of series present in the catena decreases. I agree with this as regards the more humid climates (e.g. cf. types (1) and (3) above) but have doubts regarding the arid end of the spectrum, since processes of accumulation of carbonates and salts are quite sensitive to small differences in drainage.

THE CATENA IN SOIL SURVEY

It was as a device for soil mapping in exploratory surveys that the catena was first used (Milne, 1935/6), and its usefulness for this purpose has continued. Two conventions may be employed in soil maps on a reconnaissance scale. First to show on the map only the most extensive series present, usually the freely-drained upper member; secondly, to employ the catena itself as the mapping unit, the sequence of soil types within it and their relative extent being given in accompanying text. In land systems survey (p. 333) some systems can be described in terms of one or more catenas.

The usefulness of the concept is not confined to reconnaissance work. In detailed soil surveys, a profitable plan of field observations is that of transects down slopes, carefully identifying the starting position on an air photograph, recording the direction and pacing the distance. The spacing of auger observations within the catena is adjusted following inferences from topsoil colour or vegetation. Spacing between catenary traverses varies with intensity of survey. The lines of soil type identifications produced in this way are subsequently plotted on air photographs and joined by interpolation. This arrangement of field sampling is superior to that of straight-line transects whereever there is a well-developed and repeated pattern of interfluve and valley floor.

In detailed soil maps, showing soil series, each catena has a characteristic appearance. The dendritic or other pattern of valley floor soils is the most prominent feature. This may be flanked, sometimes discontinuously, by one or more valley-floor margin series (fig. 23, p. 336). In catenas with gently-sloping or level interfluve crests (as in type (1) above) the uppermost series is usually the most extensive (e.g. Anderson, 1957). In inselberg and pediment catenas it is the third series that occupies the largest area, crests having central areas of bare rock and lithosols surrounded by narrow belts of less highly-weathered soils (e.g. Radwanski and Ollier, 1959).

In the savanna plains of Africa a common pattern of indigenous land use is one of arable cultivation (permanent or with bush fallowing) on the freely-drained part of the catena, a narrow belt of specialized cultivation, e.g. vegetables, at the valley-floor margin, and permanent grazing on the gleys of the valley floors. Hills, where present, provide firewood and wet season grazing. This natural catenary arrangement of land potential can be employed in land re-organization schemes. Each productive unit, whether a family smallholding or some form of communal tenure, is allocated land in each section of the catena (although for reasons of soil conservation works, individual fields run parallel to the contour). Many parts of Rhodesia have been re-organized in this way: viewed from the air, the cultural features and land use pattern emphasize the catenary divisions of the landscape. An interesting analogy is that medieval English parishes in scarplands were aligned in this way.

SELECTED REFERENCES

Milne, 1936, 1947, Watson, 1965; and see table 28.

PART FOUR

SOIL FERTILITY, SOIL SURVEY
AND LAND EVALUATION

CHAPTER 16

SOIL FERTILITY

It was planted in a good soil by great waters, that it might bring forth branches, and that it might bear fruit, that it might be a goodly vine.

Ezekiel xvii.8

The *fertility* of land is its productive potential, or capacity to produce crops on a sustained basis. Fertility is not a property of soil alone but of land, the totality of environmental conditions at a site (p. 383). Climate and soil are of equal importance in determining potential productivity, and slope angle can exercise an additional constraint. Moreover, the significance of properties of the soil itself varies according to the other environmental conditions with which it is associated; for example, high moisture retention is more important where rainfall is marginal for the crop concerned, and erosion resistance more significant on steeply sloping ground.

A narrower definition of fertility is sometimes employed, namely the quantities of nutrients present in the soil in forms available to plants. Its use in this way inevitably leads to a myopic view of soil, to the neglect of physical and other conditions. *Nutrient content* will be used to refer to this concept, being preferable to 'chemical fertility' in that the latter implicitly ignores the substantial source of nutrients arising from mineralization of organic matter.

There is no recognized term to refer to the relation between the actual soil nutrient content and its estimated content under natural (unfarmed) conditions; 'nutrient status' would be etymologically suitable, but is commonly used synonymously with nutrient content. The term *relative nutrient status* will be used in this sense. It is defined as the ratio, expressed as a percentage, between the existing nutrient content of a soil and its estimated content under natural (climax) vegetation. Relative nitrogen status, relative organic matter status, etc., are similarly defined.

Fertility must be assessed within the context of agricultural technology and economic conditions. The natural ('inherent') soil nutrient content is particularly important under traditional, low-technology farming; under commercial agriculture, nutrient deficiencies can be remedied with fertilizers, and the capacity of the soil to retain added nutrients becomes more significant. Fertilizer costs and market prices affect the implications of crop responses to fertilizers, and labour costs determine the practicability of many other technological inputs.

Soil fertility

Notwithstanding the links with other environmental factors, technology and economics, it is possible to assess the contribution of soil *sensu stricto* to the productive potential of land. Many soil properties are of general significance whilst for others, qualifications can be made with respect to the related conditions.

Each crop has specific soil requirements. There are, however, many soil properties which are important for most crops, excluding those with abnormal growing conditions (e.g. rice, jute). Fertility will therefore be discussed in two parts: in this chapter, general aspects, and in chapter 17, soil conditions specific to individual crops.

Table 29 gives a framework for the assessment of soil fertility, in terms of conditions affecting fertility and the morphological and analytical properties which affect these conditions. These will be discussed in four groups: physical conditions, plant nutrients, chemical conditions (other than nutrient content), and organic matter.

PHYSICAL CONDITIONS

It is a common error to assess the fertility of a soil largely in terms of its nutrient content. The physical conditions are at least of equal importance and limitations of physical properties can less easily be remedied than nutrient deficiencies.

TABLE 29 *A framework for the assessment of soil fertility. Compare table 32, p. 301.*

Fertility conditions	Relevant soil properties
Physical conditions	
A. Rooting conditions:	
effective depth	Depth to weathered rock, laterite, stone lines, fragipans
root penetration	Texture, structure, consistence
B. Moisture conditions:	
drainage	Depth of water table, permeability
moisture retention	Field capacity, wilting point, available water capacity; indirectly texture
C. Resistance to erosion	Permeability, structure; indirectly organic matter content
Plant nutrients	
A. Present nutrient status, available and reserve	Nitrogen content, carbon:nitrogen ratio, exchangeable potassium, 'available' phosphorous, content of other nutrients; weatherable minerals, total potassium and phosphorous; indirectly organic matter content
B. Capacity to retain and make available added nutrients	Cation exchange capacity, reaction; indirectly texture, organic matter content
Chemical conditions	
A. Properties of the exchange complex	Reaction, base saturation, proportions of exchangeable bases
B. Salinity or other forms of toxicity	Soluble salts, exchangeable sodium percentage, calcrete
Organic matter	Organic carbon content, carbon:nitrogen ratio

286

Rooting conditions

The rooting conditions of a soil are the properties which permit a plant to develop its full, biologically determined, rooting system or which prevent it from doing so. Limitations which may occur concern the depth to which roots can penetrate and the density of root development.

The *effective depth* of a profile is the maximum depth to which roots can penetrate. Effective depth is determined by the presence of a *limiting horizon*, one which prevents or severely inhibits downward root penetration. Limiting horizons may be formed by rock *in situ*, rock fragments, laterite, or a fragipan. Roots do not normally penetrate far into weathering rock, e.g. a whitish kaolinized horizon or one with a clear weathering mottle. Massive laterite, cemented nodular laterite, ferruginized rock and calcrete form limiting horizons. It is physically possible for roots to penetrate non-cemented nodular laterite and stone lines and those of some indigenous plants do so, but such horizons provide such a sterile soil environment that where over 30 cm thick they may be treated as limiting. Very compact horizons, fragipans, have an effect intermediate between that of a limiting horizon and that of poorly-penetrable soils. These observations apply to arable crops; tree roots have much greater powers of finding sufficient soil in more deeply-weathered pockets or between core-stones, and tree plantations (e.g. pines, eucalypts) can be established on soils too thin for cultivation.

An effective depth of 200 cm is sufficient for all crops; a limiting horizon at 150 cm has little effect on annuals, but could reduce yields for some perennials, e.g. coffee, tea, especially where dry-season moisture is critical. In the humid tropics crystalline rocks weather deeply but (contrary to common belief) shales and sandstones may give profiles only 30–100 cm deep, whilst quartzite soils in the savanna zone are frequently too shallow to cultivate. Laterite and stone lines commonly form limiting horizons, often at depths which vary substantially over short distances. In the dry tropics soil depth is a frequent limitation, restricting cultivable (irrigable) soils in the arid zone largely to areas of unconsolidated materials.

Ease of root penetration is a most important, and sometimes neglected, aspect of soil fertility. Nutrients move only short distances towards roots; average distances are 2.5 mm for phosphorus, 5 mm for calcium and magnesium, and 7.5 mm for potassium, whilst nitrates move freely in solution (Cooke, 1962). Hence, unless there is a dense network of fine root hairs, a substantial proportion of the soil volume will not be effectively utilized.

Sandy soils permit free root penetration, a feature which to some degree compensates for their chemically less favourable properties. With textures of sandy clay loam and heavier, penetrability depends on structure and consistence. Conditions inhibiting root penetration are massive or weakly-structured horizons which are compact with a firm or hard consistence.

Soil fertility

Such conditions are common in weathered ferrallitic soils of the savannas, may occur in the B horizons of shale-derived soils in the rainforest zone, and occur in some calcimorphic soils. Also unfavourable are soils which, whilst technically described as having a strong grade of structure, have coarse and internally compact peds, as is common in vertisols and gleys. Good root penetration is conferred by a moderately to strongly developed fine or medium blocky structure coupled with a friable or labile consistence; particularly favourable is the condition in which the larger peds break down into smaller blocky or crumb aggregates; these properties are associated with iron-rich profiles, and are held by some ferruginous and leached ferrallitic soils and most basisols.

Rooting conditions cannot readily be improved by treatment and are thus a fixed soil limitation. For high-value perennial crops it may be economic to dig individual plant holes through a limiting horizon and fill with surrounding topsoil. A decrease in root penetrability as a result of structural deterioration is one of the many unfavourable consequences of allowing soil organic matter to decline to a low level.

Moisture conditions

Free drainage is a requirement or desiratum for most crops. Some (e.g. maize, flue-cured tobacco, tea) suffer lowering of yields or quality from even a slight degree of drainage impedance, whilst others can tolerate seasonal impedance in the lower horizons. In depositional landscapes, site drainage impedance is a widespread limitation, whereas in erosional relief it is confined to valley floors. Profile drainage impedance, arising when rainfall intensity exceeds soil permeability, is common on clays containing an appreciable amount of 2:1 lattice minerals.

Site drainage can be substantially improved or transformed by drainage works, of which a system of open ditches 2 m deep is the most common. With the construction and regular maintenance of such a system, crops such as oil palm and sugar cane can be grown on flood-plain gleys or even peats. A rise in the water table, leading ultimately to waterlogging, can occur if irrigation is applied without associated drainage works.

It is a common practice in traditional agriculture to plant vegetables early in the dry season on low-catena sites that are waterlogged in the wet season, making use of the stored soil moisture. This is an example of how labour-intensive technology can make profitable use of a resource which development schemes usually neglect as being too small in scale.

Moisture retention is an important condition of soil fertility. It is instructive to consider water in some respects as one of the plant nutrients, in which case it displaces nitrogen as the nutrient most often deficient in the tropics. There are a number of agroclimatological studies in which a substantial measure of statistical 'explanation' of variance in crop yields is achieved in terms of moisture stress alone, disregarding soil variations (Hanna, 1974). It was formerly in dispute whether crops experience moisture stress, i.e. decline in yields, at moisture contents above wilting point. It is now established that they do; that is, that the graph of yield against moisture content begins to slope downward before wilting point is reached.

Moisture retention is important for both annual and perennial crops. For annuals, high available water capacity has the effect of extending the growing season, since temperatures and radiation are often not limiting in the tropics. It is also an insurance against dry spells. For perennials, good moisture retention limits yield losses caused by moisture stress and reduces the risk of crop loss during occasional long dry periods. The more marginal the rainfall for the crop concerned, the greater the importance of moisture-retentive soils. Near the margin between the semi-arid and savanna zones, available water capacity will determine the choice between a crop such as maize or a lower-yielding but more drought-resistant cereal; the farmer's aims, to maximize average yield or to ensure an acceptable yield in dry years, are relevant in this context. In the rainforest–savanna transition zone, heavy-textured soils with high available water capacity permit the cultivation of perennials such as coffee where sandier soils will only support annuals.

Soil moisture conditions have an added significance where irrigation is planned. Under intermittent surface flooding, the cheapest and most common form of water application, desired moisture levels can only be maintained where the soil has good moisture retention combined with adequate permeability. On permeable sandy soils with low moisture retention, water applications must be more frequent, which may entail the more expensive and energy-demanding system of spray irrigation. Unfortunately sandy soils are very extensive in dry climates.

Resistance to erosion

The erosion hazard at a given site depends on slope angle and crop as well as on soil. Properties conferring a high erosion resistance are moderate or rapid permeability combined with a well-developed and stable topsoil structure. The spacing of contour bunds and similar conservation works can be wider, and therefore the cost of such works less, on sandy than on heavy-textured soils. A sandy topsoil is not necessarily a protection against erosion; ferruginous soils with a less permeable textural B horizon have a moderate erosion hazard. The initial high erosion resistance, when newly-cleared from rainforest, of most leached ferrallitic soils declines rapidly during the first one to three years of cultivation, as a result of loss of structure consequent upon decline in organic matter. Impermeable clays can succumb to gullying on slopes as low as $\frac{1}{2}°$, as has occurred notably on vertisols.

PLANT NUTRIENTS

The supply of nutrients usually receives disproportionate emphasis by comparison with that given to soil physical conditions. Reasons include the prestige attached to chemical analysis, the appearance of precision in the quantitative data that it gives as compared with the qualitative scales by which morphological properties are assessed, and the ease of merely taking topsoil samples. A farmer or development planning organization who send away samples for laboratory analysis feel that they are getting value for money. There is also the 'proof of the pudding' that correct application of fertilizers leads to immediate and usually substantial increases in yields.

The apparent precision of soil analytical data is misleading since in some cases, notably 'available' phosphorus, they represent chemical procedures designed to approximate to soil/plant relations, whilst the natural soil environment is to a greater or lesser extent altered in the course of analysis. As argued above, soil physical conditions have as much effect on crop selection, farming practices and yields as do nutrient supplies, and are less readily altered. Hence the supply of nutrients, whilst undoubtedly contributing to an important extent to soil fertility, should not be over-emphasized at the expense of other aspects.

Full accounts of the chemistry of plant nutrients and the use of fertilizers in the tropics are available, notably those by Ignatieff and Page (1958), Ahn (1970, chapter 6), Jacob and Uexküll (1963) and National Academy of Sciences (1972). Consequently the discussion here will be relatively brief. Many governments produce their own fertilizer handbooks; these combine general principles with local experience and should be consulted in the course of soil surveys.

The sixteen chemical elements known to be essential for plant growth are shown in table 30. Carbon, hydrogen and oxygen are not normally regarded as nutrients, their adequacy being a matter of water supply and soil drainage. The *primary nutrients* are those required by plants in relatively large quantities or which are frequently deficient; they are also the nutrients supplied in most fertilizers. *Secondary nutrients* are either required in smaller amounts or are only infrequently deficient. *Trace elements*, or micronutrients, are required by plants in very small quantities. Potassium, calcium, magnesium, nitrogen in the form of ammonium, and four of the trace elements are taken up by plants as cations, whilst phosphorus, sulphur, nitrate nitrogen, and three trace elements are taken up as anions.

Levels of nutrients that are adequate for crops depend upon many factors, including type of crop, variety of seed, climate and soil texture. It can nevertheless be useful to have a standard of reference for comparison with analytical values. Generalized levels are given in table 31(*a*). 'Low' values indicate that a fertilizer response is probable and may be substantial,

TABLE 30   *Chemical elements and plant growth (after Ahn, 1970)*

	Element	Chemical symbol	Main forms in which taken up by plants
Elements obtained from air and water	Carbon	C	CO_2
	Hydrogen	H	H_2O, H^+
	Oxygen	O	CO_2, H_2O, etc.
Primary nutrients	Nitrogen	N	NH_4^+ (ammonium), NO_3^- (nitrate)
	Phosphorus	P	H_2PO_4 (orthophosphate)
	Potassium	K	K^+
Secondary nutrients	Calcium	Ca	Ca^{2+}
	Magnesium	Mg	Mg^{2+}
	Sulphur	S	SO_4^{2-}
Trace elements	Iron	Fe	Fe^{2+}, Fe^{3+}
	Boron	B	Various anions
	Zinc	Zn	Zn^{2+}
	Copper	Cu	Cu^{2+}
	Manganese	Mn	Mn^{2+}
	Molybdenum	Mo	MoO_4^{2-}
	Chlorine	Cl	Cl^-

'medium' that a response is possible, and 'high' that a substantial response is unlikely. The values given, compiled from various sources, are very generalized and should be adjusted for particular crops and countries; local fertilizer or farm management handbooks often contain such information.

TABLE 31   *(a) Generalized levels of primary nutrients in tropical soils. Values refer to topsoil analyses*

Nutrient	Unit	Low (response probable)	Medium (response possible)	High (response unlikely)
Nitrogen	N, percent	< 0.1	0.1–0.2	> 0.2
'Available' phosphorous	P, p.p.m.	< 10	10–40	> 40
Exchangeable potassium	K, m.e./100 g	< 0.2	0.2–0.4	> 0.4

(b) Conversions

$P_2O_5 \times 0.44 = P$ $P \times 2.29 = P_2O_5$
$K_2O \times 0.83 = K$ $K \times 1.20 = K_2O$

Assuming bulk density of dry soil is 1.5:

1 percent of nitrogen per centimetre horizon thickness   = 1500 kg/ha N
1 p.p.m. of phosphorous per centimetre horizon thickness   = 0.15 kg/ha P
1 m.e./100 g of potassium per centimetre horizon thickness   = 59 kg/ha K

(1 kg/ha = 0.1 g/m² = 0.89 lb/acre.   1 lb/acre = 1.12 kg/ha)

It is nowadays the preferred practice in soil chemistry to work in terms of weights of elements whereas fertilizer composition and requirements are often expressed as phosphate and potash. Conversions are given in table 31(*b*); comprehensive lists of fertilizer composition and conversion factors are given by Jacob and Uexküll (1963). Soil analytical data takes the form of percentages or other proportional measures of elements within each horizon; the calculation of plant–soil–fertilizer nutrient budgets requires data in the form of weights of nutrient per unit area. To convert soil analyses into this latter form, horizon thicknesses may be multiplied by the factors given in table 31(*b*). There are many approximations in such calculations but they enable estimates involving bulk nutrient supply and use to be made, for example the number of years in which the soil content of a nutrient would be exhausted if not replenished.

Nutrients exist in the soil in available, unavailable and reserve forms. *Available* nutrients are those present in chemical forms in which they can be taken up by plant roots. The term is also used, in the case of phosphorus, to refer to the amount extracted with a weak acid, and intended to approximate to that actually available to plants. Nutrient availability varies with soil reaction (p. 299). *Reserve* supplies of nutrients are those held in the soil combined in rock minerals, not presently available to plants but potentially convertible to available forms. *Unavailable* nutrients include both reserves and elements which, after initial release, become chemically combined ('fixed') in forms not available to roots.

Rock weathering provides a slow but continuing renewal of nutrient supplies, supplying the main input to the soil–plant cycle. In tropical soils, the presence of a supply of weathering minerals within rooting range has a highly beneficial effect on soil fertility, greater than might be supposed on the basis of estimates of the rate of release of nutrients; possibly this effect is related to the enrichment of soil–plant nutrient cycles by means of slow but regular net input.

Crop growth and yield is limited by the nutrient that is in shortest supply, relative to the biological requirements of the plant. If, for example, there is a potassium deficiency there will be little or no response to nitrogen or phosphorus fertilizer, the application of which may accentuate symptoms of potassium deficiency. If the potassium deficiency is remedied, then responses to other elements may be revealed. This 'principle of the limiting factor' is of great importance. Soil that, on the basis of analytical data, is low in some primary nutrient may give no response to this element owing to a deficiency in, for example, sulphur or occasionally a trace element. In this context it may be noted that the limiting factor is frequently water.

The matter of relative supplies of elements is complicated by the *nutrient balance*. An excessive supply of one element may inhibit plant absorbtion of another. This effect occurs particularly between the exchangeable bases; thus highly calcareous soils may exhibit potassium deficiency and sometimes iron chlorosis.

Improved cultivars of crops (including the so-called 'high-yielding varieties') require larger amounts of nutrients than lower-yielding traditional varieties. Hence it is not possible to state in absolute terms that a soil is or is not deficient in a particular nutrient. Improved varieties may give a fertilizer response on sites where local varieties do not.

The results of experiments carried out by the FAO Freedom from Hunger Campaign Programme show that, taking averages of many trials and various annual crops, the addition of a single primary nutrient increased control (unfertilized) yield by only 10–20 percent, whereas the three primary nutrients together result in increases of about 40 percent. These results, however, refer to fertilization alone; when combined with improved seed varieties and other good farming practices, increases of 200 or 300 percent are common.

Fertilizers alone are not sufficient to remedy nutrient deficiencies, and farmers frequently do not obtain potential yield increases through failure to combine fertilization with other good management practices. For example, it has been demonstrated in Kenya that whereas the application of nitrogen to maize gives a considerable economic return if planting is at the start of the rains, the yield improvement is reduced to such an extent that it becomes uneconomic in some years if planting is delayed by three weeks, and invariably if delayed by six weeks (Allan, 1971). There is a danger of regarding fertilizer as a panacea; it is a necessary condition for sustained high yields but not a sufficient one, and needs to be used in conjunction with the other components of improved management.

Nitrogen

Nitrogen is the nutrient most frequently deficient in the tropics. In the humid climates, lack of nitrogen is the greatest single cause of low crop yields, whilst even in semi-arid climates it may rank equally with shortage of water as a limiting factor (cf. Benninson and Evans, 1968). Crops of which the leaf is harvested, including tea and planted grasses, are particularly responsive to nitrogen, as also are maize, rice, cotton and sugar-cane. There are few tropical soils that will not show some response, often substantial, to the application of a nitrogenous fertilizer.

The reason for this widespread deficiency is that most soil nitrogen is derived from mineralization of organic matter, and some decline in the latter is an inevitable consequence of cultivation in the tropics. The annual net addition from atmospheric fixation is small by comparison with the amount circulating between plant and soil. Also, where vegetation is burnt prior to cultivation, the nitrogen is lost to the atmosphere whereas most other nutrients (apart from sulphur) are retained in the ash.

Nitrogen in organic forms, as a constituent of humus, is not available to plants and must first be converted to available mineral forms ('mineralized') by fungi and bacteria. It is first decomposed to ammonia (NH_3); subsequently the ammonium cation (NH_4^+) is oxidized by successive groups of nitrifying bacteria to nitrite (NO_2^-) and then nitrate (NO_3^-). Nitrate is the form in which nitrogen is mainly taken up by plants. Unfortunately nitrate is also very soluble and readily lost by leaching. It follows that it is important to have a continuing supply, released in the course of crop growth, and that there is little residual effect from nitrogen fertilizer in the second and subsequent years. The continuing supply can in part be provided by applying a second fertilizer dressing to the growing plant, but a more continuous and steady release is obtained if there is sufficient organic matter in the soil to provide a continuing source.

A carbon:nitrogen ratio of between 10:1 and 12:1 in the soil organic fraction indicates satisfactory mineralization of nitrogen. The ratio in fresh plant material is much higher, and values of over 12:1 in the soil indicate that nitrification is inhibited. The most common cause of inhibition is poor drainage.

The input to the plant–soil nitrogen cycle comes from fixation from the atmosphere, accomplished by two groups of nitrogen-fixing bacteria, free-living and symbiotic. Of the group living freely in the soil, the chief genus in temperate soils, *Azotobacter*, is rare in the humid tropics as it does not tolerate acidity below pH 6.0, but others are present (*Beijerinckie, Derxia*) (Richards, 1974). The second group consists of symbiotic bacteria of the genus *Rhizobium*, living in root nodules on leguminous plants. The latter include crops (groundnuts, beans and peas, lucerne), and legumes in natural grass fallows. It is beneficial to include an admixture of leguminous species in planted grass leys. Leguminosae are common in the vegetation of both the savannas and the semi-arid zone. Rainforest also includes leguminous species, some but not all of which carry nodule tissue (Norris, 1969). Certain leguminous shrubs are effective in nitrogen fixation and may be used in planted fallows. Both free-living and symbiotic nitrogen-fixing bacteria require small quantities of molybdenum to function efficiently. Nitrogen increases have also been recorded in grassland without legumes, apparently through non-symbiotic fixation (Moore, 1963). Microbial symbionts of termites are also capable of fixing atmospheric nitrogen (Breznak *et al.*, 1973).

In climates with a dry season there is a 'nitrogen flush' at the start of the rains. When dry soil is first moistened there is a large increase in the rates of organic matter decomposition and nitrogen mineralization; these rates then decline in a negative exponential manner, falling considerably within a few days. The longer the period of drying, the greater the flush of nitrogen mineralization. The effect can be repeated, both experimentally and in the field, by repeated drying and re-wetting. A possible explanation is that prolonged drying causes greater exposure of the surfaces of organic colloids; there follows a moisture-induced population burst of nitrifying bacteria, which declines again as soon as the readily-accessible organic surfaces decrease. This phenomenon has been termed the 'Birch effect' after its discoverer (Birch, 1958, 1960).

For many years it has been known that early planting of annual crops, especially maize, improves yields, and the nitrogen flush appears to provide an explanation of this effect. If the seed is able to germinate immediately the first rains occur it may benefit from the augmented nitrogen supply; if planting is delayed until after the onset of the rains, much of the nitrogen is lost by immediate leaching. Recent work at the East African Agriculture and Forestry Research Station has cast doubt on this explanation since a similated nitrogen flush, artificially applied to late-planted maize, does not have the same beneficial effect on yields. Weekly measurements of soil nitrogen at Samaru, Nigeria, showed that its availability did not decrease with lateness of planting (Jones, 1975). Notwithstanding, the reality of the early planting effect on yields is undoubted, even though its explanation remains uncertain.

It may be speculated whether the curious habit of *Brachystegia* tree species of producing their new leaves, picturesquely coloured red or yellow, during the height of the dry weather may be a means of enabling photosynthesis and growth to start immediately at the first rains.

In the savannas there is a small upward movement of nitrate into the topsoil during the dry season, and downward transfer to the textural B horizon during the rains (Robinson and Gacoka, 1956; Stephens, 1962; Wild, 1972). The rate of downward leaching has been estimated as 0.2–0.3 cm per centimetre of rainfall for native soil nitrogen, with the possibility that it is higher for fertilizer nitrogen (Jones, 1975).

Fixed ammonia, in the form of ions held by silicate minerals and hydrous oxides, can form up to 20 percent of the nitrogen present in subsoils, and can make some contribution to the crop nitrogen economy particularly in the dry savannas (Opuwaribo and Odu, 1974). Rainfall makes an appreciable contribution to soil nitrogen, e.g. of the order of 5 kg/ha/yr under a 1100 mm rainfall in Northern Nigeria (Jones and Bromfield, 1970).

Nitrogen levels may be increased by bush fallowing, green manuring, rotation with a leguminous crop, grass leys, compost, farmyard manure and fertilizer. Fallows usually improve nitrogen status through an increase in organic matter. An exception is fallows of tall-grass savanna dominated by the Andropogoneae family, which appears to inhibit nitrogen mineralization (p. 44). Compost and farmyard manure are highly beneficial, and nitrogenous fertilizer is most effective when applied in conjunction with either or both of these.

The commonest and cheapest nitrogenous fertilizer, sulphate of ammonia, has an acidifying effect on the soil, and may therefore reduce phosphorus availability. It should therefore be used with caution on strongly acid or sandy soils, and is increasingly being replaced by other forms such as calcium ammonium nitrate or urea. Two fertilizer applications, one on sowing and a top dressing to the young plant, help to increase the proportion utilized compared with that lost by leaching.

The supply of nitrogen in flooded padi fields is maintained through fixation by blue-green algae; avoidance of its fixation in unavailable forms, raises special problems (p. 228).

The keys to providing plants with an adequate nitrogen supply are therefore first, to maintain reasonable levels of soil organic matter; second, to supplement natural supplies by substantial additions of nitrogenous fertilizer, preferably in conjunction with manure or compost; and third, to avoid where possible planting nitrogen-demanding crops (notably maize) in two successive years.

Phosphorus

Phosphorus is required by plants, and present in soils, in very much smaller quantities than nitrogen, but is classed as a primary nutrient because it is essential to growth and often deficient. Phosphorus deficiency restricts root growth with consequent effects on growth of the plant. Softwood (coniferous) plantations on strongly acid latosols may experience restricted growth through phosphorus deficiency. Phosphorus is frequently deficient in the humid tropics, less widely so in the dry lands.

The two problems in providing plants with phosphorus are maintaining the supply in the soil solution and preventing fixation in unavailable forms. To be absorbed by root, phosphorus must be present in the form of the phosphate anion, $H_2PO_4^-$. The amounts of this anion present in the soil solution at any one time are only a small fraction of the needs of a single year's crop growth, and therefore it must be continuously replaced.

The ultimate source of phosphorus in the soil is from the weathering of the rock mineral apatite. In any one year, however, the main natural source is from the mineralization of humus. The supply can be augmented by farmyard manure and fertilizers, mainly the various forms of superphosphate.

Analytical data for 'available' phosphorus is recognized to give only rough approximation to that truly available to plants, and the three methods commonly used ('Olsen', 'Bray' and 'Troug') give differing results. It is one of the most highly variable of soil properties, with a coefficient of variation within a soil series or small area of the order of 50 percent. In most tropical soils the available phosphorus is heavily concentrated in the humic topsoil. Data are also sometimes given for *reserve* phosphorus, as estimated by solution with a strong acid; this is a useful guide for estimating the likely soil phosphorus status, especially under the rapid weathering conditions of the rainforest zone.

Phosphate *fixation* refers to processes in which phosphorus combines with other elements, chiefly iron and aluminium, forming compounds of very low solubility. Fixation is greatest under strongly acid conditions and in soils rich in iron and aluminium oxides. It is therefore an inevitable problem in soils of the rainforest zone (leached ferrallitic soils and strongly leached ferrisols). A substantial proportion of the phosphorus added in fertilizer gets fixed in insoluble forms before it can be taken up by the plant. There are again special problems in padi soils.

There is no easy solution to the problem of phosphate fixation in strongly acid soils. Liming is not usually effective on latosols, and in any case would either be impracticable or costly under conditions of intense leaching. The application of fertiliser in the form of slowly-soluble pellets helps to maintain a steady supply but is expensive. The maintenance of a good soil organic matter content is once again valuable. Firstly, it provides a continuing source of phosphorus through mineralization,; secondly, organic phosphates appear to be less readily fixed than inorganic forms; and thirdly, organic acids dissolve some 'fixed' phosphate and so render it available.

Potassium

Potassium is present in soils in larger quantities than phosphorus and is less frequently deficient. It is taken up by plants as the K^+ cation, hence the more strongly leached the soil, the lower will be its concentration in the soil solution. Deficiency is therefore common in the rainforest zone and in sandy-textured soils of the savannas. Root crops have a high potassium requirement, together with tobacco, oil palm and bananas.

Primary supply, from the weathering of feldspars and micas, can make a substantial contribution to the potassium cycle. Where weatherable minerals are absent, supplies come from breakdown of organic matter, or from ash in burnt vegetation. Fixation is not a serious problem, although 2:1 lattice clays can take up potassium in slowly-available forms.

Other nutrients

Calcium and *magnesium* are both taken up by plants as cations, hence deficiencies are possible in strongly-leached soils of the rainforest zone. However, many crops grown on strongly acid soils, such as rubber, have a low calcium requirement. Magnesium is a constituent of chlorophyll and may become deficient in crops with a large leaf area, such as tobacco. Deficiencies of calcium or magnesium are possible where the exchangeable base values of either are less than 0.2 m.e./100 g. An excess of either of these bases may inhibit uptake of the other; hence magnesium deficiency is possible in calcareous soils.

Sulphur is absorbed as the sulphate anion, SO_4^{2-}. It is mainly derived from the mineralization of organic matter, but like nitrogen is lost in bush burning. It is lost from the soil by leaching. Sulphur may be deficient in soils low in organic matter, especially sandy soils, and in areas repeatedly burnt. Groundnuts sometimes give a response to sulphur fertilizer. Deficiency has also been identified in maize (Allan, 1971). Except where a deficiency is suspected, it is not a routine determination in soil analysis.

Deficiencies of *trace elements* cannot be detected from soil analysis as the amounts involved are too small (Silanpää, 1972). Where suspected they can sometimes be diagnosed by pot experiments, in which small amounts of various elements are added. Iron, manganese, copper and zinc become less available in alkaline conditions, and calcareous soils sometimes show deficiencies. Trace element deficiencies can also occur in acid soils, e.g. zinc deficiency in cocoa and coffee. Copper is often low in organic soils. On *a priori* grounds it should be possible to detect likely deficiencies by total element analysis of parent materials, but this has not been demonstrated.

Very small amounts of *boron* are essential to plant growth, but it becomes toxic at higher levels (still very low in absolute terms). Attempts are being made to increase rates of tree growth by boron addition. Boron toxicity can occur when irrigating calcareous soils, and potential irrigation waters are analyzed for boron content. Toxicity in soils is possible at boron levels above 0.7 p.p.m. and probable above 1.5 p.p.m. (approximately equivalent to 3–6 kg/ha).

Nutrients and climatic zones

Nitrogen deficiency is widespread in all parts of the tropics, and almost ubiquitous on soils that have been under cultivation for any length of time. Cultivated soils of the *rainforest zone* are frequently deficient in both nitrogen and phosphorus, because of the inevitable lowering of organic matter, or in all three primary nutrients; potassium is especially likely to be deficient on sandy soils because of intense leaching. *Savanna zone* soils are commonly deficient in either nitrogen alone or both nitrogen and phosphorus; leguminous crops frequently require phosphorus and sometimes also sulphur. Potassium is only likely to be deficient on sandy soils. Soils of the *dry tropics* are commonly deficient in nitrogen, and sometimes also phosphorus, because of low organic matter levels. Where the reaction is alkaline there may be trace element deficiencies.

CHEMICAL CONDITIONS

Soil reaction is not the reliable guide to fertility in the tropics in the way that it is in the temperate zone. Most crops evolved in the humid tropics are adapted to acid conditions. Within a given climatic zone, however, the less acid soils will generally be more fertile than more strongly acid ones. All primary and most other nutrients have their maximum availability in the pH range 6.0–7.5. At pH values below 5.5 primary and secondary nutrients become less available, this effect being most marked in the case of phosphorus. Some trace elements become unavailable in moderately to strongly alkaline soils. (A diagram of nutrient availability against pH is given in Ignatieff and Page, 1958, fig. 1). It has repeatedly been found that liming, almost a panacea in the temperate zone, has little or no beneficial effect on latosols. Why this is so is not known; possible adverse side-effects are upsetting the nutrient balance through excess calcium, or reducing availability of trace elements. There are exceptions; for example, heavy dressings of limestone have increased sugar-cane yields on an acid latosol in the West Indies (Rodríguez *et al.*, 1968).

The *apparent base saturation* is an indication of the intensity of present-day leaching in the soil, and hence of the likely rate of loss of added nutrients. There is an approximately linear relation between pH values of 5.0–6.0 and saturation values of 25–75 percent (Greene, 1961). In strongly acid soils the apparent value of base saturation is an artificial value, reduced by the underestimate of the true value of cation exchange capacity (p. 95). Saturation values need to be judged in the context of the climatic zone and crop. As with reaction, within a given climatic zone, soil fertility tends to fall with decreasing saturation. For example, cocoa yields fall where saturation in the B horizon drops below 30 percent.

A third aspect of the soil chemical conditions is whether any form of *toxicity* is present. The most common forms of toxicity occur in arid zone soils: excess soluble salts, a high exchangeable sodium percentage, and excess boron. Sulphur toxicity can occur under the waterlogged conditions of padi fields. The presence of calcrete may cause nutrient imbalance (p. 199). Manganese dioxide concretions are common in imperfectly-drained soils of the humid tropics, but do not appear to have the toxic effect attributed to their presence in the temperate zone. The depressed yields of some crops on strongly acid soils may be caused in part by aluminium toxicity.

ORGANIC MATTER

The importance of organic matter in soil fertility has been repeatedly noted above, and its maintenance has been discussed (p. 115). A fall in the organic matter content leads to deterioration in both the soil physical conditions and the nutrient supply. The various roles of organic matter in maintaining soil fertility may be summarized as follows:

Soil fertility

1. To improve the grade of soil structure, and thereby: (a) root penetration; (b) moisture conditions: permeability coupled with good moisture retention; (c) resistance to erosion.

2. To maintain the nutrient levels, especially of nitrogen and phosphorus; acting as a store, from which nutrients are released by mineralization, slowly and in forms available to plants. The nitrogen and phosphorus cycles are thus dependent on the carbon, or total organic matter, cycle.

3. To increase the cation exchange capacity, and thereby the capacity to retain added nutrients.

The lower the mineral soil is in clay or in weatherable minerals, the more important is the role of organic matter. More than half the nutrient content and cation exchange capacity of the entire profile is frequently contained in the 20–30 cm of the humic A horizon.

The organic matter level shown by soil analysis gives an indication of how much the natural soil fertility has been lowered by cultivation. Where possible, values for cultivated land should be compared with profiles from the same soil series under mature natural vegetation. A relative organic matter status (p. 285) of over 75 percent is very satisfactory. If the value is below 50 percent the soil is being over-cultivated and remedial measures are desirable. Estimates of the relative organic matter status are at present rarely available. As a general guide, topsoil organic matter content (organic carbon $\times 1.72$) may be considered 'low' if it falls below the following percentages:

	Heavy-textured soils *(sandy clay and clay)*	*Sandy soils* *(sandy clay loam and* *sandy loam)*
Rainforest zone	3.0	2.0
Savanna zone	1.5	1.0

Corresponding values for the semi-arid zone are slightly below those of the savannas.

CONDITIONS FOR HIGH SOIL FERTILITY

High soil fertility is the capacity for sustained production, under good management, of high crop yields. The various conditions contributing to high fertility are listed in table 32. They may be summarized as a deep, freely-drained, medium to moderately heavy-textured soil, well structured, with adequate organic matter and some weatherable minerals within rooting depth. Any one of the unfavourable conditions, or limitations, can cause a soil to have low fertility, whereas for the highest fertility all the favourable conditions should be present. There are some ferruginous soils of the savannas, and ferrisols of the rainforest–savanna transition zone, which approach this optimum (e.g. the Lilongwe Series of central Malawi, Profile 1, p. 425).

300

TABLE 32 *General conditions for soil fertility*

Soil property	Conditions for high soil fertility	Conditions unfavourable to high soil fertility
Depth to limiting horizon	> 150 cm	< 100 cm
Texture	Loam, sandy clay loam, sandy clay; clay if structure and consistence favourable	Sand, loamy sand; heavy clay
Structure and consistence	Moderate or strong, fine or medium structure; friable or labile consistence	Massive, or coarsely structured, with very firm consistence
Moisture conditions	Free drainage with good moisture retention	Substantial drainage impedance; low moisture retention and rapid permeability
Plant nutrients	High levels (table 31)	Low levels
Cation exchange capacity	Medium to high levels (> 20 m.e./100 g in topsoil, > 10 in lower horizons)	Low levels
Weatherable minerals	Present within 200 cm	Absent above 200 cm
Reaction	Generally pH 5.0–8.0, but varies with crops	See previous column
Salinity	Soluble salts and exchangeable sodium low (p. 209)	Soluble salts or exchangeable sodium high
Organic matter	Adequate in relation to levels under natural vegetation	Low levels (p. 300)

Soil fertility

For a given crop and for limited areas, e.g. a country, it is possible to obtain statistically significant relationships between crop yields and soil conditions (e.g. Riquier *et al.*, 1970). The connection is simplest where one limiting factor, such as salinity, is widespread. Indices of soil productivity are discussed below in the context of land evaluation (p. 398).

The conditions for soil fertility given in table 32 apply to all normal crops, although particular properties and limitations are of more significance to some crops than others. These crop-specific aspects of soil fertility are discussed in the following chapter.

RELATING FERTILITY TO SOIL TYPES

If the results of agronomic studies of soil fertility are to be widely applied it is desirable to relate these to soil series or soil types of a higher category. In many countries the situation in this respect is unsatisfactory. Reports of agronomic experimental work frequently do not give the soil series (nor, for that matter, the rainfall in the years concerned). Agricultural recommendations, for example on fertilizer treatments, are usually given by administrative district, and rarely separately for each soil type.

An outstanding example of what can be achieved, and at the same time a demonstration of fertility differences between soil series, is the work on fertilizer responses of rubber and oil palm in West Malaysia (Pushparajah and Guha, 1968; Guha, 1969; Ng and Law, 1971). The main soil series in the country were first defined by soil survey. The means and ranges of analytical properties were determined for each of these series. A series of trials for rubber and oil palm were carried out on identified soil series. It has thereby been established for each series, firstly, whether soil nutrient levels are such that deficiencies may be expected, and with what statistical probability; secondly, whether nutrient deficiencies do in fact occur in crops, as revealed by foliar analysis; and thirdly, to which nutrients there are fertilizer responses and the magnitude of these. For example, the yield response of oil palm to potassium fertilizer is 33 percent on the granite-derived Rengam Series but only 2 percent on the shale-derived and heavier-textured Durian Series.

It is not always necessary to set up new trials; the soil series on existing experimental sites can be identified and the data reinterpreted, as was done in Malaŵi (Young and Brown, 1962, 1965). Following identification and definition of the main soil types present in a country by reconnaissance soil survey, the next stage should be to establish the fertility characteristics and agricultural behaviour of each.

302

SELECTED REFERENCES

Richardson (1963), Ignatieff and Lemos (1963), Nye (1963), Greene (1961), Ignatieff and Page (1958), Ahn (1970, ch. 6 and 9), Wrigley (1969, pp. 125–44), Meiklejohn (1955), National Academy of Sciences (1972), Ng and Law (1971). For reviews of the fertility of particular soil types see Dudal (1965) (vertisols) and van Wambeke (1974) (ferrallitic soils).

CHAPTER 17

CROPS AND SOILS

This chapter is concerned with the contribution that soil survey can make to decisions about the planting sites for individual crops and subsequent soil management. Questions of choice between types of land use, e.g. arable use, pasture or forestry, are discussed in the context of land evaluation in chapter 20.

THE NATURE OF SOIL–CROP RELATIONS

The aspects of soil–crop relations which influence crop selection are:

1. Differences between the optimal soil requirements of crops.
2. The tolerance of crops to suboptimal conditions.

With regard to the first aspect, the ideal soil conditions for high fertility, differences between individual crops are not large. Of the 34 crops treated below, 23 have as their optimal requirements the 'normal conditions for high fertility' as given above (p. 301). Nine of the remainder have a preference for what may be called 'normal but sandy' conditions, i.e. a sandy loam or sandy clay loam texture but with other soil properties similar; these crops are bulrush millet, yams, cassava, groundnuts, flue-cured tobacco and, with a less strong preference, lucerne, pineapple, coconut and citrus. Only rice and cocoyams have soil requirements which depart substantially from the normal conditions for high fertility, in both cases with respect to drainage.

There are, however, aspects of the 'normal' conditions which are of particular significance for individual crops. Examples are nitrogen status for maize, potassium supply for root crops and oil palm, and for tea a combination of a deep profile, high available water capacity and free drainage.

With regard to the second aspect, tolerance to suboptimal conditions, differences between crops are greater. Thus, some crops are sensitive, i.e. suffer a fall in yields or quality, to even a small degree of drainage impedance whilst others are little affected. Tolerance to salinity is one of the few such cases for which quantitative data are available, namely salinity levels which cause a 50 percent reduction in yields (p. 209). The general situation is that those crops which give the highest yields under conditions of high fertility, the *demanding* or eutrophic crops, suffer the more severe depression in yields as soil conditions deteriorate; maize and yams belong to this class. In contrast are the *hardy* or oligotrophic crops, such as finger millet, which are able to grow satisfactorily on shallow, sandy or infertile soils. The yields of hardy crops are higher on good soils than on poor, but on good soils they do not compete with those of the demanding crops.

The same principle is sometimes applicable to cultivars (cultivated varieties) of a single crop. Cultivars bred for maximum yield under good conditions (including the so-called 'high-yielding varieties') may have a poor tolerance of low fertility – a fact which weighs heavily with the peasant farmer considering their adoption.

By taking into account optimal soil requirements, productivity under these optimal conditions, and tolerance to suboptimal conditions it is possible to derive the *marginal advantage* of one crop over another on any given site. It is necessary to convert productivity into a common unit, which in the case of subsistence crops can be units of nutrition and for other crops gross or net cash returns per unit area. For example, the substantial marginal advantage of maize over finger millet on good soils decreases considerably, and possibly becomes reversed, on infertile soils. The size of a marginal advantage, and the point at which a reversal of advantage, if any, takes place, is affected by management practices (e.g. use of fertilizer) and economic conditions. It may be noted that in calculations of marginal advantage, climate has an equal or greater effect than soil.

DECISIONS OVER SOILS AND CROPS

The types of decision and action which are affected by soil–crop relations may be epitomized by three questions:

  (i) Shall we plant oil palm here?
  (ii) Which crop shall we plant here, oil palm or rubber?
  (iii) If we plant oil palm here, what soil management measures will be necessary to bring about conditions for high fertility?

The first question arises where the crop for a planting expansion programme has already been selected, and virgin land for settlement is being selected – a situation becoming increasingly rare. It involves both optimal requirements and tolerance to suboptimal conditions. The second question is applicable to land redevelopment as well as to new land settlement schemes and arises when a prior decision that a site is suitable for arable cropping has been made. It also involves both optimal requirements and tolerance, together with calculation of marginal advantages.

It is questions of the third type, however, which are by far the most common. The crop to be grown on any given site is frequently already settled, having been determined by climatic conditions and economic considerations. The farmer, whether smallholder or estate manager, is then faced with the problem of modifying whatever soils may be present in such ways as are necessary to obtain good yields. This aspect is frequently stressed by agriculturalists with specialist knowledge of particular crops. Thus Barnes (1974), writing of sugar-cane, notes that it is possible to grow it on all textures from sands to clays, and under a range of (initial) drainage conditions, but emphasizes that 'Each [soil] type requires its own particular form of treatment to enable it to produce profitably'.

The matter of soil management can also be brought in at the stage of site selection. The question to be asked should not be, for example, 'Is this a good banana soil?', an approach which places too much emphasis on the inherent, stored fertility which will rapidly be used up during the first cropping cycle. Rather, the question should be 'Can this soil be made to grow bananas, and if so, at what cost and return?'; that is, what is its potential capacity, under competent management, for sustained production, and what inputs are required (cf. Simmonds, 1966). It is therefore one of the functions of soil survey to provide guidance on the different soil modifications (e.g. drainage) and management practices that will be necessary on each of the soil types of the survey area if particular crops are to be grown (Young, 1975).

Crop selection involves agroclimatology, soil conditions and economics, and in determining the crops suitable for a given area, climate is frequently more important than soil. The accounts which follow are confined to soil conditions, climatic requirements being mentioned only to give a general context for the assessment of soil moisture properties. Thus the following section is not a comprehensive guide to crop ecology. Crops which are likely to remain limited to local areas, e.g. jute and beniseed, are excluded, as are most fruits and vegetables. The treatment is necessarily qualitative; information on climate–soil–crop management relations is rarely sufficient to permit yield estimates to be made, and where available applies only to limited regions. Table 33 is a summary of crop preferences and tolerances for soil conditions, and table 34 gives typical yields.

CEREALS

The normal conditions for high soil fertility apply to cereals. Most yield best on medium to heavy textures. Cereals are nitrogen-dominant, that is, they nearly always respond to nitrogenous fertilizer, and fail to respond to other nutrients unless nitrogen is also added. If grown without rotation, cereals (except rice) tend to exhaust soil organic matter; since they also necessarily expose bare soil in the early stages of growth they have a high erosion hazard.

Besides rice, which is *sui generis*, the various types of cereal differ in respect of drought resistance, and tolerance of drainage impedance, sandy textures and shallow or otherwise inherently poor soils.

Maize (Zea mays) *(Indian corn, corn)*

Maize is the major cereal crop of the savanna zone, requiring an annual rainfall of 600–1200 mm with a growing season of 120 days and upwards. It is a demanding crop, yielding higher than other cereals (rice excepted) where climate and soil are favourable but suffering severe depression of yields on poor soils. Normal conditions for high soil fertility are applicable (table 32, p. 301), with a preference for medium to heavy textures. Sandy and shallow soils depress yields both because of increased drought hazard and lower nutrient supplies. Maize is not drought-resistant, hence where rainfall is marginal (600–800 mm) a high available water capacity is particularly important. Sites with substantial drainage impedance, i.e. mottled above 100 cm, should be avoided. The preferred pH is 5.5–7.0, and strongly acid soils (pH < 5.0) are unsuitable. There is a high nutrient demand, particularly for nitrogen; a crop of 1000 kg removes 30–50 kg nitrogen. Inherent soil supplies cannot meet nutrient demands for more than 1–2 years, therefore rotation and nitrogen fertilizer is necessary. The practice of maize monoculture, although widespread, is highly undesirable as it exhausts nitrogen and organic matter. Maintenance of organic matter status is particularly important in maize cultivation, both to augment the nitrogen supply and to confer erosion resistance. It is therefore beneficial to combine artificial fertilizer with farmyard manure. Provided that other management practices are good, maize responds very well to nitrogen fertilizer on nearly all soils and to phosphorous on some; dressings as high as 150 kg/ha of nitrogen may be profitable.

Ferruginous soils, ferrisols and well-drained alluvial soils are suitable, and on such sites maize will frequently possess a marginal advantage over other cereals. Latosols which on analysis are clays or sandy clays but which, owing to dominance by kaolinitic clay minerals, are well-structured and friable are well suited to rotations based on maize with groundnuts. The approximate magnitudes of yields that may be expected are shown in table 35 (p. 310).

Sorghum (Sorghum *spp.*) *(guinea corn, jowar)*

Sorghum has the fourth largest world production of cereals, after wheat, rice and maize; it is preferred to maize in much of West Africa. There are many cultivars, with growing seasons ranging from 110 to 180 days. It is a more hardy crop than maize, although with lower yields where conditions are

TABLE 33 *Soil requirements of tropical crops. L, M, H = Low, Medium, High. For soil textures, 'Moderately heavy' = sandy clays, heavy loams and well-structured clays; 'Moderately sandy' = sandy clay loams and other light loams, with sandy loams only slightly suboptimal. Cf. also Vink (1975, Table 36).*

Crop	Texture				Drainage			Depth		Moisture			Erosion hazard	Reaction		Nutrient content		Salinity tolerance class
	Moderately heavy preferred	Moderately sandy preferred	Moderately sandy well tolerated	Very sandy tolerated	Free essential or very desirable	Imperfect well tolerated	Poor tolerated or needed	Deep soil very desirable	Moderately shallow well tolerated	Drought resistance	High AWC important	Low AWC well tolerated		Optimum pH	Range of pH tolerance	General level of requirements	Specific requirements	
Cereals																		
Maize	+					+		+		L	+		H	5.5–7.0	5.0–8.0	H	High N	M
Sorghum	+		+	+	+				+	M		+	H	5.5–6.5	5.0–8.5	M	High N	M
Finger millet			+		+					H		+	M			M		
Bulrush millet			+		+					H		+	H			M		
Panicum millet	+		+		+					L		+	M			M		
Rice			+				+		+		+		L	5.0–6.5	4.0–8.0	M	High N	M
Upland rice					+			+		L			M		Tσ < 40	M		
Hungry rice					+					M		+	H			L		
Root crops																		
Yams			+		+					M	+		M			H	High K	
Sweet potato			+	+						L			M	5.8–6.0		H	High K	
Potato		+	+					+					M	5.0–5.8	4.5–7.0	M	High K	
Cassava		+					+	+		H		+	M			M	Low tolerated	M
Cocoyam	+								+	L			L			H	High K	
Legumes																		
Groundnuts			+	+	+					M		+	M	5.3–6.6	5.0–7.0	M	Balanced	
Soya beans			+		+	+							M	6.0–7.0	4.5–7.5	M	Balanced	
Field beans		+	+							L	+		M	6.0–7.0		M		L
Lucerne						+		+		M	+		L	6.2–7.8		H	Low N, high Ca, S	M

Crop							Salinity tol.	pH optimum	pH range		Nutrient requirements	
Annual cash crops												
Cotton				+		+	H	5.2–6.0	5.0–7.0	M	Balanced	H
Tobacco, flue-cured		+	+	+	+		M	5.5–6.0	5.0–7.5	L	Low N, high K	
Tobacco, fire-cured	+			+	+		M	5.5–6.0	5.0–7.5	L	High N, K	
Tobacco, light air-cured			+	++	++	++	M	5.5–6.0	5.0–7.5	L	High N, K	
Sunflower				++	++	++	M	6.0–7.5		M		M
Field perennials												
Sugar-cane			+	++	++	+	H	6.0–7.5		L	High N	
Sisal	++		+	++	++		M			M		
Pineapple		+					H	5.0–6.5		M		
Tree and shrub perennials												
Tea	++	+	++	++	++	+	H	4.0–5.5	4.0–6.0	L	High N	M
Coffee (arabica)	++		++	++	++	+	M	5.0–6.0	4.5–7.0	L		M
Cocoa	+	+	++	++	++	+	M	6.0–7.0	4.5–8.0	L		M
Oil palm			+				M	5.5–6.0	4.0–8.0	L	Very high K	H
Rubber		+	++	+			L	4.0–6.5	3.5–8.0	L	N when young	L
Coconuts	+	+	++	++	++	+	L	6.0–7.5	5.0–8.0	L		
Bananas	+	++	++	++	++	+	H	6.0–7.5	4.0–8.0	M	High N, K	H
Citrus fruits	+	+	+	++	++		L	5.5–6.5	5.0–8.0		High N, K	L

Notes:

1. Cation exchange capacity omitted because information available for only a few crops; in general, the higher the better. Organic matter omitted because invariably the higher the better, up to at least 5 percent.

2. Compiled from various sources. Optimum pH ranges mainly from Jacob and Uexküll (1963). Salinity tolerance classes from Richards (1954).

TABLE 34 *Yields of tropical crops. Owing to the predominance of unimproved methods, the mean yields for less developed countries are fairly representative of traditional farming. Yields given for improved management are estimates.*

Crop	Mean yields for all less developed countries 1961–5 (kg/ha)	Mean yields for all less developed countries 1970–2 (kg/ha)	Approximate yields, favourable climate and soil, improved management (kg/ha)	Unit
Maize	1130	1280	3000–4000	Grain
Sorghum	640	730	2000–3000	Grain
Finger millet			1500–2000	Grain
Bulrush millet	520	590	600–1200	Grain
Panicum millet			600–800	Grain
Rice	1630	1830	4000–6000	Grain
Yams	9130	9280	10000–15000	Tuber
Sweet potato	6480	6280	10000–15000	Tuber
Cassava	8690	9490	20000–30000	Tuber
Cocoyam	4930	4960	15000–30000	Tuber
Groundnuts	810	790	1200–2400	Nuts in shell
			700–1400	Shelled nuts
Soya beans	750	1120	800–1200	Dried beans
Field beans	430	450	800–1500	Dried beans
Lucerne*			100000	Green forage
Cotton	670	780	600–1200	Seed cotton
			200–400	Lint
Tobacco	810	850	800–1200	Green leaf
Sunflower	690	790	1500–2000	Seed
Sugar-cane	46900	50700	100000–200000	Cane
Sisal	780	740	1000–1200	Fibre
Pineapple			50000–60000	Fruit
Tea	870	970	1000–2000	Made tea
Coffee	440	500	1200–1500	Green beans
Cocoa	280	320	600–2000	Dry beans
Oil palm			10000–20000	Fresh fruit bunches
Rubber			1000–2000	Dry rubber
Coconuts			500–1000	Copra
Bananas	12000	13300	40000–50000	Fruit

*Irrigated.

TABLE 35 *Typical yields of maize in relation to soil type and management level. For definition of management level, see p. 370.*

Management level	Maize yields, kg/ha grain — Heavy-textured soils of moderate to high fertility (ferruginous soils, ferrisols)	Maize yields, kg/ha grain — Moderately sandy soils of low fertility (sandy ferruginous and weathered ferrallitic soils)
1. Traditional farming	800–1200	500–1000
2. Improved management	3000–4000	1500–3000
3. Advanced technology	5000–7000	3000–5000

favourable. Sorghum is moderately drought-resistant, with the ability to remain dormant through dry periods and then resume growth. At the same time it will tolerate a moderate degree of drainage impedance, including temporary waterlogging. It can be grown on a wide range of textures, from sandy loams to moderately heavy clays. Satisfactory yields are obtained on soils exhausted by previous cropping. Sorghum will tolerate an alkaline reaction, calcareous soils and moderate salt concentrations, but not strong acidity. It nevertheless gives highest yields on freely-drained moderately heavy-textured soils, and responds well to nitrogen fertilizer.

Sorghum may have a marginal advantage over maize where there is a substantial drought risk, at rainfalls of 600–800 mm and on soils of low moisture retention; in such circumstances it is certainly the better 'insurance' crop, that which gives the better yield in a dry year, and hence preferable for subsistence. Under traditional farming with bush fallowing it is sometimes planted late in the rotation, and suffers rather less than maize under the undesirable practice of cereal monoculture. On irrigated land with calcrete, salinity or risk of interruption in water supply, sorghum will often be the preferred cereal.

Finger millet (Eleusine coracana)

Finger millet is a hardy cereal. It is not especially drought-resistant, but will give satisfactory yields when soils are moderately shallow, stony or low in nutrients. It tolerates weathered ferrallitic soils with massive structure, and is grown in shifting cultivation of the *citemene* type. On ferruginous soils it is clearly outyielded by maize. On weathered ferrallitic soils ('plateau soils') it will give 400–800 kg/ha under traditional farming and about 1800 kg/ha under good husbandry. There may be a place for finger millet in improved agriculture on such soils, and for this reason it warrants inclusion in plant breeding programmes.

Bulrush millet (Pennisetum typhoides) *(Pearl millet, bajra)*

Bulrush millet is a drought-resistant cereal which grows well on sandy soils of low organic matter and nutrient content. It prefers medium to light textures and will tolerate very sandy soils, including loamy sands or even sands. It is intolerant of poor drainage. Notwithstanding its hardy characteristics it gives a good response to nitrogen and potash fertilizer (since the latter is frequently deficient in sandy soils). At the transition between the dry savanna and semi-arid zones, with an annual rainfall of 500–600 mm, sandy soils with low available water capacity (arenosols and sandy ferruginous soils) happen to be extensive, and bulrush millet is well adapted to these conditions.

Crops and soils

Common millet (Panicum miliaceum)

This is not a tropical crop, but its cultivation extends into subtropical semi-arid and arid regions. It has the lowest water requirement of any cereal, coupled with a short growing season. It will grow satisfactorily on poor, sandy soils, either during a short rainy season or under irrigation. Typical yields are 450–800 kg/ha under rain-fed cultivation, rising to a maximum of 2000–3000 kg/ha where irrigated.

Rice (Oryza sativa) *(swamp rice, padi)*

Rice is a cereal with abnormal growing requirements, arising from its physiological need for standing water for part of the growth cycle. Hence it is grown predominantly on alluvial soils and gleys, although where land pressure has enforced terracing it climbs up the hillsides onto latosols. Rice cultivation itself causes substantial modifications to the soil profile, and conditions favouring high yields are discussed in the context of padi soils in chapter 12 (p. 230).

Upland rice (Oryza sativa) *(hill rice)*

This crop is widely grown in Brazil and south-east Asia, in part under shifting cultivation. It appears to tolerate strong acidity (pH $c.4.0$), with its associated high aluminium saturation, better than other cereals. It is therefore grown on leached ferrallitic soils where rainfall is seasonal but high, $c.1500$–2000 mm. Highest yields are obtained from soils of medium to heavy texture, but it is tolerant of moderately sandy or stony soils. It cannot tolerate dessication.

'Hungry rice' (Digitaria exilis) *(acha)*

This minor crop (which is not truly a rice), grown in West Africa, is included as an example of an extremely hardy crop. It gives some yield (400–600 kg/ha) on shallow or rocky soils where most other crops would totally fail. Under improved agriculture, the land on which hungry rice (possibly) possesses a marginal advantage would not be recommended for arable use.

Cereals: comparison

The requirements of rice for level land with standing water set it apart from all other cereals, as a crop primarily of the rainforest zone although also with a potential on irrigated alluvial lands of the savannas. Leaving these special conditions aside, the highest-yielding cereal of the savanna zone, given adequate rainfall (>750 mm) and normal conditions for high soil

312

fertility is maize. Other cereals may possess a marginal advantage over maize under the following circumstances:

Sorghum: Sites with a moderate drought risk, arising from low rainfall and/or low available water capacity; sites with imperfect drainage; alkaline irrigated soils.

Finger millet: Massive or weakly-structured weathered ferrallitic soils.

Bulrush millet: Sandy to very sandy soils of the dry savanna to semi-arid zone transition.

Common millet: Semi-arid zone, including on sandy soils.

Upland rice: Strongly acid soils, high rainfall.

ROOT CROPS

The features of root crops that lead to distinctive soil preferences are that the tuber develops within the soil and has a high starch content. The underground tubers mean that shallow soils and heavy clays are unsuitable, the preferred textures being either moderately sandy or, if clays, those with good structure and friable consistence. The starch content causes a high demand for potassium, the deficiency of which is often increased because sandy soils are the more strongly leached. Crude yields of root crops are high, of the order of 10–30 t/ha, but this is because of the large amounts of moisture present in the harvested tuber; in terms of calorific value the productivity does not greatly exceed that from cereals.

Yam (Dioscoria *spp.*)

Yams are a highly-prized food in some regions, notably West Africa, whilst little grown in others. Whilst not entering international trade, in countries where they are favoured there is a good demand from towns. With a long growing season and high moisture requirements they are best suited to the rainforest–savanna transition zone, with a rainfall over 1200 mm and 2–5 dry months.

Yams are a demanding crop, only giving high yields where the soil is moderately deep, freely drained, neither very sandy nor a heavy clay, and well-structured. A profile depth of over 100 cm coupled with friable consistence are necessary for unhindered growth of the large tuber. Textures as light as sandy loam are satisfactory provided the soil has a good organic matter content. There is a high demand for both nitrogen and potassium, and therefore sandy soils which are physically suitable do not provide sufficient nutrients except after a long fallow. Soils which are moderately heavy-textured but at the same time well-structured and high in nutrients are optimal; this, together with the climatic requirements, indicates ferrisols as the most suitable soil type. Whilst the crop will survive in drier climates and sandy soils, yields are severely depressed. In shifting cultivation systems, yams are usually planted first in the cropping cycle, and may be replaced by a less demanding crop if fertility is lowered by a shortening of the bush fallow.

Sweet potato (Ipomoea batatas)

This delicious vegetable can be grown throughout the savanna zone. Soil requirements are generally similar to yams, including free drainage and good structure and consistence, but it prefers rather more sandy textures and does not grow well on clays. A sandy loam or sandy clay loam topsoil with a heavier-textured B horizon is ideal. Nutrient demands for high yields are large, especially for potassium of which a 6000 kg crop removes about 45 kg. The crop is more tolerant than yams of adverse conditions, and can be grown on loamy sands. Sweet potato is a suitable component of savanna-zone rotations on medium-textured to moderately sandy soils.

Potato (Solanum tuberosum)

The potato, called in the tropics Irish or 'European' potato, whilst primarily a temperate crop, can be a useful cash crop in the savanna zone at altitudes above about 1500 m; it is grown, for example, in the Andes and on high plateau areas of Africa. Freely-drained soils of medium to light texture are suitable, and sandy soils with some gravel can be used provided that rainfall is adequate. Heavy clays make harvesting difficult. The soil should be well-structured and free-working. There are moderate demands for all primary nutrients, especially for potassium on sandy soils.

Cassava (Manihot esculenta) *(manioc, tapioca plant)*

Cassava is one of the highest-yielding of all crops in terms of crude weight, giving upwards of 25 t/ha under good conditions. It is an unusual plant in that whilst preferring ample rain it can withstand periods of severe drought, and whilst yielding best on medium-textured soils it will grow satisfactorily on almost pure sands. Because of the high starch content there is a large potassium requirement and response to potash fertilizer. Nevertheless, a harvest can be obtained from soils very low in nutrients. High nutrient levels may produce excessive vegetative growth at the expense of the tubers.

The question of whether to plant cassava is linked to dietary preferences and other non-agronomic considerations. Preparation as food is lengthy, *inter alia* to remove the prussic acid content of bitter varieties, and maize-eating people despise cassava. Hence on the fertile sandy clay loams on which it yields well it is often not the preferred crop. It is grown as late in the cropping period in bush fallowing systems. Cassava may have a place in improved farming under two circumstances. Firstly, it is a suitable subsistence crop for very sandy soils (arenosols and regosols), for example soils derived from raised beaches, and also on the most sandy kinds of weathered ferrallitic soils on which cereals grow poorly. Secondly, small parts of farms may usefully be planted with cassava where there is a risk of cereal crop failure through drought.

Cocoyam (Colocasia esculenta) (taro)

Cocoyam is a crop with abnormal requirements. It needs a large water supply and can tolerate poor drainage, hence it is grown on swampy ground, poorly-drained alluvial soils and near river banks. It requires moderately fertile soils, and can tolerate clays but not sands. Deep alluvial silts and loams are best. Yields are high, 15–30 t/ha under good conditions, although it is often planted in small patches as a horticultural rather than a major crop.

LEGUMES

The distinctive feature of leguminous crops is their ability to fix atmospheric nitrogen. This does not necessarily mean that they will not respond to nitrogen fertilizer. It does, however, give them a measure of independence of soil nutrients, and tolerance of a wide range of conditions. Free or only slightly impeded drainage is always preferable, and for high yields there must be good rooting conditions.

Groundnuts (Arachis hypogaea) (peanuts, monkey nuts)

The optimal soil conditions for groundnuts are the 'normal but sandy' type, i.e. the normal conditions for high fertility but with textures on the sandy side rather than heavy. All accounts agree that a sandy loam is the best. Loamy sands are acceptable. The lower moisture retention of more sandy soils is not critical as the crop is moderately drought-resistant. Heavy textures are only satisfactory if the soil is permeable, well aerated and friable. Plastic clays and soils of very firm consistence should be avoided. Reasons for the preference for sandy textures are to permit peg penetration and facilitate harvesting. The nut develops from a shoot (peg) which grows downwards from the plant, and must be able to penetrate the soil surface. Clayey soils adhere to the nuts during harvest. Waterlogging depresses yields. Excess lime is harmful, but there must be adequate calcium. Neither alkaline nor strongly acid soils are suitable, and moderately acid soils may need added lime to meet calcium demands. There is some apparent conflict between the generally acknowledged beneficial effect of groundnuts in rotations, especially with maize, and statements that it is a soil-depleting crop. A possible resolution is that there is a beneficial effect on the succeeding crop providing that the groundnuts themselves are well fertilized. Yields on favourable soils under good husbandry are of the order of 700–1400 kg/ha (shelled nuts), about double those of the unimproved forming level. Under irrigation, yields over 2000 kg/ha can be attained.

Soya beans (Glycine max)

The soya bean is a crop of the subtropical to warm temperate zone; cultivation extends into parts of the tropics although with lower yields. It grows best under normal conditions for high fertility, but is not very sensitive to soil conditions; it will grow on both sandy and heavy soils, tolerates slight drainage impedance and a wide pH range. The demand for primary nutrients is balanced, usually responding to potassium and phosphorus, and sometimes to nitrogen. An adequate supply of calcium is necessary. Soya beans can be grown on a wide range of soil types, including organic soils.

Field beans (Phaseolus vulgaris *and other spp.*)

Beans are a valuable dietary source of vegetable protein. They are widely grown as a mixed crop, interplanted with cereals, a practice which gives about half the yields of a pure stand. A moderately wide range of soil textures is permissible, but good structure and friability are desirable, coupled with free drainage. Beans are not drought-resistant, hence sandy soils should be avoided in areas of marginal rainfall.

Lucerne (Medicago sativa) *(alfalfa)*

Lucerne 'occupies pride of place among the forage crops that are grown in the dry regions' (Arnon, 1972). For high production it requires abundant water and a deep soil with good rooting conditions. The water table must be kept down, by drainage, to at least 100 cm in sandy soils and 180 cm in clays. If the texture is sandy, irrigation applications must be frequent. A reaction within one pH-unit of neutral is best. It is very sensitive to salinity when young, but moderately tolerant when mature. Lucerne requires little nitrogen but a good supply of phosphorus and potassium, and has an unusually high requirement for calcium and sulphur. Given these soil conditions coupled with an abundant supply of water, very large quantities of forage can be obtained. Lucerne improves soil structure and nitrogen status, and has a beneficial effect on succeeding crops in an irrigated rotation.

Cotton (Gossypium hirsutum *var.* latifolium) *(short-staple cotton, American cotton)*

Normal conditions for high fertility produce the best yields for cotton, although there is a fairly wide tolerance of suboptimal conditions. There should not be any substantial degree of drainage impedance. Good root penetration is important. Nutrient demands are balanced and there is a wide range of reaction tolerance. Ferruginous soils and freely-drained alluvial soils are suitable. Cotton monoculture exhausts soil nitrogen and lowers organic matter with a consequent deterioration in structure, and this all too common practice is one reason for the very low yields common under traditional farming. Farm-yard manure coupled with rotation or fallow periods is beneficial. When irrigated, a good drainage system is necessary. The gap between yields under traditional farming, improved management (including spraying) and advanced technology is very large, being respectively of the order of 100, 400 and over 1000 kg/ha (lint).

Cotton is also widely grown on vertisols, for example in Sudan and the Indian Deccan. Although chemically favourable, the stiff clay and coarse grade of structure of these soils has adverse effects on the root system, both tap root and laterals, with substantially depressed yields. This problem is not solved. Deep ploughing is expensive and its effects only last for one year. Lowering of organic matter aggravates the structural problems. Grass roots are effective in producing smaller peds, and if a place in the farming system could be found for productive grass leys this could be a solution.

Tobacco (Nicotiana tabacum)

In tobacco production leaf quality is as important as yield, and greatly influences the price. Leaf quality is affected by the properties and condition of the soil, especially structure and organic matter content. 'No crop is more sensitive than tobacco to small variations in soil' (Purseglove, 1968).* Some soil requirements are common to all types of tobacco whilst others are related to varieties and to leaf intended for different methods of curing.

All types of tobacco require adequate profile depth (preferably over 120 cm), a texture neither excessively sandy nor a heavy clay, good structure and friability, and adequate levels of organic matter and nutrients. Free drainage, including both a permeable soil profile and absence of a high water table, is essential. An open structure is particularly important, even for varieties suited to heavier textures. The preferred reaction is weakly to

* But the grape vine?

moderately acid. There is a high demand for both nitrogen and potassium, and supply by fertilizers is essential. Fertilizers are the more necessary when, as is often the case, the crop is grown on soils of low inherent fertility.

Flue-cured tobacco has the special requirement that nitrogen must be adequate during the early stages of growth but low in the later stages. It is achieved by using a sandy soil and fertilizing early. Sandy loams are the best, particularly when there is a B horizon of heavier texture to give adequate moisture retention. Sandy clay loams and even sandy clays can be used provided they are friable and freely drained. The crop is also grown on very light sands but yields are low. Heavy clays, strongly acid soils, alkaline soils and even a slight degree of drainage impedance should be avoided.

Fire-cured and dark types of air-cured tobacco both require heavier textured soils, ranging from medium loams to sandy clays, or well-structured, friable clays. Steady supplies of moisture and of nitrogen throughout the season are needed. Free drainage is still essential.

Light types of air-cured tobacco, including Burley are suited to soil textures intermediate between those for flue-cured and fire-cured, that is, clay loams, sandy clay loams and sandy clays. Other requirements are as given above for all types.

Soil types. Ferruginous soils of heavy texture will usually have desirable properties for fire-cured tobacco; sandy ferruginous soils, weakly ferrallitic soils and some weathered ferrallitic soils are suitable for the flue-cured variety. The growth of flue-cured tobacco is a good way of obtaining a high cash output from soils of low fertility.

Soil management and leaf quality. High-quality tobacco can only be obtained if the soil has a good organic matter status. It is commonly recognized that newly-cleared land produces good quality tobacco. Once soil conditions have deteriorated through over-use recovery is slow and there is also an erosion hazard. In Malaŵi, a decline in average leaf quality over the period 1950–1970 has been attributed to the replacement of former cultivation on newly-cleared land by cultivation without fallows as land has become fully used (Akehurst, 1968). The recommendation in Rhodesia is that tobacco should be preceded by at least two years of grass or fallow.

Sunflower (Helianthus annuus)

Sunflower has a highly efficient root system and is drought-resistant. It is a crop suited to the dry savannas, for example in association with sorghum. It cannot tolerate drainage impedance or strong acidity. Ferruginous soils and brown calcimorphic soils with a weakly acid to very slightly alkaline reaction are suitable.

FIELD PERENNIALS

Field perennials are crops which remain in the ground for more than one year but which do not have a tree or shrub habit. They lack the period of several years to maturity characteristic of tree and shrub perennials, coming into bearing in the first year and continuing for one or more subsequent years. Few generalizations about soil requirements can be made, other than that sufficient soil depth and moisture-holding capacity to carry the plant over a dry season is important.

Sugar-cane (Saccharum officinarum *and other spp.*)

In terms of calories per unit area, sugar-cane is the most productive of all crops. A feature of its pedology is that, rather than being restricted to specific soil conditions, each of the wide range of soil types on which it is grown requires different management practices in order to bring about the rooting environment necessary for good yields. These soils include various types of latosols (notably limestone-derived latosols in the West Indies), andosols, vertisols, riverine and marine alluvial soils, and organic soils (including the *pegasse* soils of the Guyana coastal plain). Yields under good management are of the order of 100–200 t/ha cane (10–20 t/ha sucrose) in the first year, falling slightly in the first ratoon crop and to about half this level in the second.

The essential requirements of the plant are firstly, an abundant moisture supply coupled with free drainage at least in the upper 100 cm; and secondly, a large supply of the primary nutrients, particularly nitrogen. Deep soils are essential, and the existence and maintenance of a high organic matter content is desirable. Both moisture and nitrogen are likely to be more abundant in heavy-textured soils, which in general are to be preferred. If grown on sandy loams, heavy and continuing fertilizer applications are needed. Level ground facilitates cultivation activities as well as irrigation, hence the widespread use of alluvial soils. These usually require a system of drains to control the water table, the more closely spaced the lower the soil permeability; on coastal plains pumping may be necessary. The drainage problem is most acute on vertisols owing to their very low permeability. Where light-textured alluvial soils are used, the necessary continuous moisture supply can be maintained by spray irrigation. Sugar-cane exhausts soil nitrogen and organic matter, and where grown on latosols on sloping ground there is a severe erosion hazard. The optimum pH is close to neutral, but moderately or even strongly acid latosols can be tolerated; on strongly acid soils there may be a response to liming, an exception to the general rule that liming is not beneficial in the tropics.

Crops and soils

Sisal (Agave sisalana)

Sisal is a fibre crop grown in cycles of 7–12 years. The xeromorphic nature of the leaf surface confers a high drought resistance, enabling it to survive the dry season of savanna climates. The optimum rainfall is 1200–1800 mm but it can be grown, with reduced fibre yields, down to 750 mm rainfall. Deep soils with high moisture retention help to offset the effects of a moisture deficit and hence are particularly important where rainfall is low. Moderately heavy-textured latosols with a stable structure are the best. The crop is intolerant of waterlogging, and attempts to use vertisols and gleys, by drainage and ridging, have not been successful. On light-textured, acid latosols the crop will grow but yields are lower and the soil becomes exhausted. Maintenance of a cover crop between rows is an essential erosion protection measure.

The species grown in the Yucatan Peninsula of Mexico, *A. fourcroydes* or henequen, is a lower-yielding but more hardy plant, surviving on thin and stony limestone soils.

Pineapple (Ananas comosus)

Although drought-resistant, a regular moisture supply throughout the year is necessary for good pineapple yields. Medium to moderately light textures are preferred, on sandy clays if well-structured. Where grown on leached ferrallitic soils there is a severe erosion hazard. In Malaya, early cultivation was on steeply-sloping land, and erosion ensued. It was then found that the peats of the coastal plain could be made suitable by deep drains, and cultivation has been largely transferred to these.

Pyrethrum (Chrysanthemum cinerariaefolium)

The location of areas growing pyrethrum is largely determined by its unusual climatic requirements. It is a crop of the high-altitude tropics, yields being highest at 2500 m and above, falling considerably below 2000 m. Free drainage, good soil structure and a pH above 5.5 are required.

TREE AND SHRUB PERENNIALS

The tree and shrub perennials include many of the most profitable cash crops of the tropics. With the exception of bananas, they do not produce the carbohydrates needed in a basic subsistence crop, and hence under traditional farming they are grown in conjunction with rice or a root crop. In part because of their relatively complex agronomy and processing technology

320

and in part because of profitability, they have been widely developed as plantation crops. The establishment of the young crop and its maintenance over several years until bearing involves a capital investment that is substantial not only on a plantation but also, relative to resources, on a small-holding. Hence the choice of productive environmental conditions for new planting areas acquires additional importance. General aspects of the soil requirements of tree and shrub perennials are as follows:

1. Deep profiles are desirable, to permit tap root development.
2. Where the climate has a dry season (e.g. Köppen Am climates) a high available water capacity is important.
3. Perennials (rubber and coconuts excepted) have a high nutrient requirement, they are necessarily grown in climates where leaching is strong, and unlike annual crops, rotation is not available as a means of restoring soil fertility. Consequently heavy and continuing fertilizer application is necessary, and the existence of a sufficient cation exchange capacity to hold added nutrients is important.
4. Some perennials (tea, rubber, oil palm and to a lesser degree coffee) are able to tolerate strong acidity (pH 4.0–5.0).

Where well managed, the mature crop protects the soil against erosion and therefore moderate or even steep slopes can be used. There is, however, a severe erosion hazard on such slopes if conservation measures are not followed. The widely-spaced planting of the young crop leaves the soil exposed, and establishment and maintenance of a cover crop is essential. Terracing is desirable on steep slopes and, because of the high value of the production, is frequently economic.

Tea (Camellia sinensis)

The essential features of good tea soils are that they should be deep, have good moisture retention and be freely drained. In the upland landscapes climatically suited to the crop such conditions are frequently found on moderate to steep slopes. The plant will survive dry spells providing that the soil does not dry to wilting point down to the depth reached by the tap root, but yields suffer as it does not give flush growth unless the soil is moist. There should be at least 180 cm of soil free from a limiting horizon. Friable kaolinitic clays are the most suitable, including leached ferrallitic soils, strongly-leached ferrisols and humic latosols. Very coarse textures are unacceptable owing to low moisture retention. Whilst giving highest production with an all-year supply of moisture, tea is widely grown in climates with up to four, occasionally six, dry months. Moisture retention then becomes of particular importance. Basin ridging (contour ridges with cross ties) is a useful means of eliminating runoff and so ensuring that all rainfall in excess of field capacity infiltrates to the lower layers. Supplementary spray irrigation during dry spells can give a substantial response, sometimes sufficient to be justifiable in economic terms.

As it is the leaf which is harvested, tea is a nitrogen-dominant crop. Except on virgin land high in organic matter it rarely fails to respond to nitrogen fertilizer. The plant is a calcifuge and an aluminium accumulator. Strong acidity can be tolerated, although there are difficulties with phosphorus fixation below pH 4.5.

The steep slopes under a high rainfall that are characteristic of tea-growing areas offer a serious danger. Under good management, the nature of the tea plant is such that this can be averted, since the dense leaves of a correctly pruned stand provide an almost complete ground cover. The soil organic matter must be maintained, and terracing is necessary on steep slopes. Poor management can lead to severe erosion, as has happened on some smallholder tea plots in Sri Lanka.

Coffee (Coffea arabica, C.canephora)

The better quality *arabica* type of coffee is grown in climates with a longer dry season, up to 5–6 months, than that of any other tree or shrub perennial. A critical feature of its soils is therefore a sufficiently high available water capacity to supply moisture to the tap root through this period. A deep profile, preferably over 180 cm, and medium to moderately heavy textures are desirable. At the same time there must be free drainage. Stiff clays are unsuitable because of poor root aeration, and sandy soils because they dry out quickly. Land with a limiting horizon of laterite should be avoided (unless it is economic to dig planting holes). A high organic matter content when the trees are planted is particularly desirable. Coffee has a balanced but high nutrient demand, and the natural fertility of virgin soils is soon exhausted. Thus, apart from having a wide range of pH tolerance, the soil requirements for sustained high-yielding cultivation of *arabica* coffee are relatively specific and demanding. They are met *par excellence* by ferrisols derived from basic volcanic rocks, as for example in the Wa-Chagga areas on the southern slopes of Mount Kilimanjaro.

The 'robusta' type (now botanically *C. canephora*) has higher yields but lower quality. It is a more hardy plant, with the same optimal requirements as *arabica* coffee but tolerant of a wider range of soil conditions provided there is free drainage.

Cocoa (Theobroma cacao)

There is a measure of incompatibility between the climatic requirements of cocoa and its optimum soil conditions. It needs a high rainfall with few or no dry months, yet does not yield well on soils with a low base status. Heavy-textured soils are preferable provided they are well aggregated and permeable. Good site drainage is important, and profile depth should be at least 150 cm. Ferrisols derived from basic volcanic rocks are highly suitable, as they can maintain base status despite strong leaching.

When grown under shade, cocoa does not usually respond to fertilizer but yields are moderately low. If unshaded it responds to nitrogen, and to magnesium where deficient. The higher yields that are possible with unshaded conditions can only be sustained by continuous fertilizing (Hartley, 1968).

The suitability of soils for cocoa is the subject of one of the most detailed studies of soil–crop relations to be carried out (Smyth, 1966). It is based on consideration of the moisture and air requirements, rooting system and nutrient requirements of the plant. The findings are summarized by dividing cocoa soils into four classes: I good, II fairly good, III poor, and IV unsuitable. The soil properties necessary for inclusion in classes I and II are as follows:

Depth:	> 150 cm
Drainage:	Free
Organic matter:	> 3 percent in top 15 cm
pH:	6.0–7.5 in topsoil; no horizon < 4.0 nor > 8.0
Cation exchange capacity:	> 12 m.e./100 g in topsoil, > 5 m.e./100 g in subsoil
Base saturation	> 35 percent in lower horizons
Exchangeable bases:	In top 15 cm: > 8 m.e./100 g calcium > 2 m.e./100 g magnesium > 0.24 m.e./100 g potassium

There is no specific textural requirement but many sandy soils would be excluded on grounds of the properties of the exchange complex.

Oil palm (Elaeis guineensis)

The oil palm has the highest yield of oil per unit area of any crop, and in terms of calorific production is exceeded only by sugar-cane. It has a correspondingly high rate of nutrient removal, especially of potassium and nitrogen. These large nutrient demands must be met by fertilizers. Consequently it is the physical conditions of the soil that are important in the selection of planting sites. The key feature is root penetration. The palm has an extensive system of feeder roots extending well beyond the diameter of the crown. Depth to a limiting horizon should be over 150 cm, coupled with a well-developed structure and friable or labile consistence. Strongly leached ferrisols are suitable, such as those derived from andesite in Malaysia (Segamat series) which although technically clays have such strong micro-aggregation that they remain friable and permeable. Moderately sandy soils can be used, such as the leached ferrallitic soils derived from granites and sandstones in West Africa, but yields are substantially lower than on soils of heavy texture. Free drainage is needed, but riverine and marine alluvial soils are extensively used, with lowering of the water table below the rooting zone by deep open ditches where necessary. A dry season, such as is found in areas of cultivation in West Africa, reduces yields, and this reduction is aggravated by sandy soils with low moisture retention. Sites with massive laterite close to the surface should be avoided, and such is the investment involved that a very intensive special-purpose soil survey to locate areas of laterite may be justified. Soils with up to one metre of peat overlying mineral soil can be used if drained, but deeper peat fails to provide a sufficiently firm root anchorage.

The palms in semi-wild groves of West Africa are unfertilized yet do not show nutrient deficiency symptoms; these, however, are low yielding. With the higher-yielding varieties grown commercially, substantial fertilizer applications are necessary, not only to the young plant but continuing through the bearing period. Nitrogen deficiency is common in the young plant and potassium deficiency in mature palms. Potassium removal is of the order of 100–200 kg/ha per annum, which represents one quarter or more of the total content in many of the soils used. Because of the strong leaching, inevitable in the climates in which it is grown, magnesium may also become deficient. The reserves of nutrients in newly-cleared jungle are of great advantage in the first 2–3 years of the initial planting cycle, but are soon used up and thereafter balanced nutrition is important. Both in Malaysia and Nigeria detailed studies have been made of the relations between soil and leaf nutrient content and fertilizer response; yields are found to vary with quite small differences in nutrient status.

A cover crop, e.g. *Pueraria*, a rapidly-growing creeping legume, must be established soon after forest clearance. Terracing is desirable on moderately steep slopes. Production costs (notably harvest roads) are considerably higher on sloping land but the higher return obtainable from the crop may justify its use.

Rubber (Hevea brasiliensis)

Rubber is a less demanding crop than any other perennial with the possible exception of coconuts. Whilst the highest yields are obtainable on deep and fertile soils, satisfactory growth and economically adequate yields can be obtained on fairly shallow soils of low nutrient status. Some stands in Malaysia, for example, are on the Malacca series, a leached ferrallitic soil with a thick horizon of massive laterite commencing at 50–100 cm depth. Texture may be anything from a light sand (provided there are no dry months). to a heavy clay. The young plant requires good supplies of the primary nutrients; these may well be present if planted on cleared jungle, but must be supplied by fertilizers where old rubber stands are being replanted. Once the tree is mature its main need is for water, the amounts of nutrients removed in the latex being small. There are responses to nitrogen fertilizer but only of about 5–20 percent increase in yield of which the lower range is not economic. Correctly spaced and maintained mature rubber trees provide, through leaf and litter cover, a complete protection against erosion. Rubber in Malaysia is one of the few crops for which specific information on soil series in relation to crop response is available (p. 302).

Coconuts (Cocos nucifera)

The picturesque sight of coconut palms on a coastal sand bar is indicative of their tolerance rather than optimal requirements. The palm will grow on almost pure sand that would support few other crops, provided that an abundant supply of groundwater is available in depth. It has a relatively high salinity tolerance, although yields are higher where the water beneath dunes is kept fresh by a high rainfall and lateral underground flow toward the sea. Yields are depressed to about half by groundwater with 1500 p.p.m soluble salts and to nil by 3500 p.p.m. (Jenkin and Foale, 1968). Thus, on present and former coastal dune sands with a high rainfall, coconuts have a considerable marginal advantage over all other crops.

This does not mean that the palm will only grow on coastal sites nor that it is a halophyte. Besides dune regosols, alluvial and organic soils of coastal plains are suitable if the water table can be maintained at some 150–200 cm depth. Latosols and andosols can also be used provided there is a plentiful supply of water; physiological drought reduces the number and size of nuts and the copra out-turn per nut. Normally some other crop will possess a marginal advantage on erosional relief, but coconuts may be considered on such sites in island economies geared to copra. Nutrient requirements are not large, but on regosols potassium is likely to be deficient.

*Bananas and plantains (*Musa *spp.)*

Bananas require free drainage and a deep profile with a high available water capacity. Sandy soils are unsuitable owing to their low moisture retention. The water table must be below 150 cm, or must be lowered to this depth by drainage. They can be grown on alluvial soils or even former swamps if these are drained by deep trenching. The physical properties of the soil, good structure and porosity, are of particular importance, as bananas have a delicate root system which cannot penetrate compact clays. The productivity of newly-cleared land is high but falls rapidly as organic matter decreases. Soils with a low nutrient content can be used if fertilized; the crop usually responds to nitrogen and sometimes to potassium. Ferrisols and latosols derived from limestones are ideal. The banana tree depletes soil organic matter and does not provide a very complete vegetative protection against erosion, therefore conservation works are important on sloping land.

Citrus fruits (Citrus *spp.*)

Citrus fruits are mainly Mediterranean to subtropical crops, commonly grown under irrigation; the fruits obtained in the tropics are often of lower quality. The trees suffer if root aeration is inadequate, and soils should therefore either be of moderately light texture or well-structured and porous. A problem in arid areas under irrigation is that citrus are particularly sensitive to salinity and to boron toxicity. They have a high zinc requirement and deficiency is common.

SELECTED REFERENCES

General

Purseglove (1968, 1972), Ignatieff and Page (1958), Arnon (1972), Jacob and Uexküll (1963), Irvine (1969), Acland (1971), Richardson (1963).

Specific crops

Maize	Berger (1962), Allan (1971)
Sorghum	Doggett (1970)
Rice	Grist (1965)
Yams	Coursey (1967), Irving (1956)
Cassava	Jones (1959)
Soya bean	Norman (1963)
Lucerne	Bolton (1962)
Cotton	Prentice (1972)
Tobacco	Akehurst (1968)
Sugar-cane	Barnes (1974)
Sisal	Lock (1969)
Tea	Eden (1965), Child (1953)
Cocoa	Smyth (1966), Urquhart (1961), Hartley (1968)

Oil palm	Hartley (1967, 1968), Hill (1969), Ng (1966), Coulter (1972) Tinkler (1964), Ng and Law (1971).
Rubber	Coulter (1972), Ng and Law (1971), Guha (1969)
Coconuts	Child (1964), Jenkin and Foale (1968)
Bananas	Simmonds (1966), Baeyans (1949)

CHAPTER 18

SOIL SURVEY AND SOIL MAPS

Soil survey is one of a group of activities collectively known as natural resource surveys. These are studies of the natural environment with special reference to its resource potential. Resource surveys may cover each of the eight factors of the physical environment: geology landforms, climate, hydrology (surface water and groundwater), soils, vegetation, fauna and disease. Geological survey is the longest established branch of resource survey. At the other extreme, surveys of the incidence and areal distribution of disease, whilst undoubtedly a factor in development, are as yet infrequent. The production of maps is a major element in geological, landform, soil and vegetation surveys, and plays a lesser role, as compared with statistical and descriptive matter, in other types of survey (Young, 1968a, 1974b).

Of these branches of natural resource survey, soil survey is that which is most widely used in development planning. A soil map is one of the primary documents on which land development projects are based.

On theoretical grounds there is no reason why soil surveys in the tropics, or in less developed countries, should be different from those in temperate or advanced countries. In practice distinctive approaches and methods are used. This distinction arises in part from differences in the physical environment, such as the importance in temperate latitudes of glacial and periglacial drift deposits as soil parent materials, and in part from differences in the conditions for survey – the available topographic map base, accessibility, the absence of permanent field boundaries over much of the tropics, climatic conditions for fieldwork, availability of labour, and even the contrasted attitudes of local farmers to the appearance of surveyors. To an important degree, however, the distinction results from differences in the types and scales of survey and the purposes to which they are put; these in turn are consequent upon the stage reached in the technological use of land resources and the role of government agencies in their development. Thus the approaches and methods characteristic of soil survey in the tropics are in large part applicable to less developed countries of subtropical and temperate latitudes, including North Africa, the Middle East and perhaps, although we have little information, China.

THE GROWTH OF SOIL SURVEY IN TROPICS

Soil maps of the tropics dating from before the 1930s are rare. There is a soil map of Cuba at 1 : 800000 by H. H. Bennett and R. V. Allison of the US

Department of Agriculture of 1928; a survey of the island of Tobago published in 1929 by the Imperial College of Tropical Agriculture; and a map of
1930 on 1:30000 scale by R. L. Pendleton of the La Carlota area of the
Phillipines (18° 20′N, 123° 00′E). This last shows 30 mapping units consisting of soil series with textural phases, e.g. Guimbalaon clay loam. There
may be other surveys prior to 1930 in the libraries or filing cabinets of
Departments of Agriculture of former Colonial territories.*

These achievements pale, however, before the work carried out by Dutch
scientists in the East Indies. In the Library of the Royal Tropical Institute.
Amsterdam, there is a survey of Sumatra at 1:800000 published in 1920,
and two undated maps of parts of Java (one at 1:10000) believed to date from
about 1910. The earliest, however, is the *Groondsoortenkart van een gedeelte
van Deli* by D. J. Hissink, published at Buitenzorg in 1901. The area lies near
Medan on the north-east coast of Sumatra (3° 45′N, 98° 45′E). This remarkable map, on a scale of 1:100000, shows five soil types, which may be translated as peat, clay- sand- and mixed-soil, black humus-rich soil [vertisol?],
chocolate-coloured soil, and red hill soil [ferrisol?]. The accompanying 18-
page memoir includes a table giving for each type the density, water capacity,
ignition loss, nitrogen, phosphorus and potassium, and cation exchange
capacity. It is noteworthy that this earliest survey should include the most
fertile of tropical soils, those derived from basic rocks.

There were a number of surveys carried out during the 1930s and the succeeding War years, often on an exiguous budget and with a staffing of one
man per country (a situation occupied by the author as late as 1962). Among
the more noteworthy to reach published form were a reconnaissance account of the soils of the Belgian Congo by J. Baeyans (1938); soil-vegetation
surveys of Northern Rhodesia by C. G. Trapnell and others (Trapnell and
Clothier, 1937; Trapnell, 1943; Trapnell *et al.*, 1948); the eleven 'grey
books' covering the British West Indies produced by the Imperial College
of Tropical Agriculture (see Hardy and Ahmad, 1974); and a printed soil
map of part of Malaŵi by A. J. W. Hornby (1938) which, considering the
poor topographic base-maps and the lack of aerial photographs, is astonish-
ing in its accuracy. Isolated from the rest of the scientific world, R. L.
Pendleton conducted surveys in Thailand. But the outstanding survey of this
period, and possibly of all time, is the *Provisional Soil Map of East Africa*
by Geoffrey Milne (1935/6; cf. also 1947). This was responsible for the introduction of the catena as a soil mapping unit, and shows a remarkable grasp
of the nature and origins of soil distribution patterns in the African savanna
zone.

The period 1945–55 was one of rapid growth, borne on the Western
world's new awareness of the need for tropical land development. It saw

---

*The author would be grateful for details of any such maps, published or in manuscript.

the establishment of soil survey as a branch (usually small) of many Government Departments of Agriculture, and the production of the first soil maps of most countries (e.g. Ceylon in 1945, Thailand in 1949). By 1949 C. F. Charter was in a position to publish a paper on 'Methods of soil survey in use in the Gold Coast', and systematic surveys were under way in French African territories.

An important stimulus to the use of soil surveys in development projects was the failure of the 1947 East African Groundnuts Scheme, an attempted large land settlement project in Tanganyika. There were faults of many kinds in the planning of this scheme, but among them was the fact that two of the three areas selected for planting proved to be environmentally unsuitable; at Kongwa the rainfall was too low and the soils excessively compact and abrasive, whilst at Urambo there was much low-lying land subject to waterlogging (Phillips, 1959): Two *post mortem* detailed soil surveys of these areas, by Anderson (1957) and Charter (1958), are of historical as well as much scientific interest. Out of the costly failure of this scheme there arose the recognition that land development should be preceded by soil and other natural resource surveys, a practice which subsequently became a standard requirement of international planning and aid-giving agencies.

From 1955 onwards soil surveys have been carried out in growing numbers. A distinction has arisen between soil mapping for general resource inventory purposes, analogous with routine geological mapping, and soil surveys at larger scales carried out for specific development projects. Surveys of the former kind are usually carried out by governments, sometimes with the help of an international aid subvention, whilst the latter are frequently done by consultant firms and international agencies. One of the most extensive single mapping projects was a landform and soil survey of the vast area of the Indus Plains, carried out by a consultant firm under Colombo Plan aid, which has regrettably remained difficult of access (Fraser *et al.*, 1958). Noteworthy contributors to soil survey in the post-war period include H. Brammer in West Africa and subsequently Bangladesh, C. F. Charter in West Africa, R. Dudal in the Far East, R. Maignen in French African territories, C. Sys in the Belgian Congo and H. Vine in the West Indies and Nigeria.

In the field of integrated surveys (p. 335) the earliest work could be taken as the *Vegetation–soil map of Northern Rhodesia* by C. G. Trapnell *et al.* (1948) already noted. Whilst nominally showing vegetation–soil units, the map legend in fact also incorporates features of landforms. Priority is usually, however, accorded to the *General report on survey of Katherine-Darwin region, 1946*, by C. S. Christian and G. A. Stewart of the Commonwealth Scientific and Industrial research Organization, Australia (published in 1953). This was responsible for the introduction of the land systems method, which has since been widely used for reconnaissance surveys.

Both Trapnell's vegetation–soil maps and the land systems surveys are essentially resource inventories, and do not proceed far in the direction of evaluation of resource potential and development possibilities. The pioneer works in this latter respect are reports of the 1950s by consultant firms, not generally available. Apart from such reports, an outstanding work is the two volumes of the *Agricultural survey of Southern Rhodesia* (Vincent *et al.*, 1961). Part I, the *Agro-ecological survey*, covers physiography, climate, soils and vegetation. These are synthesized into natural regions, based on climatic and soil moisture conditions as indicated by the plant response, and sub-divided into natural areas based on landforms and soils. The farming systems and agricultural potential are assessed for each natural area. Part II, the *Agroeconomic survey*, carries the evaluation forward into economic terms. It gives, for samples of farms from each natural area, the percentages of land under each crop, yields, production, sales and gross income; both large estates and smallholdings are covered. The integration of resource survey, land evaluation and economic analysis achieved in this remarkable and little-known survey has never been repeated for a survey of a whole country.

SOIL MAPS

Someting of the order of 10 percent of the tropics has been covered by reconnaissance soil or integrated surveys, and probably less than one percent by semi-detailed and detailed surveys. Maps of the remaining area are based on exploratory studies or on inference from climate and vegetation. For South America and proportions are 13 percent systematic soil surveys (mostly in coastal areas), 48 percent exploratory studies and 39 percent (mostly in the Amazon Basin) inferential (FAO–Unesco, 1971). Coverage of this order refers to surveys on very small scales, often 1:250000 or 1:500000. At larger scales the proportional areas covered fall off very rapidly. A recent soil map inventory of Tanzania shows that only one percent of the country is covered by surveys at 1:100000 or smaller, and less than 0.4 percent at 1:25000 (Cook, 1974).

Countries with complete coverage by reconnaissance surveys include Cameroun, Sri Lanka, Guyana, Jamaica and other islands of the former British West Indies, Lesotho, Malaŵi and Uganda. Substantial parts of Chad, Ghana (78 percent), Malagasy, Malaysia, Niger, Nigeria and the Phillippines have been surveyed. A survey is in progress in Kenya which will cover the whole country at 1:100000 (humid areas) and 1:250000 (dry areas). Reconnaissance surveys cover approximately 16 percent of Bangladesh, 10 percent of Pakistan and 2 percent of India. Nearly all countries have produced national soil maps at scales of between 1:500000 and 1:2000000, often with several revisions, but in many cases substantial parts of these are based on exploratory observations only.

Land systems atlases are available for the whole of Lesotho (Bawden and Carroll, 1967), Swaziland (Murdoch *et al.*, 1971) and Uganda (Ollier *et al.*, 1969), and one-third of Kenya (Scott *et al.*, 1971). For tropical Australia there are soil-landscape maps at 1:2000000 (*Atlas of Australian soils*) whilst substantial parts, together with part of Papua New Guinea, are covered by land systems surveys (see below).

The starting point for soil map information is the 19 sheets and 10 explanatory volumes of the *FAO–Unesco Soil map of the world 1:5000000*. For Africa, the *Soil map of Africa* (D'Hoore, 1964) is still of value. The most comprehensive world lists of soil maps are the *Catalogue of maps* (FAO, 1965/73), which includes most national compilations and many detailed surveys, and the map sections of the bibliography by Orvedal (1975). A more extensive list for countries of humid tropical Asia is given by Dudal *et al.* (1974). Other sources are the catalogues issued by the Office de la Recherche Scientifique et Technique Outre-Mer (ORSTOM), Paris, and the Land Resources Division of the Directorate of Overseas Surveys, Britain.

Soil maps and memoirs are generally more difficult of access than much other scientific publication, and the final stages in any search are to write to, and ultimately visit, the Soil Survey Branch of the Ministry of Agriculture of the country concerned.

The following are the more noteworthy series of tropical soil surveys and natural resource surveys that include soils:

1. FAO/Special Fund and FAO/UNDP project reports, which include surveys of parts of Afghanistan, Colombia, Ethiopia, Ghana, Somalia, Sudan and Zambia. These multi-volume reports vary in content from soil surveys as such to development surveys.

2. Soil maps and memoirs published by ORSTOM, Paris, covering many territories of former French Africa and Malagasy. These are purely pedological surveys, not covering resource potential.

3. The *Land Resource Studies* and *Technical Bulletins* of the Land Resources Division, Britain. These constitute a non-uniform series which includes land systems surveys (of parts of Botswana, Lesotho and Nigeria), a development survey of a small part of Tanzania, and two surveys of crop–soil– environment relations (coconuts on Christmas Island and the oil palm potential of Gambia).

4. The *Land Research Series* of the Division of Land Research and Regional Survey, CSIRO Australia. With a few exceptions these are land

systems surveys, covering parts of tropical and interior Australia and Papua New Guinea (CSIRO, 1953–75).

5. Twenty-six soil survey volumes ('green books') covering Jamaica and the former British West Indies, produced by the Regional Research Centre of the British Carribean, Trinidad.

Mention should also be made of soil surveys carried out by consultant firms in land development planning. They are difficult of access, being obtainable only through the clients (usually national or regional governments) and not, except with clients' permission, from the consultants. These surveys are frequently of high standard and possess particular interest in being directed towards specific development projects, some of which have been implemented. It is unfortunate that this considerable body of scientific information is not more widely available.

TYPES AND SCALES OF SOIL SURVEY

Integrated survey and soil survey

A distinction may be drawn between integrated survey and soil survey. The method of integrated survey, also known as the land systems method, is based on mapping the total physical environment. In practice landforms are the main basis for the definition of mapping units and, especially, for the drawing of boundaries between them. Soils are an important but not usually a defining property of the mapping units, although it is possible by supplementary field survey to derive a soil map from a land systems map, as has been done for Lesotho (Carroll and Bascomb, 1967).

In soil survey *sensu stricto* the essential and defining feature of the mapping units is the soil itself, mapped as soil series, soil associations or other units of classification. There is, however, a reservation to the apparently tautologous statement that a soil map is a map of soils. Some are indeed of this nature, showing only base map detail and soil information. In many cases, however, information on landforms is incorporated in the soil map; this is done partly because landforms are much used as a means of delineating soil boundaries, and partly because it is apparent in the course of survey that landforms as well as soils will affect the uses to which the survey is to be put. The degree of inclusion of landform data varies from an overprint showing slope angle class to the use of landform descriptors in the definitions of mapping units, e.g. 'shallow red soils on moderately dissected terrain'.

Thus many of what are nominally soil surveys are in fact soil–landform surveys. The extent to which landform information is likely to be included in the map legend varies with scale. At the 1:250000 scale combined soil–

landform definitions are frequently employed. At scales of 1:100000–1:50000 the practice varies, whilst at scales larger than 1:50000 the detailed mapping units are likely to be soil types, landforms being used if at all as a a higher category grouping.

Scales of soil maps and soil survey

There is an hierarchy of soil maps and surveys at different scales, each using different methods and intensities of survey and having different purposes. The terminology of different survey intensities and scales is well established but not standardized. The FAO has proposed a new set of terms on the grounds that they lack the varying connotations of the older ones. In the following list the established terms are used, the FAO equivalents being given in brackets. The scales given refer to the published map; field soil survey is usually carried out on base maps 2–2½ times the scale of the intended final map. The first three types listed are distinguished from each other primarily on the basis of how they are produced and only secondarily on scale (fig. 22):

Compilations (FAO: syntheses) are soil maps based on abstraction from other surveys plus, where gaps in coverage render it necessary, inference. Scales are usually 1:1000000 or smaller. The national soil maps of most countries belong to this type.

Exploratory surveys (FAO: same) are not surveys in the strict sense in that they do not attempt full coverage of the area. They are rapid road (oc-

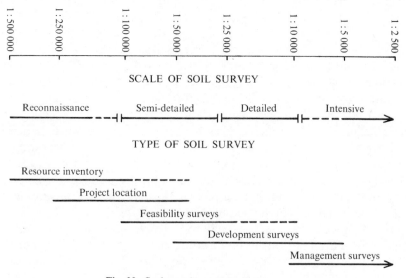

Fig. 22. Scales and types of soil survey.

casionally air) traverses made to provide a modicum of information about otherwise unknown regions. Scales vary from 1:2000000 to 1:500000. A number of exploratory surveys were carried out to provide information for the FAO–UNESCO soil map of the world.

Reconnaissance surveys (FAO: low intensity) are the smallest scale of survey to achieve coverage of the whole survey area. The usual scale is 1:250000, but maps at from 1:500000 to 1:120000 are included. They are frequently integrated surveys, and much use is made of air photograph interpretation. Some smaller countries are completely covered by reconnaissance surveys. Examples: Sri Lanka (de Alwis and Panabokke, 1972–3), Uganda (Uganda Government, 1958–61), Malaŵi (Young and Stobbs 1965/71), the Zambian Copperbelt (Wilson, 1956).

Semi-detailed surveys (FAO: medium intensity) cover the scale range 1:100000 to 1:30000, typically 1:50000. Survey is by air photograph interpretation combined with a substantial amount of field survey. Mapping units vary from soil–landform classes to soil associations and series. Examples: Semporna Peninsula, North Borneo (Paton, 1953–9), Western Samoa (Wright, 1963).

Detailed surveys (FAO: high intensity) cover the scale range 1:25000 to 1:10000 inclusive. They are produced mainly by field survey. The usual mapping units are soil series and phases of series. Examples: Kongwa/Nachingwea, Tanganyika (Anderson, 1957) (see fig. 23), St. Vincent (Watson, 1958).

Intensive surveys (FAO: very high intensity) are at scales larger than 1:10000. Grid or systematic traverse methods of field survey are commonly employed. Mapping may be soil series and phases or, additionally, parametric maps of individual soil properties. Examples Kpong, Ghana (Brammer, 1952), Dwangwa, Malaŵi (Stobbs, 1970, 1971).

INTEGRATED SURVEY

The method of integrated survey, or *land systems method*, is based on mapping units of the total physical environment that can be distinguished on air photographs. It is an outcome of the use of air photograph interpretation and was devised independently by several different organizations employing this technique (Brink *et al.*, 1966; Christian and Stewart, 1968). The primary mapping unit is the *land system*, defined originally as 'an area with a recurring pattern of topography, soils and vegetation' (Christian and Stewart, 1953) and subsequently as 'an area with a recurring pattern of genetically linked land facets' (Brink *et al.*, 1966). Both are *post hoc* definitions, the working practice being that they are the areal divisions that become apparent in the course of photo-interpretation. *Land facets* are areas within which, for practical purposes, environmental conditions are uniform; in practice they

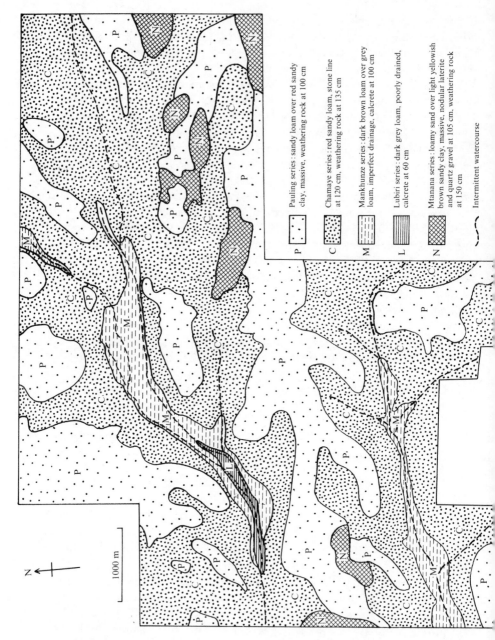

Fig. 23. Example of a detailed soil map, showing distribution dominated by a catenary pattern. Kongwa area, Tanzania (6° S., 36½° E.). Scale of original, 1:25000. Pauling, Chamaye, Mankhunze and Lubiri series form the main catena; Pauling and Chamaye series are weathered ferrallitic soils, Mtanana series a pallid ferrallitic soil. Felsic gneiss, rainfall *c.* 600 mm, *Commiphora* thicket. Based on Anderson (1957).

are the smallest areas that can be distinguished as separate but internally homogeneous areas on air photographs. (Smaller units, called *land elements*, are sometimes initially distinguished, and subsequently amalgamated into land facets; the distinction between the two, however, is not clear.) The land system is not simply a collection of contiguous facets; the component facets are causally linked, for example by geomorphological origin or groundwater movement. Each land system is known by a local name, e.g. the Masaka Land System, and is unique.

The land systems method is largely used in reconnaissance surveys, with mapping at scales of $1:1000000$ to $1:250000$. Examples have been noted above (p. 332). The method becomes less useful at larger scales, although the extent to which it might still be employed at semi-detailed and detailed scales has not been systematically investigated. There is a technical scale limit to its potential utility; this occurs where it no longer becomes possible, by means of air photograph interpretation, to subdivide the landscape into land facets which are significantly more homogeneous than the areas which they subdivide.

Survey is carried out by a combination of air photograph interpretation and field observation, but there are differences from the methods used in soil survey *sensu stricto*. Firstly, the soil receives no special emphasis in comparison with other environmental factors. Secondly, field survey is usually a team project, with each land system being studied by a geomorphologist, hydrologist, soil surveyor, ecologist and possibly agronomist. Thirdly, the ratio of time spent on field survey to air photograph interpretation is lower than that in soil survey.

The main advantage of the method is the speed and cheapness with which large areas can be covered at the reconnaissance scale. This arises from the economies of delineating boundaries by air photograph interpretation, and confining field survey to a form of stratified sampling of the areas so delineated. There is a further economy in that some areas, such as mountains, can be eliminated from field survey entirely. The amount of information and accuracy of boundaries is more than twice as much as could be achieved at the same cost by field traverses alone. A second advantage lies in the manner of data presentation using a regional basis with two scales, those of the land system and the land facet; by this means the essentials of the physical landscape can be assimilated more readily than from separate maps of each factor.

It is not self-evident that the natural landscape is ordered into units primarily at two levels of scale. The size of land systems varies widely between different surveys, from the order of 10 m² to 1000 km² (Young, 1969a). This is in part a true reflection of the nature of landscapes, but to what extent is difficult to test owing to the lack of specification of the amount of internal variation permitted within a system. Some surveys have employed *land units* as a mapping unit at intermediate level (Bawden, 1965). With respect to landforms alone there is an hierarchy of areal units at different scales (p. 23). There is a danger that the system/facet framework may be treated as a Procrustean bed, into which natural landscape units of differing size, nature and internal homogeneity are forced.

There is a lack of logical rigour in the way that every environmental factor is itself part of the definition of the units of total landscape. Since no single factor will exhibit complete covariation with the mapping units, the latter are a loosely defined base from which to proceed. It is possible to find the extent of correlation of any one variable with land systems or facets, or correlations between parameters characterizing land systems (King, 1975). However, as a basis for statistical analysis, units defined in terms of the totality of environment are not very satisfactory. Some of these difficulties could be met if the *defining characteristics* of each land system were made explicit, and separated from the *associated characteristics* in the map legend. In a high proportion of cases land systems are initially distinguished on the basis of landforms, and it would provide a firmer logical basis to the map if this practice were explicitly recognized and landforms normally taken as the defining characteristics. Such a practice would in any case be necessitated if the method of land system definition by parametric descriptions of landforms advocated by Speight (1968, 1974) were to be adopted. The difficulty with making land systems synonymous with landform regions, however, is that there are 'flatland' areas, especially coastal plains, where physiognomic vegetation types are the only means of recognizing areal divisions on air photographs. In such cases the defining characteristics would need to be in terms of vegetation.

The most substantial criticism of the land systems methods is its failure, at least in most examples to date, to collect the type of information needed for land use planning. This is particularly apparent in the Land Research Series of Australia, in which the chapters and maps on land evaluation are considerably less strong than those concerned with environmental description. The fact that much non-useful information is collected is not in itself a fault, for it may be necessary to define the mapping units themselves. But because of the emphasis on landforms the approach is a static one, based on present conditions and treating the environment as a fixed quantity. To meet the needs of development planning it needs to be supplemented–Moss(1969) suggests it should be replaced – by studies of the dynamic relations between land use and the less static factors of hydrology, soils and vegetation.

PURPOSES OF SOIL SURVEY

The purpose of a soil survey, and of its product the soil map and associated text, may be viewed in two ways. The first, or proximate, purpose is to provide the user with information about the soil and landform conditions at any site of interest. The second, or ultimate, purpose is to supply information which will assist in decisions about land use and land development planning. Such decisions range from allocation to a particular type of use, e.g. to agriculture or forestry, to details of crop management practices. It is a common misconception to regard soil surveys as fulfilling only the first of these purposes – i.e. that the function of soil survey is to produce a soil map. The latter is a means to an end, and the measure of a successful survey is not only how accurate is the map but how far has the survey been able to assist farmers, agricultural advisory staff, agronomists, economists, planners and other potential users in making decisions concerning the land.

Types of survey

Within the broad field of decisions concerning land use there are different types of survey with a range of aims, corresponding in part to different scales although with some degree of overlap (fig. 22).

Surveys for *resource inventory* are national or regional surveys intended as a guide to the varying resource potential and problems possessed by different areas, and more specifically as a guide to areas with development potential; they also provide basic scientific information of use in other disciplines, e.g. engineering, plant ecology, human geography. These are analogous to systematic geological surveys. Resource inventory in less developed countries is initially at the reconnaissance scale, although systematic mapping may subsequently be extended to semi-detailed survey.

Not clearly separate from the above are surveys intended for *project location*. Their intention is to locate and define a range of alternative development schemes, e.g. land settlement or irrigation projects. They are undertaken in areas which appear to have more promising development potential, possibly as indicated by previous reconnaissance surveys. Surveys for project location are generally at semi-detailed scales.

Feasibility surveys are intended to assess the technical and economic feasibility of specific development projects. They do not provide a complete plan, but only an estimate of whether the project is sufficiently likely to be successful to warrant detailed investigation. Both resource inventory and feasibility surveys are sometimes termed 'pre-investment surveys', meaning that no commitment to the investment of capital has yet been made.

Surveys undertaken as part of the actual planning of a project are *development surveys*. Both feasibility and development surveys may be at either semi-detailed or detailed scales, extending in some irrigation projects to the range of intensive surveys.

Management surveys are normally at the intensive scale. They are concerned with specific management questions, such as the application of irrigation water, precise planting locations for individual crops, or fertilizer application.

There is a further distinction, which has implications for survey method and the type of soil classification used, between general-purpose and special-purpose surveys (Mulcahy and Humphries, 1967). At one extreme, surveys of the resource inventory type are *general-purpose* in nature. They are intended to provide information relevant to all types of potential land use, both agricultural and non-agricultural. For surveys of this type the soil, or soil–landform, classification used must be a general classification; it may be natural or artificial (p. 236), but should aim to provide general information on the soils without omission or over-emphasis of particular properties. At the opposite extreme, management surveys are *special-purpose* in nature. The aims are more narrowly defined and hence the information to be collected may be closely specified, e.g. effective soil depth, salinity levels or presence of drainage impedance; a specialized soil classification may be employed, which selectively emphasizes the soil properties relevant to the purpose required.

Surveys for project location, feasibility and development may be of either type: general-purpose where the anticipated or planned development scheme covers a variety of terrain and requires the allocation of areas to different types of land use, or special-purpose where concerned with development of a specific type, e.g. irrigation, or flue-cured tobacco schemes. It has even been known for reconnaissance surveys to be special-purpose, in that they may be directed only towards locating land suitable for one type of development, e.g. tea planting; this narrow approach to soil survey stands forever condemned by the words of Charter (1957*a*): 'This is nothing less than turning soil scientists into pedological procurers who are given the task of finding attractive virgin lands for agricultural rape.'

Planning decisions in land development

The purposes of surveys directed towards specific land development projects, namely feasibility and development surveys, may be examined in more detail. The planning recommendations on such surveys are concerned *inter alia* with the following.:

1. Possible hazards that may arise from development, e.g. soil erosion, drought, salinization.
2. Allocation of areas to land use of different types: agriculture, grazing, forestry, water catchment conservation, settlement sites, etc.
3. Types of crops to be grown on the cultivated areas.
4. Agricultural practices, e.g. rotations, fertilizers.
5. Size and layout of landholdings.

It is common practice to make use of soil survey information for the first two types of recommendation, and to a lesser extent for the third. The function of hazard avoidance provided the original stimulus for the use of soil surveys in development planning, and it remains an important safeguard against wasteful investment. It is however, the second purpose, land use allocation, which is the primary function of soil survey, and towards which the information collected and the systems of land evaluation used are direcrected. With regard to the third set of recommendations, types of crops, soil information is combined with data from agroclimatology and economics in arriving at decisions. Recommendations concerning these three aspects are *soil-specific*, that is, they refer separately to each type of soil mapped.

It is, however, often the case that recommendations on detailed agricultural methods and on size of landholdings are *soil non-specific*; that is, they are the same for all soil types within the area recommended as suitable for arable use. This common practice results in under-utilization of the results of soil survey. In countries with a technologically advanced system of farming, such as the United Kingdom, the primary purpose of soil survey is to provide guidance to agricultural advisory staff over recommendations on such matters as fertilizers, drainage works and rotations. There is certainly a potential for applying such soil-specific recommendations to tropical areas, notwithstanding lower average standards of farming. It has rarely been found acceptable to planners to recommend that a landholding of a given size on one soil series is equivalent to a larger holding on another, yet this is a fact which geographical analysis can readily demonstrate.

Another need in development planning, not yet widely recognized, is to predict the response of the soil to changes in land use. Development will cause alterations in the type and intensity of land use, which in turn will affect the soil. The most important requirement is to ensure that under the proposed land use, soil organic matter will be maintained at an acceptable level. In irrigation schemes it is necessary to predict the soil behaviour under an altered moisture regime, including the salt balance.

There is a further class of information which might be obtained from soil survey but at present is rarely given, namely estimates of crop yields under specified management. Soil surveyors and agriculturalists are reluctant to make these owing to the many sources of uncertainty. In particular, in the normal farming situation in less developed countries, differences in standards of farm management have a greater effect on yields than differences between soil type. Yet for the purpose of economic evaluation of a development scheme, such estimates must be made. In cost/benefit analysis, crop yield is a variable on which much subsequent calculation is based, and the sensitivity to variations in estimated crop yield (that is, the amount by which the outcome of the calculations is affected by changes in its value) is high. If those technically qualified to provide realistic estimates of yields do not do so, then economists are forced to make rough approximations which will be soil non-specific.

Yield estimates may be made initially in terms of two management levels, traditional farming and improved management, as defined on p. 370. In each case they are given in statistical terms that take rainfall variability into account. Alternative estimates can then be made of the proportion of farmers adopting improved management at different periods, and the crop yields for each soil type on the scheme derived from these. It would also be possible to demonstrate the consequences for crop production of variations in the proportion farmers using improved management methods, and so provide a basis for evaluating the effects of different intensities of agricultural extension services.

Estimates of production can also be made for other types of land use. In the case of forestry plantations, information is required in the form of mean annual increment, and annual offtake, of merchantable timber. Pastoral productivity requires intitially quantitative estimates of grass growth, under unimproved and improved pastures, and estimates of carrying capacity, as livestock units per unit area, from which other parameters of production can subsequently be derived. The extent to which estimates of these two types of production can be made soil-specific is probably less than is the case with arable production.

Soil surveys for land development projects should therefore provide soil-specific information of the following kinds:

1. Indications of environmental hazards which may arise from development.
2. Land suitability for various types of land use.
3. Predictions of the soil response to changes in land use.
4. Land management recommendations.
5. Crop yield estimates under specified management levels.
6. Production estimates for other types of land use.

Information of types 1–3 is obtained largely from soil survey alone. Types 4–6 will require collaboration with agriculturalists and other applied scientists. It is suggested, however, that such collaboration should not be a matter of the soil survey providing maps of soil types and leaving subsequent matters to others; rather, soil survey should be actively concerned, including its field operations period, with land management and crop yields.

A reservation concerning these more detailed types of information concerns the degree of subdivision of soil types employed. It will not usually be possible to make land management recommendations or crop yield estimates separately for each of the soil series or other detailed soil mapping units present; moreover, such refinement would neither be practicable for agricultural advisory purposes nor, in view of the approximate nature of yield estimates, realistic. The cultivable soils series should be grouped into a smaller number of *soil management types*, that is, groups of soil series for which the management recommendations are for practical purposes the same. The fewer such classes there are, the better. They can be quite broad, and should preferably be defined in terms easily understood by local extension workers, e.g. 'red and yellow clays', 'light sandy soils'.

The survey activities that are necessary in information of these types is to be provided are considered below (p. 370).

Soil survey and land management

The role of soil surveys in less developed countries has usually been confined to the pre-investment and planning stages of projects, and their function tacitly regarded as completed when development has been physically implemented, that is, when the production phase of the project is under way. This attitude is in marked contrast to countries such as the United Kingdom, where the main function of soil survey is seen as that of aiding the farming advisory service. The tendency in less developed countries to neglect soil survey in the productive period of projects is accentuated by the fact that many such surveys are carried out by external consultants who are no longer on hand.

This situation is undesirable. The problems encountered, and the decisions to be made, in land management cannot all be anticipated at the planning stage; many will arise in the course of growing crops (or trees). Indeed, whereas in developed countries there is long-term experience of the 'improved management' type of farming, projects in less developed countries may involve abrupt transformations in types or intensities of land use, and problems are particularly likely to arise during the first 5–10 years of such projects.

Whilst many such problems will be ones of practical farming advice, and within the competence of extension workers, there are respects in which soil survey may have a bearing. If a management problem, say a suspected nutrient deficiency, arises at a particular location, it is a matter of skill and judgement to decide on the extent of the area over which it is likely to be encountered. The initial assumption is that it is co-extensive with the mapped extent of the soil series on which the problem has arisen. However, to rely entirely on such a solution is asking too much of the original survey. This is not only because of its limitations in accuracy but also from the fact that there is a considerable degree of variability within soil mapping units, and the soil properties used for mapping may not correspond to those which subsequently prove significant in agriculture.

A newly-identified and substantial soil management problem should therefore be considered by reference to, and reinterpretation of, the original survey. In many instances this will prove to be sufficient. In others, it may

343

be necessary to carry out a special-purpose management survey, at the detailed or intensive scale, of selected areas; the decision on whether to undertake such a survey can itself be arrived at by a short reconnaissance of the affected area, by soil surveyor and farm advisory worker, with the original map in hand.

A further more general role of soil survey during the productive phase is to monitor soil changes. It has been argued above that the original survey should attempt to predict soil response to changed land use. Such predictions will only be approximate, for the methods of making them are as yet little developed. A continuing record of the conditions is valuable both for local and more general purposes. Firstly, it may provide early warning of soil management problems, for example salinization, depletion of organic matter or erosion, enabling remedial measures to be taken whilst the cost is relatively low. Secondly, such soil monitoring could add to a store of experience on soil response to land use practices and so assist future development planning.

It is therefore desirable that soil survey should play a continuing role in land management, in association with the agricultural extension service, and not be confined to the planning stage.

SOIL DISTRIBUTION AND LIMITATIONS OF SOIL MAPS

Soil distribution patterns and mapping units

The degree to which soil distribution patterns are governed by different soil-forming factors at different scales, discussed above in a general context (p. 60), has consequences for soil survey. Climatic influence (including climatic differences caused by altitude) needs to be taken into account in reconnaissance survey, and is frequently used as a means of higher-category grouping; examples are the 'Wet Zone' and 'Dry Zone' classes of the reconnaissance soil survey of Sri Lanka (de Alwis and Panabokke, 1972–3), and the 'natural regions' of the agro-ecological survey of Rhodesia (Vincent *et al.*, 1961). Most detailed and some semi-detailed surveys cover areas within which climatic differences are not great enough to have a significant effect on soil differentiation.

Parent material dominates soil distribution patterns over a substantial range of scale, from 1:1000000 maps of smaller countries (e.g. Malaysia, Rhodesia) down to local surveys at 1:50000. Soil survey at the semi-detailed and detailed scales is to a large, sometimes predominant, degree concerned with ascertaining variations in parent material and the resulting soil types; this is not synonymous with geological mapping owing to the widespread occurrence of superficial materials. Much of the soil variation over short

344

distances which seems impossible to rationalize is probably governed by minor variations in parent material.

The factor of relief affects soil distribution over a wide range of scales, and becomes especially important in soil survey at the two extremes. In reconnaissance survey, landform regions are frequently the basis for mapping units, in part for practical reasons, because they can be seen on air photographs, but in part also because they truly differentiate soil associations at scales of about 1:250000. At the opposite extreme, that of intensive surveys, soil distribution on erosional relief is dominated by the catenary pattern, and on depositional relief by the pattern of alluvial landforms (cf. figs. 14 and 23, pp. 223 and 336). At intermediate scales the importance of relief varies with the area concerned; individual soil series within a catena are likely to be mapped at scales larger than 1:50000 to 1:25000, although poorly-drained valley floors can sometimes be delineated from air photographs at smaller scales.

These distribution patterns are reflected in the mapping units used at different scales. In reconnaissance surveys, composite soil–landform or total environment units are usually employed, including *land systems* (p. 335) and *soil landscapes*; the latter are defined as associations of soils delineated by landform regions (*Atlas of Australian soils*, 1960–68; Pullan, 1970). Where there is a regular variation over a large area, as on many erosion surfaces, *catenary associations* may be mapped at the reconnaissance scale. In semi-detailed surveys, *associations of soil series* are usually mapped. These may either be shown by naming the series present, e.g. 'Munchong–Serdang–Kedah Association', or by mapping *physiographic units enclosing soil series*, e.g. 'gently-sloping pediments with Kamenya and Jalira Series'. At detailed scales *soil series* are the principal mapping units, together with *soil complexes* where two or more series cannot be separated. At intensive scales, soil series, *phases* of soil series, and single *soil parameters*, e.g. depth, clay percentage, may be employed.

Mapping units are divided into simple and compound units. *Simple* mapping units contain only one class in a classification system. Soil series and phases of series, mapped as such, are examples. *Compound* mapping units contain two or more taxonomic classes. The soil landscape, association and complex are all compound units, containing more than one soil series.

It would be far from the truth to suppose that the whole of the spatial variation in soil types and individual soil properties can be explained in terms of a rational and ascertainable distribution of causative factors. It has been shown that more than half the total variability in soil properties, with respect to a large region, commonly occurs within areas as small as one hectare or even 10 m^2. Within areas or mapping units of this size, coefficients of variation (standard deviation as a percentage of the mean) for individual soil properties are rarely less than 20 percent and sometimes as high as 70 percent. Variability tends to be highest (50–70 percent) for chemical properties (including exchangeable bases and available phosphorous), and lowest (20–30 percent) for physical properties (including texture). A similar order of variability is found within areas mapped as belonging to a single soil series (Beckett and Webster, 1971).

The limitations of soil maps

There are limitations to the accuracy with which soil types and individual soil properties can be predicted from soil maps. At small scales these arise through the type of mapping units employed, and at large scales through the nature of soil variability and the minimum management area.

The mapping units employed in reconnaissance and semi-detailed surveys are mainly compound. They have been adopted for pragmatic reasons, based on the nature of soil distribution patterns and the methods of soil survey. When considered from a user's viewpoint, however, such maps suffer from considerable limitations. It is axiomatic that: (i) homogeneous mapping units are more useful than heterogeneous units; (ii) units with regular and specified variation (e.g. catenas) are more useful than those with random or unspecified variations; (iii) units within which the relative proportions of the contained soil types are specified are more useful than those in which they are not.

Judged by these criteria, the information content of many medium and small scale soil maps is fairly low. Even making the assumption that the survey is 'perfect', that is, the mapping units contain no inclusions of soil types other than those specified, a unit of which the legend consists of two soil series with no specification of either their proportions or their mode of variation with the landscape has a predictive accuracy of only 50 percent, and one with three series of 33 percent. The inclusion of mapping errors, probably between 15 and 30 percent, lowers these figures. Judged on the nature of the legend alone, exclusive of mapping errors, many soil maps achieved overall a predictive accuracy of less than 50 percent. Moreover the most 'useful' units, in terms of information content, are often those on uncultivable land or infertile soils such as rocky hills and marshes, whilst the fertile areas are more frequently mapped as composite units (Bie and Beckett, 1971; Tomlinson, 1970). Specifically, that widespread tropical landform, the gently undulating plain, often has soils of high and low fertility that are indistinguishable on a landform basis, and by means of vegetation indicators can be distinguished only by field survey and not from air photographs.

There is no simple way out of this conflict between the ideal requirements of soil map users and the practical exigencies of survey, but two aspects are of help; consideration of requirements prior to the design of the survey, and attention to the specification of manner of variability.

346

In commissioning a soil survey, whether by a national survey organization or by consultants, the potential users may wish to specify the nature of the legend, taking into account their potential requirements. The next step is to commission a reconnaissance directed towards finding how much time and money would be required to achieve a map of the desired information content and accuracy, or alternatively, what predictive ability could be secured for a given expenditure. An honest answer to these questions might well result in reconsidering the nature and scale of the survey.

The second means is to increase the utility of the maps by giving more attention to specifying the manner of variation of sub-units within mapping units. Two types of information may be given, the relative proportions of the sub-units and their position within elements of the landscape recognizable by visual inspection, notably catenary position. Much information of these types is collected in the course of survey, but it is often omitted from the legend printed on the map itself, and not always given fully in the accompanying memoir. Systematic description of the proportions of soil series, or other sub-units, and their relations to each other and to the landscape should be a standard part of any map containing composite mapping units.

At the detailed and intensive scales composite mapping units are less often employed, and limitations to the predictive accuracy of soil maps arise from other causes. Much of the work on which the following conclusions are based refers to temperate areas, but there is no reason to suppose that it is not applicable to the tropics. The main findings concern soil survey costs and the prediction of soil properties (Beckett, 1971; Beckett *et al.*, 1971).

The time and cost of soil surveys varies mainly with the length of boundary mapped and the extent to which these are coincident with visible features of relief and vegetation. There is a strong tendency to increase the number of mapping units distinguished, and hence the length of boundary mapped, with increase in scale. Hence soil survey costs vary mainly with scale, rising, as expected from the two dimensions involved, approximately as the square of the map scale.

The *purity* of mapped soil series, that is, the proportion of the mapped areas actually occupied by the series shown, is usually assumed to be about 85 percent, although some follow-up studies have shown that it may fall as low as 50–65 percent.

The accuracy with which a single soil property, e.g. clay percentage, nitrogen content, can be predicted varies with map scale. The increase in predictive accuracy obtained by the use of a soil map is calculated as $(1 - RV)$, where RV is relative variance as given by:

$$RV = \frac{\text{variance of soil property pooled within mapping units}}{\text{variance of soil property over the whole survey area}}$$

The expression 'pooled' indicates that the variances within each of the mapping units are combined in proportion to the areas occupied by these units. On a 'perfect' map, $RV = 0$ therefore $(1 - RV) = 1.0$. If $(1 - RV) = 0$, the map is no use at all (Beckett *et al.*, 1971).

Comparing predictive accuracy, $(1 - RV)$, with map scale, there is initially a rapid increase; that is, there is a great gain in information from reconnaissance surveys. There is then a less rapid but still substantial increase in predictive accuracy through the range of semi-detailed and detailed surveys. A point is reached, however, at which in general-purpose surveys, showing soil series or similar units, any further increase in scale brings little or no further improvement in accuracy of prediction. This point is typically reached at a scale of about 1:25000, and the maximum accuracy of prediction is of the order of 30–50 percent, i.e. $(1 - RV) = 0.3$–0.5, varying with the property concerned. This ceiling can only be penetrated, and high predictive accuracies obtained, by intensive special-purpose surveys which directly record the property about which information is required.

The nature and amount of soil variability within a given area, and the predictive accuracy likely to be attained by survey at various scales, could be ascertained by reconnaissance field studies carried out prior to the design and specification of the full survey.

There is a further limitation to the potential usefulness of large-scale soil maps, arising from the uses to which they may be put. For any system of farming, or other land use, there is a *minimum management area*, the smallest area to which it is both practicable and economic to apply a distinct management system. Under mechanized farming this may be as large as 2–5 hectares. With peasant farming using animal power and manual methods it is perhaps as low as one or half a hectare for general arable farming. (For horticultural enterprises it is of course considerably smaller.) There is thus no practical purpose served by increasing the detail of soil maps to a point at which substantial parts are occupied by mapping units smaller than the minimum management area.

Thus, there are distinct limits to the precision with which soil maps, both at small and large scales, can provide information on the spatial distribution of soil types and soil properties. These limits arise from the nature of soil distribution patterns in conjunction with the available methods of soil mapping, and are not easily surpassed. The minimum management area sets a further ceiling of utility. However, to conclude from this that soil surveys are of very limited value is to take a restricted, mechanistic, view, namely that the only purpose of soil survey is to produce a soil map. What these results do indicate is that the usefulness of soil surveys in development planning can be increased as much by widening the range of information provided as by improvements in methods of mapping.

NON-AGRICULTURAL APPLICATIONS OF SOIL SURVEY

There has recently been a growth in the application of soil survey to types of land use and environmental management other than agriculture and forestry. This broadening of potential applicability was developed particularly in the Netherlands, and has since been employed largely in temperate countries. The applications include road engineering, foundation engineering for buildings, site layout of building complexes (e.g. hospitals), waste disposal, general aspects of regional planning and in particular the location of urban expansion (Edelman, 1963; Simonson, 1974).

With one exception, less use has been made of non-agricultural applications in less developed countries. This is in part because competition between agricultural and non-agricultural types of land use is not the key problem in land use planning that it is in Western countries. The major exception is the use of soil survey in road engineering.

Soil survey for road engineering

It is the usual practice of engineers to take numerous soil (i.e. regolith) samples along all possible alignments of a projected road and analyze them for engineering properties. This may be acceptable for high-cost roads in countries such as Britain, but in less developed countries the greater distances and cheaper types of new road that are required make it an inefficient and expensive procedure. Work in South Africa showed that it was possible to produce soil engineering maps by dividing the land surface into units on the basis of landforms and other features visible on air photographs, and then taking samples for analysis from each of the mapped units (Kantey and Williams, 1962; Brink and Williams, 1964). It has subsequently been demonstrated that integrated surveys employing the land systems method can be used as the basis for soil engineering maps.

The environmental properties relevant to road alignment fall into four groups:

(i) *Landforms.* Slope steepness and relief pattern (arrangement, orientation and continuity of slopes), which affect alignment, length of route and amounts of cut-and-fill; drainage density, pattern and stream flow regimes, which affect bridge construction and maintenance.

(ii) *General properties of the regolith.* Depth of weathering, surface and subsurface drainage, which affect foundation requirements, expense of cuttings, stability of cuttings and verge erosion.

(iii) *Engineering properties of the regolith.* These include Atterberg limits, plasticity index, bulk density, linear shrinkage, unconfined compressive strength, expansion pressure and California bearing ratio.

349

(iv) *Availability and nature of constructional materials.* These include crushable rock, gravel, sand and laterite.

The document required for these purposes is a *soil engineering map* (geotechnical map) (Kantey and Williams, 1962; Clare, 1967). This is a map constructed by air photograph interpretation combined with special-purpose field sampling and testing of samples for engineering properties. For each mapping unit, the legend shows the mean value and range of each engineering property. Drainage conditions and sources of aggregate may also be indicated.

It has been demonstrated that land systems and facets form a reliable basis for stratified sampling of the regolith for this purpose. The variance of engineering properties within land facets is very much less than within the landscape as a whole, and falls within a range acceptable for provisional selection of road alignments (Dowling and Williams, 1964; Dowling, 1968; Beaven *et al.*, 1971). Thus the mapping units and boundaries shown on land systems maps, surveyed originally for general purposes, probably form as good a basis for soil engineering maps as those obtained by special-purpose photo interpretation. This is only to be expected, since everyone can see the same areal land divisions on air photographs.

Whether soil surveys *sensu stricto*, surveyed for agricultural purposes only and showing soil series, can be used in the same way is less firmly established. On *prima facie* grounds they are not necessarily suitable, since agricultural soil series are defined on the basis of properties of the upper two metres of regolith at most. However, such is the dependence of (agricultural) soil types on parent material that there may well be as good or better a correlation of engineering properties with soil series as with land facets. Such a relation has been formally demonstrated in Britain (McGown and Iley, 1973) and New Zealand (Northey, 1966) and soil maps have certainly been made use of by road engineers in some tropical development projects. In Ethiopia, a clear distinction has been shown between the engineering properties of 'red soils' (probably ferrisols) and 'black soils' (vertisols) (Morin and Parry, 1971), but in view of the appalling properties of the latter from a geotechnical viewpoint this is hardly surprising.

Geomorphological mapping may also have a contribution to make. It can indicate areas of superficial deposits, probable deep weathering, high landslide risk and groundwater seepage. Such mapping has been applied to road siting in a steeply dissected area with much landsliding in Nepal.

A soil engineering map, whether constructed as such or based on a land systems or soils map, is not of itself sufficient to take final decisions on road construction. The manner of use is approximately as follows. First, land clearly unsuitable for roads is eliminated or reduced in length as far as possible; this will include steep slopes, closely-dissected land, land subject to landsliding, and swamps. Next, the soil engineering map is used to classify land according to its suitability for roads. Then, using this classification together with re-examination of the air photographs, in conjunction with the user requirements, i.e. the routes required, one or a small number of alternative alignments are selected. A detailed topographic and engineering field survey (analagous to the development survey in agricultural planning) is still required for costing purposes, but it is confined to the small number of alignments selected by means of the soil engineering map.

It is worth adding a simple but golden rule for the alignment of earth roads in the tropics: wherever possible use watersheds. Earth roads which descend or run low on valley sides or which cross valley floors are subject to gullying, rutting by lorries and, where slopes are steep, landslides. They frequently become impassable in the wet season and require much regrading every year. In contrast, roads sited close to watersheds experience no runoff other than the relatively small amount of rain falling on their surface, dry out rapidly after rain, and remain in much better condition. It is surprising how often this simple rule has been ignored, presumably because of reasoning on a construction cost per unit length basis. Both maintenance costs and costs to users (as seasonally closed routes or stranded vehicles) are very much higher if routes lie along lower parts of slopes. On a subjective estimate it is certainly better to make an earth road 50 percent longer, and perhaps 100 percent, if by so doing watershed routes can be followed.

BENEFITS OF SOIL SURVEY

It is sometimes suggested that the costs of soil survey are not justified by the returns. This proposition can be examined by a consideration of the orders of magnitude involved. Suppose that a soil survey is to be carried out in an area of annual cropping and that the main anticipated cash crop is groundnuts. A detailed soil survey, on a scale of 1:25000, can be completed for a cost of the order of £0.50 per hectare (Bie and Beckett, 1970, adjusted to 1975 money values). One hectare of groundnuts yields about 1000 kg, valued at £80. It is reasonable to assume as a conservative estimate that gains of all kinds from soil survey, such as correct selection of planting areas and variation of fertilizer treatment with soil type, will raise net production by at least 10 percent. This gives a return of £8 in the first year alone, considerably above the cost of survey. Perennial crops have a higher value of output per unit area and so give still greater returns. Thus even at high discount rates, the benefit: cost ratio of soil surveys is high, Klingebiel (1966) gives examples with the benefit: cost ratios ranging from 45:1 upwards.

Potential economic benefits from soil maps

A method for calculating the potential economic benefit from a soil map of a given standard of accuracy has been devised by Bie *et al.* (1973). The

351

method rests on the assumption of a high standard of farming, in which each soil series (or other mapping unit) is given the method of land management which produces from it the greatest net profit. The greater the accuracy of the soil map, the more closely can management be matched to soil type.

The principle of the method may be illustrated by a simplified example. Suppose there are two soil series present in an area, Series 1 consisting of loams and Series 2 of sandy soils. For each soil series, an optimum management system has been found, e.g.:

System A: Tobacco–maize–grass rotation, with tobacco and maize each given 100 kg/ha of fertilizer

System B: Maize–groundnuts–grass rotation, with maize only given 200 gk/ha of fertilizer

System A is the most profitable known for the heavier-textured Series 1, and System B best for the sandier and less fertile Series 2. Assume that the net cash returns of each system applied to each series, in £ per hectare averaged over the rotation, are as follows:

Management system		
Soil type	A	B
Series 1	20	8
Series 2	10	15

Suppose that the farm has an area of 10 ha, and in fact contains 5 ha of Series 1 and 5 ha of Series 2. With no soil survey (i.e. no knowledge of soil differences) the farm will produce $(5 \times 20) + (5 \times 10) = £150$ under management System A, and $(5 \times 8) + (5 \times 15) = £115$ under System B, and will therefore be farmed under the former. With a 'perfect' soil map, and if it were practicable to farm each soil series under its optimum system, the farm could yield $(5 \times 20) + (5 \times 15) = £175$. There could thus be a potential gain of £25, of £2.50 per hectare, from the map.

A soil map may be defined as possessing 90 percent purity if 10 percent of the area mapped as Series 1 in fact contains other soil types, in this case Series 2, and vice versa. The farm income would then be:

System A on Series 1: $4.5 \times 20 = £90$
System A on Series 2: $0.5 \times 10 = £5$
System B on Series 2: $4.5 \times 15 = £67.5$
System B on Series 1: $0.5 \times 8 = £4$
$£166.5$

giving a gain of £16.5, of £1.65 per hectare, from soil survey.

This principle can be extended to any number of soil series and management systems. The information needed is: (i) the purity of the soil mapping units, defined as the percentage of the area mapped as possing each soil type which in fact consists of that soil type; (ii) the net profit from a number of different management systems applied to each soil type. It is then possible to calculate the potential benefit from additional soil survey that will produce given increments in the purity of the soil mapping units; a computer programme is available for this calculation (Bie *et al.*, 1973). If the cost of soil survey necessary to produce successive increments in purity can be estimated, it becomes possible in theory to calculate the point at which the cost of additional survey is less than the benefits to be derived from it. There is a ceiling set by the point at which an increase in detail of soil survey produces substantial areas of mapped soil units smaller than the minimum management area.

It may be doubted whether the precise calculations implied in the original statement of this method can be realistically applied in practice, the more particularly under the management standards prevailing in less developed countries. The basic concept, however, is useful in assessing the order of magnitude of probable gains from soil surveys at any given scale.

Increasing the range of application of soil surveys

Some ways in which the utility of feasibility and development surveys could be increased by greater attention to the land evaluation aspects have been noted above. In the case of surveys forming part of a routine mapping (resource inventory) programme, there is a need to give greater publicity to the existence and potential uses of the map. Measures of this kind taken in Barbados were to give a presentation of the map to a meeting of the Agricultural Society; to make sections of it available to farmers in the form of air photograph enlargements with the soil boundaries superimposed; and to issue a 'Soil map bulletin' giving accounts of recent experimental work on each soil type and also relating previous experimental results to soil type (Hudson, 1965). This example could usefully be followed. The first step is to indoctrinate a captive audience, the staff of the government agricultural advisory service. Malaysia is a country in which this publicity has been successfully accomplished, and where planting sites, land management measures and experimental results are discussed in terms of a nationally recognised classification of soil series.

On more general grounds there are opportunities for widening the range of application of soil survey not only in respect of the information which they provide but also the types of decision to which they contribute. There is an analogy with the position of economics; the role of economics in planning has never been confined to economic analysis *sensu stricto*, the assessment of costs and returns, but has always been conducted as an approach to the total planning process (some might say as if it *was* the total planning process). Similarly there are many aspects of planning in less developed countries on which considerations of soil and other resource management, including both conservation and profitable utilization, have

Soil survey and soil maps

a bearing. It is desirable to broaden the scope of soil survey and to incorporate it more fully into the processes of planning and development.

SELECTED REFERENCES

Types and scales of soil survey. Robertson *et al.* (1968), Stobbs (1970).

Integrated survey. Haantjens (1965*b*), Brink *et al.* (1966), Christian and Stewart (1968), UNESCO (1968), Mabbutt (1968), Beckett *et al.* (1972).

Soil variability. Beckett and Webster (1971).

Costs and limitations of soil mapping. Webster and Beckett (1968), Bie and Beckett (1970), Beckett (1971), Beckett *et al.* (1971).

Purposes of soil survey. Robertson (1970), Young (1973*a*).

Soil survey for road engineering. Dowling and Williams (1964), Dowling (1968), Aitchison and Grant (1968), Dowling and Beaven (1969).

CHAPTER 19

METHODS OF SOIL SURVEY

And Moses sent them to spy out the land of Canaan, and said unto them, Get you up this way southward, and go up into the mountain; and see the land what it is . . . whether it be good or bad . . . whether it be fat or lean.

Numbers xiii. 17–20

In discussing methods of soil survey in the tropics it is convenient to consider as the general case feasibility and development surveys, that is, surveys of limited areas at semi-detailed or detailed scales directed towards proposed land development projects. Except where otherwise stated, the methods discussed in this chapter refer to surveys of this type.

PLANNING AND ORGANIZATION

There are three main stages to a soil survey; pre-field operations, field survey and post-field operations. These are preceded by the planning of the survey, and may be followed by continuation activities during project implementation. In more detail, the sequence of activities is as follows:

Planning of survey ← (1)

Stage I. Pre-field operations   Study of existing data
General field reconnaissance
Main air photograph interpretation
Design and planning of field survey
←(2)

Stage II. Field survey   A. Soil mapping:
Survey reconnaissance   ←(3)
Mapping
Representative profile description
B. Evaluation and other field activities:
Land evaluation
Soil response

Stage III. Post-field operations   Revision of air photograph interpretation
Laboratory analysis of soil samples
Data analysis: Soils
Evaluation material

355

Presentation: Map preparation
Report writing
Participation in project planning
Survey activities in land management

The design of the survey is initially drawn up at the point marked (1). It may be modified subsequently at either of points (2) and (3).

The *planning* of the survey involves discussion of its purposes, the nature of the information required and the capacity of surveys of different kinds to supply such information. This will lead to provisional decisions on the scale and intensity of soil mapping and the types of maps and other information to be provided. The duration of the survey is estimated and phased with that of the project as a whole, possibly by network analysis. Decisions taken during initial planning of the survey can be critical, and yet from the nature of the case they are made with very limited knowledge of the region to be covered and the problems likely to be encountered. It is desirable that someone with experience of soil surveys in similar types of environment should take part in the planning.

These planning decisions lead in the case of a survey by a government organization to the allocation of the necessary time, manpower and funds, and in that of a survey by an outside body to the drawing up of a contract specifying how the survey shall be conducted and what information will be provided. In the latter situation there is some conflict between the scientific need to retain flexibility in the survey operations and the client's need for some guarantee about the standards of survey, particularly the intensity of field observations. In this respect the least desirable position is where the contract specifies the precise location of soil observations, e.g. 'spaced at 500 m intervals along traverse lines 1 km apart'. On the other hand it is probably unacceptable simply to specify the information that will be supplied and state that the operations necessary to achieve this will be carried out. The best that can be done to resolve this conflict is to ensure that in the survey specifications, and where necessary the contract:

(i) The intensity of field observations is described as an average figure for the whole survey area, e.g. 'at an overall density of not less than 5 observations per square kilometre'. This permits observation to be concentrated, as found necessary in the course of survey, in areas where they can be of most value.

(ii) Provision is made for revising the specifications (and hence the costs) as the survey proceeds.

The main part of the *pre-field operations* is air photograph interpretation. It is desirable that this should be preceded by a general field reconnaissance, made with the air photographs in hand. This serves two purposes; it assists photo interpretation, and it provides information on accessibility which will aid the planning of field survey. Photo interpretation is followed by design and planning of field survey, including phasing of areas to be covered, selection of some or all of the traverse routes to be followed in mapping, and possibly some provisional selection of observation sites.

The *field survey* stage commences with a second and more thorough reconnaissance, following which the detailed design and possibly specifications of fieldwork may again be modified. There are then two main groups of field activities, of equal importance: activities directed towards soil mapping, and activities directed towards land evaluation and other purposes.

In the stage of post-field operations the air photograph interpretation is revised in the light of field survey. The soil samples collected are analyzed. The two main sets of data from field survey, on the soils themselves and on their productive potential, are collated and analyzed. Soil and land evaluation maps are prepared and the soils section of the survey report written. There is then an important phase in which the information derived from survey is combined with that from other sources and incorporated in the project plan. It is sometimes the practice for project planning to commence before completion of resource survey, with the preparation of a series of interim plans based on incomplete information; such plans serve as a guide to the need to collect further field information.

Following completion of the development plan, soil survey may play a continuing part in contributing to land management (p. 343).

Field survey occupies roughly half the total time in a feasibility or development survey at the semi-detailed scale, less than half for reconnaissance surveys and more for detailed and intensive surveys. A detailed survey of 10000 hectares (10 × 10 km) at 1:25000 scale requires about 3–6 man-months of fieldwork, and might be accomplished by one surveyor with 1 month of pre-field and 1–2 months of post-field operations. A semi-detailed survey of 100000 hectares (32 × 32 km) at the 1:50000 scale requires 6–12 man-months of fieldwork, and with a two-man team might be planned as 2 months of pre-field operations, 6 months (12 man-months) of field survey and 4 months of post-field operations. These timings vary widely with type of terrain and survey specifications.

AIR PHOTOGRAPH INTERPRETATION AND REMOTE SENSING

The time has long passed when it was necessary to justify the use of air photograph interpretation in soil survey. It has become as much a part of standard practice as the use of the auger.

Remote sensing techniques are divided on a technical basis into the use of photographic and non-photographic sensors, and into imagery taken from aircraft and from satellites. From a viewpoint of utility in soil survey

357

a more pragmatic division is into two: *air photograph interpretation* (API), using photographs taken from aircraft (both black-and-white and true colour), and *other remote sensing techniques*, comprising the use of other types of sensors from aircraft together with satellite imagery. Of these divisions, conventional air photograph interpretation has been in the past, and is at present, of considerably more value in soil survey than all other techniques together.

Air photograph interpretation

Not least among the purposes of air photographs in soil survey is their use as a means of location during fieldwork. In the first place, in nearly all parts of the tropics recent air photographs are a better guide to roads and tracks than are maps. Secondly, the position of a soil observation site can be located on a photograph, e.g. by reference to a group of huts, clump of vegetation or a single large tree, in a way that it is impossible with even the best of maps. Thus, besides their use for interpretation, photographs are of considerable value in planning and following traverse routes, and locating and subsequently plotting soil observations.

The two most useful photograph scales in less developed countries are about 1:40000 and about 1:25000. The 1:40000 photo-scale is suitable for interpretation in reconnaissance and some semi-detailed surveys, directed towards the production of maps at scales from 1:250000 to 1:100000. The landscape patterns of the major regions (e.g. land systems) can be seen as a whole on a single pair of photographs, whilst at the same time binocular viewing at ×3 magnification enables boundaries to be delineated to an accuracy of about 100 m. 1:40000 is superior to *c.* 1:50000, at which tracks and other cultural features begin to become inconveniently small. The 1:25000 or 1:20000 photo-scale is suitable for use in semi-detailed to detailed surveys. The detailed subdivisions of the landscape (e.g. land facets) can be more clearly distinguished, and delineated to an accuracy of about 50 m. Individual huts and trees show clearly, making it the more suitable scale to use as a field base-map. It is less easy, as compared with the 1:40000 scale, to gain an appreciation of the major regions as the pattern of their component parts is not repeated sufficiently on a single pair of photographs. Moreover, if a large area is to be covered, interpretation on a 1:25000 scale becomes very tedious; with the added problems of handling and transferring boundaries between larger numbers of photographs it may take 3–4 times as long as on 1:40000. Larger photo scales such as 1:10000 do not add very much that cannot be seen equally well at 1:25000.

A normal mirror stereoscope with a ×3 magnification binocular attachment is adequate for the kind of interpretation called for in soil survey. It is important to have an instrument in which the field of vision covers the whole of the photo-pair. A parallel guidance mechanism for the photographs is useful, enabling a boundary located under binocular mangification to be followed across the full extent of the photo-pair. The 'zoom' type of instrument, which permits continuous change in magnification without losing vision, is a luxury rather than a necessity, but a very pleasant one to have. Where full photogrammetric facilities are not available, a sketchmaster or other dual-image instrument is necessary for transference of glazed-surface crayon markings from photograph to base map.

The methods used in photo interpretation for soil survey are similar irrespective of scale. *Pre-fieldwork interpretation* is the main operation. The first stage is to gain an appreciation of the primary land units, e.g. hill masses, plains, alluvial areas, from non-stereoscopic viewing of the survey area as a whole. This can be done from a print lay-down or photo-mosaic, the latter having the advantage of eliminating boundaries between photographs although not differences in tone. Satellite imagery, where available, is superior for this purpose. The next stage consists of detailed delineation of the primary units by stereoscopic examination of successive photo-pairs. This is followed by subdivision into secondary units, e.g. level crest areas, pediments, valley floors. At all stages the recognition and delineation of mapping units is largely on the basis of landform differences, and to a subsidiary extent on vegetation; direct interpretation of soils is neither necessary nor possible. It is also not necessary to be able to interpret all the phenomena employed in map unit differentiation; and photographic difference of tone, texture or pattern can be employed. It is best to follow the rule of subdividing areas wherever any consistent difference can be detected; the additional time taken is small, and boundaries which prove superfluous can be removed subsequently. The mapping units distinguished in pre-fieldwork interpretation will not correspond exactly to those of the final soil map, and may be called *API mapping units*. An interim map legend is drawn up, giving for each unit the landforms, vegetation and land use, and the photo-diagnostic features by which it has been identified. It is useful to list photo-pairs on which each unit is typically represented.

A provisional map is constructed, which need not be photogrammetrically accurate, showing the distribution of the API mapping units. This map and the photographs are used to plan field survey. The actual prints on which boundaries are marked in glazed-surface crayon are the best base for fieldwork, and so a tracing should be made to guard against loss.

Post-fieldwork interpretation occupies considerably less time. A high proportion of the boundaries between API mapping units survive, notably boundaries of hill areas and poorly-drained valley floors. It is at this stage that the relations between API mapping units and soil types, investigated during fieldwork, are finally collated, with the following possible results:

(i) one API mapping unit corresponds with one soil type; (ii) two or more API mapping units possess the same soil type; (iii) one API mapping unit contains two or more soil types.

Case (i) calls for no further action other than adding soils to the map legend. Case (ii) requires amalgamation of units. Case (iii) calls for re-inspection to see whether any difference in the photographic images associated with the soil types can be detected. If so, boundaries are inserted; if not, then either approximate boundaries are drawn by simple interpolation between field observations, or the unit is mapped as a soil association. The working rule at this stage is the opposite to that before fieldwork; when in doubt about whether a difference exists between two mapping units it is better to amalgamate them. The final boundaries are then transferred to a base map, frequently one which has in the meantime been constructed photogrammetrically from the same flight of photographs.

The value of API varies with the type and scale of survey and the kind of landscape. It is greater for integrated surveys and less, but still often considerable, for soil surveys *sensu stricto*. The benefit obtained is very large in reconnaissance surveys and least in detailed and intensive surveys. Vink (1968) estimated savings in time and cost, or gain in accuracy, from using API as compared with ground survey alone as 70–80 percent for scales of 1:250000 to 1:50000, and 10–20 percent for 1:20000 to 1:10000, but a *post facto* analysis of actual surveys has shown that those using API required on average 55 percent of the time of those not using it, irrespective of scale (Bie and Beckett, 1971). With respect to type of landscape, API is of most value in regions with clear topographic contrasts and least on plains; of highest value in deserts, where the ground surface is fully exposed, intermediate in semi-arid to savanna landscapes, and substantially less in uncleared rainforest; and of greater value in areas of natural vegetation or sparse settlement than in densely-occupied regions. Even in the latter, however, the numerous, small and in some regions changing fields characteristic of tropical regions do not obscure natural landscape patterns to the same extent as do the fixed field boundaries of the temperate zone.

The time saved and accuracy gained from photo interpretation is not in identifying soil types but in drawing boundaries between them. It should be clearly recognized that soil mapping from air photograph interpretation is largely a matter of correlation with the visible features of landforms and vegetation. Even where the topsoil is not obscured by vegetation, as in deserts or recently tilled land, it is self-evident that only the topsoil is visible – and most topsoils are dark greyish brown and tend to be sandy. Thus, attempts directly to identify soil series from photographs will inevitably give low returns. Even the extrapolation approach, of identifying soil types in the field and then seeking a characteristic photo image, fails in some areas. On the typical gently undulating relief of the savanna zone, ferruginous soils of high fertility and infertile weathered ferrallitic soils may occupy virtually identical relief; still more undetectable are the substantial soil differences that are caused by parent rock within dissected rainforest relief.

True colour air photographs are employed in the same way as black-and-white prints, and may be considered part of conventional API rather than one of the more esoteric forms of remote sensing. The cost of obtaining the photography and an initial set or prints is only 10–20 percent more for colour, but subsequent prints are considerably more expensive. It is more difficult with colour to obtain consistency between flights on different days. Whilst most photo-interpreters enjoy using colour the gain in information is usually small. The reasons are first, that colour suffers from the same limitations as black-and-white photography, that only the soil surface is visible; secondly, as regards the indirect interpretation which is the main use of air photographs, the landforms on which such interpretation is largely based can be seen equally well in black-and-white. There may be special regions in which important soil differences can be detected more readily in colour, for example detection of slight depressions and hence the possibility of saline soils on desert plains (Simakova, 1964). In the general case, however, whilst there is a slight gain in information from the use of colour photography rather than black-and-white, this gain is relatively small, perhaps no more than 10 percent.

Other remote sensing techniques

The principal remote sensing techniques, in addition to monochrome and true colour photography, are near infra-red and false colour photography, multi-spectral scanning, thermal infra-red sensing and side-looking airborne radar. All can be operated from aircraft, and in addition the photographic sensors and multi-spectral scanning are used from satellites. A review of the application of these techniques to soil survey is given by Carroll (1973).

Monochrome images from *near infra-red photography* closely resemble normal air photographs. It is possible, owing to the absorbtion of near infra-red radiation by water, that areas of damp ground should show more clearly as darker tones by this method, but the difference from normal photography is not large. The striking images of *false colour photography* from aircraft have a high potential for vegetation and land use mapping, but have not proved to have demonstrable advantages for soils. *Thermal infra-red* has a theoretical potential to detect, through heat differences, the texture or moisture of topsoil, but is unlikely to be of use in normal soil survey. *Side-looking airborne radar* has been used for topographic mapping in areas of permanent cloud cover, but has otherwise little non-military value.

The main useful addition to conventional photography has come from *satellite imagery*. The multi-spectral scanning (MSS) images from Earth Resources Technology Satellite No. 1 (Landsat 1, formerly ERTS 1) can be used separately, or combined to produce a false-colour image.

The MSS image in the near infra-red waveband (0.7–0.9 μm) produces a sharp and well-contrasted image which is useful in the initial stage of photo-interpretation. Its advantage over a photo-mozaic is that an area of 50 × 50 km is covered by a single frame of uniform tone. Most of the features shown can be identified also on conventional air photographs, but with considerably more effort. There is value in seeing large landscape regions, and their relations with adjacent regions, as a whole.

It is possible to use ERTS imagery alone, combined with field observation, in reconnaissance natural resource surveys of large areas at small scales of mapping, and many studies of the potential of this method are in progress. It is early to anticipate the outcome of these, but from experience to date, coupled with consideration of the technical nature of the imagery available, one thing is clear. Satellite imagery is unlikely to compete with conventional air photography in aspects for which the latter is technically suited and currently used, namely medium to large scale interpretation involving the stereoscopic viewing of landforms; for soil surveys of this type, satellite images are a supplement to, but not a substitute for, air photographs. Satellite imagery is likely to be of greatest value by making use of its main technical advantage, that of being able to cover very large areas. The immediate possibility that has been opened is for resource surveys, at very small scales but with uniform coverage, of the large countries which it is unlikely would ever be covered by older methods of survey. For example, a resource inventory of Brazil, India or Tanzania might be accomplished by this means. If such surveys could be financed and successfully accomplished, the way would be open for the first time for uniform resource inventory surveys first of continents and ultimately of the earth's land surface.

FIELD SURVEY

The activities carried out in the course of field soil survey comprise two parts, the soil mapping operation and the land evaluation operation. The *soil mapping operation* involves identification and classification of the soil types present in an area, and surveying their distribution – in short, the production of a soil map. The *land evaluation operation* includes field activities directed towards assessing the potential of the various soils for a range of alternative types of land use, and the identification of possible development hazards (Young, 1973*a*).

Soil survey is sometimes considered synonymous with soil mapping. Following from this assumption, field survey activities are confined to the determination of soil distribution, and land evaluation is undertaken as a post-fieldwork office exercise. This view is erroneous, and has frequently in the past resulted in the production of soil maps which have been little used for planning and other practical purposes. There is a range of field activities related to determination of the resource potential of the soils mapped and these should form an integral part of soil survey.

Although not necessarily successive in time, the mapping and evaluation operations in field survey will be discussed separately.

SOIL MAPPING

Phases and methods

There are three phases in the soil mapping operation: reconnaissance, mapping and representative profile description.

The functions of the *reconnaissance phase* are to identify the soil types present in the area and ascertain the general nature of their distribution pattern. The first aspect involves discovering the natural groupings of soil profiles, including their diagnostic characteristics and range of variability. This is followed by drawing up provisional soil series (or other soil type) descriptions, as a first draft of the map legend; some 80–90 percent of the soil types ultimately mapped are likely to be identified during this phase. The second aspect, that of investigating the nature of the soil distribution pattern, consists of ascertaining relations of the soils with landforms, vegetation, and with the mapping units identified by air photograph interpretation.

Included in this latter aspect is the study of the scale and complexity of the soil distribution pattern. There are some areas within which soils are relatively uniform, or bear a simple catenary relation with landforms, whilst in others the pattern is complex with substantial changes in soils over short distances, not visibly detectable by landform or vegetation inspection. The difference between these types of area is frequently related to broad groups of parent materials. In a large survey it may be worth making a systematic investigation of the scale of soil variability by means of a nested sampling design or by surveys of small sample areas (Nortcliff, 1974). A simpler means is to make short linear traverses with closely-spaced observations. The density of observations during the subsequent mapping phase is planned according to the results of this reconnaissance study into distribution patterns; by varying the density in accordance with complexity of pattern a considerable gain in information, for a given period of fieldwork, can be made. In cases of surveys under contract, it may be necessary at this point to agree on modifications to the specifications.

There are therefore two groups of observations during the reconnaissance

363

stage of a large survey; widely-spaced sites throughout the survey region, and detailed surveys of or traverses in small sample areas representative of the main physiographic units.

Reconnaissance is an important phase in soil mapping, and the most skilled. The saving in time and gain in accuracy during subsequent mapping that can come from good reconnaissance is considerable. The time allocated to reconnaissance, sometimes as little as two weeks in a six-month survey, is frequently inadequate. A case can be argued for allocating as much as a third of the total soil mapping time to a really thorough reconnaissance, following which the subsequent mapping phase becomes relatively straightforward. Where clients or superiors have to be appeased this extreme view may be unrealistic, but it is reasonable to suggest that in drawing up a field-work schedule, at least 10–15 percent of the time spent on the soil mapping operation should be allocated to reconnaissance.

The *mapping phase* is the largest single activity in the course of soil survey. The emphasis placed here on field activities directed towards land evaluation, and within the soil mapping operation on the importance of reconnaissance, should not obscure the fact that mapping the distribution of soil types is the foundation on which all other aspects of soil survey rest.

Field soil mapping procedures may comprise free survey or grid survey. In *free survey* the observation sites are not regularly spaced but are chosen as representative of areas identified on the basis of landforms and vegetation; additional sites are selected concurrently with survey, as the apparent pattern of soil distribution becomes revealed. In *grid survey* the sites are regularly spaced on a predetermined rectangular grid.

Free survey consists essentially of a form of stratified sampling, in which the strata are land units identified by air photograph interpretation or landform–vegetation units identified in the field at the time of survey. On statistical grounds it is desirable that the soil observation sites should be randomly located, but practical considerations dictate that they should be near tracks or other traverse routes. The sites are usually chosen at points which appear, on the air photographs and on the ground, to be 'typical' of the land units, a procedure which may be reprehensible on statistical grounds but which works in practice.

Soil observations during the mapping stage are usually by auger, supplemented by road cuttings, rubbish pits and other artificial exposures. The post-hole type of auger that brings up a cylinder of soil 10–15 cm in diameter is usually preferable in the tropics, as compared with the screw auger normal in temperate lands. Even better is the 'Dutch' pattern of auger with a head 6–8 cm in diameter, less cumbersome than the post-hole type but large enough to permit observation of mottling, consistence, concretions and weathered rock patches. It is useful to carry a supplementary screw auger which will, with difficulty, penetrate non-cemented nodular laterite and stone lines.

A soil observation during mapping may be an auger description or a series identification. In an *auger description* the horizon depths, Munsell colour, texture and other properties observable by auger are recorded. In a *series identification* the record is confined to that of the presence of a previously identified soil series, perhaps with a single parameter of significance, e.g. 'Jalira series (laterite 40 cm)'. Practice varies between different surveyors and there are arguments either way. It saves much time to make series identifications only, since many series have obvious diagnostic features and can be identified in 5 minutes or less, whereas to record colour, texture and other features down to 120 cm takes 10–20 minutes. Series identifications, however, permit of no subsequent reinterpretation if a series is subdivided, or a change made in some limiting value separating it from another series. The modern tendency is to record a full description, and there is particular reason for doing so in the tropics where more than half the time in the field is spent on travel between sites. Series identifications should only be made if it is quite clear that the profile falls into an established soil type, and if it is further believed that users will only require the soil type name and not individual properties.

In *grid survey* the method is to set out a series of parallel traverse lines and take soil observations at fixed intervals along them. The method is common at the intensive scale, and less frequently employed in detailed and semi-detailed surveys. It is necessary in surveys of uncleared rainforest, since you can only find out where you are by cutting trace lines through the undergrowth and measuring along these from a baseline; a herringbone or gridiron pattern is suitable. Another circumstance is that of flat alluvial plains of dry regions which appear uniform on air photographs but which have agriculturally significant variations in soil texture and salinity; large areas of the Sudan have been surveyed by this rather soul-destroying method.

Most free surveys and many grid surveys are general-purpose in type, recording all observable soil features and mapping soil series or other soil type classes. Grid surveys may alternatively be special-purpose, recording only soil properties relevant to a previously determined type of land use. Suppose, for example, it has been found that an area of 20 km² contains a substantial area of soils suitable for flue-cured tobacco, and farms are to be laid out; it may be worth making a grid survey with observations spaced at 50 m or less in order to map precisely the land suitable for planting, and omit that which is unsuitable on grounds, say, of impeded drainage or laterite close to the surface.

The third phase of soil mapping, that of *representative profile description* with sampling for analysis, may either succeed mapping or be carried out simultaneously. Methods of description are standardized, but two points call for comment. Firstly, it is not yet common to take undisturbed cores for pressure membrane testing of soil moisture characteristics. Such data are at least as important for land evaluation purposes as chemical analyses, and core sampling should become standard practice. Secondly, it is usual to take a single sample from each horizon of a soil pit supposedly 'typical' of a series, or a small number of such pits, and to assume tacitly that the values found from analysis are representative of the whole of the area covered by that series. As the coefficient of variation of most chemical characteristics within a soil series is known to be at least 30–50 percent this assumption is clearly unjustified. To take samples from a sufficiently large number of pits to obtain means and confidence limits for all horizons would involve considerable time and expense. A simple measure, which only increases the number of analyses by some 25 percent, is to take a *composite sample* of the topsoil around each pit; this is a sample of the type common for agricultural advisory purposes, consisting of about ten trowel-fuls from random points around the pit, mixed together in the sample bag. Whilst giving no indication of the range of variation it provides a better estimate of mean values than the single sample from the pit itself.

If the modifications to customary soil mapping procedures that are suggested here are to be made, it is necessary to allow for them when specifying the terms of reference of the survey. In particular, provision must be made for:

(i) a longer period of reconnaissance – field activities preceding routine mapping – than is at present usual;

(ii) decisions on intensities of survey in different parts of the area to be made subsequent to reconnaissance, with the possible inclusion of special-purpose surveys of limited areas.

Taking into account the land evaluation activities discussed below, these modifications could mean that out of the total time spent on field survey, the proportion occupied by the main routine mapping phase might fall to 50 percent or less.

Survey method and map scale

A schematic view of the relation between survey method, map scale and the cost-effectiveness of soil mapping has been suggested by Beckett (1971; also Beckett *et al.*, 1971). In this view it is taken as axiomatic that the primary purpose of a soil map is to equip its user to inform himself about the properties of the soil at any site in the surveyed area. Given this assumption, the *predictive ability** of a soil map is the extent to which it enables a user to make more precise statements about the soil conditions at any site of interest than could have been made without the map. Predictive ability

---

* The original references use 'utility' to refer to this concept. As the utility of a soil map in the normal sense of the word depends to an appreciable extent on non-mapping aspects, the term 'predictive ability' is substituted here.

depends on the nature of the legend, namely the proportions of simple and compound mapping units, and the purity of these units (p. 347).

The following three methods of survey are considered:

(i) *General purpose API survey.* This refers to integrated surveys and soil surveys in which the boundaries are largely or entirely drawn from air photograph interpretation, with a relatively low density of field soil observations. There is a high proportion of compound mapping units such as land systems or soil landscapes.

(ii) *General purpose free survey.* This is survey which, whilst it may make use of air photograph interpretation, has a relatively high density of field soil observations. The majority of mapping units are simple and are general-purpose, showing soil series or other soil types.

(iii) *Special purpose grid survey.* In this method individual soil properties are recorded, on a grid pattern, and may be mapped parametrically.

The predictive ability of each method in relation to map scale is shown in fig. 24(*a*). In surveys at reconnaissance scales the highest predictive ability is obtained by use of general purpose API survey. Although the predictive ability is fairly low, arising from the high proportion of compound mapping units, attempts to use simple mapping units would result in an unacceptably low purity. As the map scale becomes larger, the increase in predictive ability from API survey falls off, whilst that from general purpose free survey increases rapidly. In particular, it becomes possible by free survey with a relatively larger amount of field observation to map simple soil mapping units with a purity of over 65 percent. With further increase in scale, the ceiling on the predictive ability of general purpose surveys noted above (p. 348) is reached. At scales larger than this, higher predictive abilities can be obtained only from special purpose grid surveys.

The relative cost per unit area of surveys of these types is shown in fig. 24(*b*). Surveys with a high dependence on air photograph interpretation

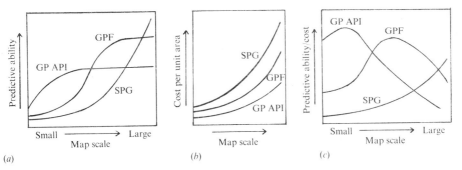

Fig. 24. Relations between scale, survey method and cost, and predictive ability of soil maps. Based on Beckett (1971).

are cheaper than those more heavily dependent on field work. Since, however, there is a general consensus that the density of field observations per square centimetre of map should remain constant irrespective of scale, survey costs vary mainly with map scale. Field soil mapping is the most expensive single activity in soil survey, and if it were the only one then survey costs would rise with the square of map scale; taking into account other activities the rise is still exponential but at a somewhat lower rate, the exponent being about 1.5 (Bie and Beckett, 1970, 1971).

Dividing predictive ability by survey cost gives the *cost-effectiveness* of different methods of soil survey in relation to map scale (fig. 24(*c*)). General-purpose API surveys have the highest cost-effectiveness at small scales, with a moderate predictive ability coupled to a relatively low cost. General-purpose free surveys become superior at intermediate scales, since although they remain more expensive than API surveys their predictive ability overtakes that of the latter. Special-purpose grid surveys only achieve a higher cost-effectiveness at very large scales and where there is a requirement for more precise soil data than can be achieved by general-purpose survey. The scales corresponding to the crossing points in fig. 24(*c*) are not specified by Beckett. API survey is undoubtedly superior at 1:250000, possibly giving place to free survey at about 1:100000. Free survey has the highest cost-effectiveness from 1:50000 to at least 1:25000 and probably 1:10000. Grid surveys only become superior somewhere in the range of intensive scales, perhaps at about 1:5000.

It may be repeated that the above discussion rests on its axiom, that the (only) purpose of a soil survey is to supply information on the distribution of soil types and soil properties. The conclusions thus refer to the soil mapping operation and the methods used therein. This does not affect the validity of the conclusions within this limited field. However, the cost-effectiveness of soil survey using these terms in their wider senses, that is, the economic gain to be obtained from expenditure on soil survey, is also considerably influenced by the land evaluation aspects of the survey and by the extent to which practical use is made of the information which it provides.

Survey to give areas of soil types or properties

There are circumstances in which the information required by users concerns the areas or proportions of soils or soil properties within a given region and not their distribution; that is, information is required in statistical form and not as a map. An example is the amount of land within a projected irrigation scheme that is saline and will require reclamation; cost estimates could be prepared from a figure of the area of such land, without needing to know its distribution until and unless it is decided to proceed with the project. There may be circumstances in the early stage of pre-investment surveys where estimates of the extent of uncultivable lands, swamps, and potential arable soils are required. In such cases the information needed is an estimate of area covered together with confidence limits.

It may be quicker and cheaper to obtain such estimates by sampling rather than mapping. If point sampling is used, the confidence limits can be obtained from the Standard Error of the Proportion (*SEP*), given by:

$$SEP = \sqrt{\left(\frac{\hat{p}\hat{q}}{n}\right)}$$

where $\hat{p}$ = the percentage of the sample points occupied by a given soil property, $\hat{q} = (100 - \hat{p})$, and n = the number of points observed. The 95 percent confidence limits lie within two Standard Errors above and below the mean percentage area as estimated from the sample. The *SEP* falls rapidly over the first 100 observations and more slowly between 100 and 200, above which the increase in confidence produced by additional sampling points is very slow; curves are given by Robertson and Stoner (1971, Fig. 1). With 200 sampling points, *SEP* = 4 percent in the least favourable circumstance, that of $p = 50$ percent. This means that the 95 percent confidence limits for p are 42 to 58 percent, which may be acceptable for broad planning purposes.

The cost of such a sampling programme could be considerably less than that of mapping, particularly in circumstances where soil boundaries cannot be detected by air photograph interpretation. Estimates of texture and salinity for flat plains of dry regions are an example. In place of point sampling, systematically or randomly spaced traverses, such as are used in forest inventory, are a possible alternative.

EVALUATION AND OTHER FIELD ACTIVITIES

The second part of field soil survey comprises all those aspects other than mapping, principal among which are activities concerned with land evaluation. Some years ago Coulter (1964) argued for the need to supplement mapping by studies directed towards the estimation of soil productivity. In the past these have often been neglected and evaluation left as a post-fieldwork office exercise. As discussed in the previous chapter, however, the nature of the information needed from soil survey for planning purposes requires that there should be a group of field activities directed toward this purpose. It is immaterial whether this is carried out wholly subsequent to soil mapping or interspersed with it. It is necessary, however, to allot an appreciable proportion of field survey time, perhaps, 25 percent, to non-mapping activities.

Field survey activities for land evaluation

As discussed above (p. 340), the principal requirement of planners and other users from soil surveys is for soil-specific estimates of the productive potential of land and the inputs required. In the case of arable land this means estimates of crop yield under given management conditions, and of the inputs necessary to obtain such yields. Both production and inputs are required initially in physical terms, i.e. as estimates of crop yields, labour requirements, fertilizer amounts, etc., and can subsequently by translated into economic costs and returns.

The most frequent case to consider is that of crop yield estimates for arable land. These need to be made with reference to specified management levels or practices. The following are useful management levels:

Level 1. *Traditional farming.* The most common existing farming practices.

Level 2. *Improved management.* The methods adopted by the more progressive farmers at the present day, such as 'master farmers'; including standard 'better farming' practices, as recommended by the agricultural advisory service, but excluding methods requiring a high capital input or sophisticated skills.

Level 3. *Advanced technology.* As used on experiment stations; including all known methods of obtaining high yields.

For any local area and crop, the crop varieties, rotations, fertilizer treatments and disease protection measures adopted at any level may be specified. It is appropriate for development schemes if yield estimates are made in terms of Level 2. Comparable estimates at Level 1 are instructive as an indication of the magnitude of benefits to be expected from expansion of agricultural education and advisory services.

The following are possible methods of making yield estimates:

(i) By reference to the physiology of growth of the crops, in relation to the physical and chemical properties of the soils.

(ii) By comparison with yields on experimental sites situated both within the survey area and on similar soil series outside the area.

(iii) By agronomic field experiments conducted within the project specifically for this purpose.

(iv) By the study of village farm production.

Comparison between crop physiology and soil properties is largely a non-field activity, although general observations of crop growth made in the course of mapping, e.g. the extent to which they appear to suffer from impeded drainage or a limiting horizon, are of assistance. The main source of information is the second. This requires visits to all sites where there are records of crop yields; these may include experiment stations, district crop variety or fertilizer trials, demonstration farms and some of the more progressive farmers. Crop marketing boards or co-operatives may possess records of area cultivated and production. Such records are not confined to the survey area, but are extended to such adjacent areas as are climatically and pedologically similar. At each site a soil description is made, preferably from a pit. It will be found that some of these sites are pedologically unsatisfactory, e.g. transitional between two series, untypical, or low in the catena. Others, however, can be used as type profiles of soil series; this is in some ways a more satisfactory manner of proceeding than the usual one of choosing locations typical of soil series and then wondering where to get agronomic data.

At each site a *crop yield record* is made. This contains:

(i) *Location*. Although this may appear self-evident, it is important in the case of farmers and minor demonstration sites to determine the location precisely at the time of the visit, for no one may know where it is on the map.

(ii) *Crop yield*. Including quality where appropriate, e.g. grade of tobacco.

(iii) *Management practices*. Including crop variety, date of planting (annuals), age of crop (perennials), fertilizer treatment, disease protection.

(iv) *Site description*. As in soil survey; including preceding crop and land use history. Subsequent to the field visit it is important to make a best estimate of the amount and timing of the rainfall for the site.

(v) *Soil description*. With samples for analysis.

The data obtained by making use of past yield records and relating them to soils may be quite considerable in quantity but will be non-standardized. It is not likely to admit of precise statistical treatment and the calculation of confidence limits, but allows a general assessment of the level of yields, e.g. 2500–3000 kg/ha. This, however, is sufficient for most practical purposes.

The third method of obtaining yield estimates is that of agronomic experimental plots established specifically for the purpose on sites representative of the main arable soil types of the area. The timing of planning or investment decisions in relation to the soil survey will frequently mean that this method is not available; it is necessarily impossible for perennial crops, and difficulties of obtaining a small water supply at isolated sites may make it impracticable in irrigation schemes. Where annual crops are concerned, however, it may be possible to obtain at least one year's results in the course of the survey. The plots may need to be established after the reconnaissance stage of soil mapping, on provisionally identified soil series. The standard experimental design for a small variety/fertilizer trial is suitable, except that only a small number of crop varieties and treatments, those most frequently used in the area or known to be best for it, are used. There remains an element of chance in that it may turn out to be an abnormally wet or dry year. If results for even one year under standardized conditions can be obtained, it will usefully complement the *ad hoc* information from yield data previously available.

Whether or not results can be obtained in time to be of use in project planning, recommendations for agronomic experimental sites should be a part of any soil survey. Rather frequently the locations of existing trials are pedologically unsatisfactory, the soil being either transitional or untypical. Sites should be established at locations typical of the main arable soil series of the area. These are known as *bench-mark sites*, and can subsequently provide information of value during project implementation.

The fourth method is that of making sample yield studies on village farms. A crop yield record similar to that from an experimental site is made, with information obtained by questioning. This data will be less reliable than that from controlled experimental plots, but has the advantage of revealing what yields are really like and not what the 'should be'. This is a better means of estimating yields under unimproved farming than are the 'unfertilized local seed variety' plots on experimental sites, which do not take account of possible late planting, inadequate weeding, or problems caused by labour bottlenecks.

Of equal importance to crop yield estimates are specifications of the inputs required to produce these, that is, the agricultural operations and their requirements in labour and purchased materials. The difference between good and less good land is often not so much in respect of yields as of inputs.

A convenient method of approach is first to find out the inputs necessary for the best land in the area, i.e. almost level, well-drained soils of high fertility; then to determine the additional inputs necessary on each type of poorer land, such as heavier fertilizer applications, soil conservation works or drainage works. Data on labour requirements may be available from *unit farms.* These are smallholdings established by the Department of Agriculture, on which all farming operations are carried out by a local farmer and his family but with a full record kept of labour, costs and output (Jolly, 1952, 1955; Cox *et al.*, 1972).

Where types of land use other than arable farming appear possible, comparable data on these is obtained. For forestry plantations the main source is visits to growth plots of the government Forestry Department. These are rated according to *site index*, a measure of tree growth from which estimates of output can be derived. As with agricultural experiment sites, it is not necessary that the growth plots should be within the area of the soil survey; a comparison can be made between soil and land factors on growth plots outside the area and the conditions found on potential forestry land within it. Information for potential output from pastoral types of use is only indirectly related to soils, and where it is likely to be an important form of production there may be a separate ecological survey for this purpose.

Soil response

It is more difficult to specify procedures for estimating soil response to changes in land use, as this technique has been little studied. For non-irrigated farming, the main prediction needed is that of organic matter levels, with consequences for soil structure, erosion resistance and fertility. One simple and practicable method is to compare organic matter levels on cultivated and uncultivated land. When representative profile pits are sampled, the land use history of the site is sought by enquiry – specifically, if cultivated, when it was last under fallow, or *vice versa*, and when it was first taken into cultivation from bush. In the case of pits on cultivated land, a composite topsoil sample is taken from a nearby and pedologically similar site under mature vegetation. This data enables an estimate to be made of the relative soil organic matter status under differing intensities of cultivation. It is also possible that data of this type may be available from agronomic experiments, or can be obtained by taking soil samples from such sites, since the cropping history on such sites is known.

Where development plans involve a change from natural vegetation to cultivation there may be changes in the soil moisture regime (Moss and Morgan, 1970). More substantial soil changes follow the introduction of irrigation, including changes in organic matter and possibly salinity under the altered moisture regime.

Soil survey will normally include outline recommendations for erosion protection measures, particularly soil series requiring special attention in this respect. They should also contain an account of the extent and severity of existing erosion. If circumstances require there may be a recommendation for a more comprehensive survey of erosion in part of the area to be undertaken as a special exercise. For example, a resource inventory of part of northern Nigeria includes a separate account of the erosion problem caused by over-grazing on high altitude grasslands, with recommendations for immediate control measures (Bawden and Tuley, 1967).

THE ALLOCATION OF TIME IN FIELD SURVEY

It has been customary in the past to spend a rather short time on reconnaissance, upwards of 65 percent of the total field survey time on soil mapping, some 10–15 percent on representative profile description, and to devote relatively little field time to activities directed towards evaluation. It is desirable that more time should be spent on both reconnaissance and evaluation activities, with a consequent reduction in the relative time spent on mapping, although this will remain the largest single activity. The proportions of field survey time spent on different activities varies considerably with the purpose and scale of survey and the type of landscape. For typical circumstances, say a feasibility survey at 1:100000 of an area of 500 km² in the savanna zone, an appropriate allocation of time might be as follows:

Survey reconnaissance	15 percent
Mapping	50 percent
Representative profile description	10 percent
Evaluation activities	25 percent

No review of soil survey in the tropics would be complete without mention of the aspect which in practice causes most problems, namely transport. Not much can be written about it, except to emphasize the considerable loss of time that can result from vehicle breakdowns and the consequent need to include regular and comprehensive servicing in the planned fieldwork schedule. Money spent on preventive maintenance, i.e. replacing worn parts before they collapse, is well spent. It is further desirable either that the soil surveyor himself should have some mechanical ability or that he should recruit a driver who has.

MAP AND REPORT PREPARATION

Maps

The soil survey report may be a document in its own right or part of a project plan. The maps and report are interdependent, the one an integral part of the other. In particular, part of the report consists essentially of expanded legends to the maps. The maps will normally include a soil or soil–landform map together with one or more evaluation maps, e.g. land capability or land suitability. General aspects of the cartography of thematic maps will not be considered, but certain points concerning soil maps, some of which have been discussed previously, may be noted.

The map scale will have been determined prior to the survey, but there may be pressures to reduce it in publication, and retain the original scale only as maps reproduced by a dieline process. These pressures should be resisted. Dieline maps have a local distribution and a short life. Economies in printing costs are possible by producing one good colour map, the basic soil map, together with a series of black-and-white maps, printed but using the same base and the same boundaries, with a different legend superimposed. It is preferable, however, that at least one land evaluation map should also be colour printed, giving it equal standing with the soil map.

The temptation towards complexity should be resisted. At the time of map preparation the field surveyor has fresh in his mind very many subtle, although real, distinctions between soil and land types. By no means all of these can ever be of practical use in planning and land management. Where there is doubt whether to split or lump together, it is better at this stage to do the latter, if necessary adding a short note in the report on the distinction that has been combined.

The legend to a soil map presents particular problems, arising from the fact that the same map has to serve for different types of users. It is therefore necessary to include both pedological information for the specialist and information in descriptive terms comprehensible to agriculturalists and other applied scientists. Fortunately the specialist information can be given briefly through use of technical terms, namely a soil series name and a taxonomic class. The latter will be in the national system of soil classification or in some foreign or international classification previously specified. The equivalents in the FAO classification should be shown on the map itself; a table of equivalents in other systems may be given in the report. The descriptive information should avoid the use of words known only to the specialist (e.g. argillic horizon) and should emphasize soil properties of two kinds: first, those that enable the non-specialist to recognize the soil in the field, especially colour and texture, and second, those that are of agricultural significance. Examples are as follows:

Lilongwe Series	Ferruginous soil (FAO Ferric luvisol)	A deep, dark reddish brown to dark red sandy clay or clay, with good structure and no limiting horizon; high fertility
Jalira Series	Weathered ferallitic soil with laterite (FAO Plinthic ferralsol, petric phase)	A yellowish brown sandy loam or sandy clay loam, structureless, with a layer of nodular laterite commencing at 30–100 cm depth; very low fertility

It is the usual practice when grouping the mapping units into classes in a higher category, and ordering them in the legend, to do so in terms of pedological classification, e.g. putting all soil series that are ferruginous soils together. It is worth considering the alternative of grouping together units of similar agricultural significance, e.g. good arable soils, arable soils of low to moderate potential, non-arable soils; regrouping on the basis of pedological taxonomy would then be shown as a table in the report.

Where compound mapping units are shown, the legend should if possible contain some information of proportions of the included soil series and their relative position in the landscape. Where it is difficult to include such information on the face of the map, it should be given in the expanded legend in the report.

375

Methods of soil survey

Non-cartographic presentation

Information on soil properties and land evaluation lends itself well to computer-based data handling. Each observation of each soil factor is stored with its locational co-ordinates. Maps can then be produced on demand, specifying the combinations of properties on which they are to be based and the classes to be mapped. Thus, a map showing topsoil texture, depth to a limiting horizon and nitrogen status could be called for. There are also possibilities for the analysis of data prior to printing. For example equations relating maize yield to climatic and soil parameters and to management practices could be obtained from agronomic data. These could then be applied to the survey area to produce maps of yield estimates under various specified management practices; means and confidence limits could be given, taking into account climatic variability, soil variability (i.e. variation of properties over small areas) and the error of estimate in the equations. If, subsequently, some additional property were shown to be limiting yields, revised maps incorporating this could be produced.

This method supplements rather than replaces maps of the conventional type. It will not always be practicable, besides which there is the danger of creating a spurious impression of precision. It is probably of most value as a means of running a series of data-processing trials, with rapid production of a corresponding series of maps, only one or a few of which will be selected for subsequent printing and use in planning.

The soil survey report

The form of soil survey reports has become fairly standardized and calls for no substantial modification. Assuming the soil survey report to be an independent document, rather than a section of a project report, the arrangement of chapters is approximately as follows:

INTRODUCTION. Including origin of the survey, its aims and terms of reference.

SUMMARY OF CONCLUSIONS AND RECOMMENDATIONS.

Part I. Environment

GENERAL DESCRIPTION OF THE AREA. Including location, climate, main relief features, drainage, vegetation, land use, and a brief account of population and agriculture.

Part II. Soils

METHODS OF SURVEY.

LANDFORMS. A more detailed account than that in Part I, including features of significance to land use potential.

SOIL TYPES. General properties, classification, main soil types, properties of agricultural significance.

SOIL MAPPING UNITS. An expanded legend to the soil or soil–landform map.

Part III. Land evaluation

SUITABILITY FOR MAJOR KINDS OF LAND USE. Including land capability or land suitability classifications.

POTENTIAL PRODUCTIVITY. Estimates of output under specified management conditions from agriculture and other kinds of land use.

Appendices

PEDOLOGY. Technical discussion of soil properties, genesis, classification and correlation.

SOIL TYPES. Morphological descriptions and analytical data.

LAND EVALUATION. Technical discussion of the derivation of production estimates and inputs.

Where the soil survey forms an integral part of a project report, part of the material in the general description of the area may be given elsewhere, as more detailed chapters on climate, hydrology, vegetation and present land use.

Different parts of the report are directed at different classes of reader and should be written in appropriate style. The Summary of conclusions and recommendations is intended for the administrator, planner or politician, possibly at a high level in government, who wishes to know how the recommendations could affect national or regional development planning. This should therefore be written in simple language, and confine itself to giving one or a small number of alternative recommendations without outline reasons, omitting or making only brief reference to alternatives discussed in the text but rejected. The main body of the text, Parts I–III in the above outline, is intended in part for other soil scientists but also for agriculturalists, economists, planners and administrators who are closely concerned with the area, and who may have no specialist knowledge of soil science but have time to read and absorb the detail of the report. It should therefore set out the data and reasoning fully but without unnecessary use of technical terms, omitting material that is not directly relevant to the conclusions reached, including landform and soil genesis and details of methods used; for example the results of statistical calculations (e.g. of confidence limits) will be given but not the basis for calculation. The Appendices are intended first for professional soil scientists, and second as a source of reference on technical aspects of the report. Project managers are usually kind enough to allow the soil surveyor a few pages of discussion on soil genesis or other matters of academic interest.

Matters of arrangement, style and presentation are discussed in more detail in the FAO handbook *The preparation of soil survey reports* (Smyth, 1970).

SOIL SURVEY FOR IRRIGATION PROJECTS

In a feasibility survey for a proposed irrigation project there are two fundamental questions to be answered prior to economic analysis: how much water is available, and how much land? The former is answered by studies of river discharge, the storage capacity of potential dam sites or rates of water yield from underground aquifers, the latter by soil survey. Having estimated the water duty, i.e. the depth and hence volume per unit area irrigated, and the probable water losses in storage, transmission and distribution, the water required to irrigate the land available can be calculated. Comparison of these two sets of studies will show whether the potential size of the scheme is limited by the amount of water available or the amount of irrigable land. On extensive alluvial plains water is usually the limiting factor on size, but where there are valley floors of limited width amid erosional relief the limit may be set by the amount of suitable land.

Many of the approaches and methods of normal soil survey are applicable in irrigation investigations, but certain aspects are given special emphasis:

1. Field survey is usually confined to relatively flat land, often to depositional relief.

2. Soil properties related to the control of water receive particular attention.

3. The proposed change in moisture regime requires prediction of the soil response to this changed environment.

4. Investment costs are considerably larger than in other forms of agricultural development, therefore particular attention is given to providing soil survey information in a form suitable for economic analysis.

In most irrigation schemes in less developed countries the water distribution is by gravity irrigation, the flooding of flat fields enclosed by bunds. Where this method is intended, land sloping at more than some given angle will be identified by air photograph interpretation and omitted from field survey; the angle varies with climate and economic circumstances, but is usually less than 6° (12 percent). Often, areas of erosional relief are identified and eliminated by photo-interpretation, and field survey confined to depositional landforms. In addition, an early estimate of the location of possible dam sites and their height may result in apparently suitable land being eliminated because it cannot be commanded by a potential irrigation supply. Where there proves to be much more sloping land than flat it may be necessary to recommend investigation of the more expensive method of sprinkler irrigation. This can happen, for example, in semi-arid regions where there are relict sand dunes. Soil survey is sometimes complemented by photogrammetric contouring at a close vertical interval, and it is desirable to schedule field soil survey so that these contour maps become available prior to the preparation of land evaluation maps.

The importance of the study of soil moisture characteristics becomes equal to or even greater than that of fertility. It requires determination of the following properties, either at sites of representative profiles or on a regularly-spaced grid: soil moisture retained at $\frac{1}{3}$ atmosphere ($\simeq$ field capacity) and 15 atmosphere ($\simeq$ wilting point) tensions, and hence calculation of the available water capacity of the profile, permeability (hydraulic conductivity), infiltration rate, and transmission capacity in depth. Infiltration rate is usually measured by the double cylinder infiltrometer method, tests with which occupy an appreciable part of field survey time (US Bureau of Reclamation, 1953). Hydraulic conductivity must exceed 0.1–0.2 cm/hr or there will be difficulties in obtaining an even distribution of water; very heavy textures or high exchangeable sodium percentages can cause unacceptably low values. Infiltration rates in excess of about 12 cm/hr coupled with a low available water capacity, both caused by very sandy textures, make the soil unsuitable for gravity methods and necessitate sprinkler irrigation. Deep borings, beyond the 2 m limit of normal soil survey observations, are necessary at intervals to check that the deeper layers of the regolith have adequate transmission capacity to drain away excess water.

Irrigation involves substantial changes to the soil itself and its environment. Grading (levelling) by bulldozer physically alters the profile; it should not be undertaken without consideration of the probable effects upon agricultural potential. The physical, chemical and biological properties of the soil will be altered by the change in moisture regime. In arid regions the increase in plant growth may improve soil organic matter levels. Of particular importance is to predict the salt balance, taking into account the present salinity and exchangeable sodium percentages of the soils and the salinity of the irrigation water. Thus prediction of soil change is a major element in soil survey for irrigation. In the fields of moisture and salinity estimation there are standard methods available (Thorne and Peterson, 1954; Richards, 1954; FAO, 1973*a*). Other aspects are not standardized, and soil inspection visits to areas of comparable land already under irrigation should be a regular part of the soil survey.

The financial loss involved in supplying water to land which subsequently proves unsuitable for irrigation is so high that survey on a detailed scale is justified. In developed countries surveys at 1:10000 or 1:5000 may be justifiable, but in less developed countries the size of schemes and proposed intensity of development is usually such that 1:25000 or 1:20000 is a sufficiently large scale.

The need to justify investment in terms, *inter alia*, of cost/benefit analysis means that the land evaluation data supplied by soil survey must permit calculation of management costs for each class of land as well as providing crop yield estimates. The system for comparing management costs is first to obtain costs of farming operations on the best class of land, and then specify the additional measures necessary on each type of poorer land, e.g. levelling, stone clearance, desalinization or more frequent water applications.

Methods and specifications for soil survey irrigation projects are discussed in detail in the manual of the US Bureau of Reclamation (1953), and an FAO manual is in preparation.

The cost-orientation of irrigation projects has a beneficial effect on soil surveys. The user requirements for information are more closely specified which gives a clear purpose-orientation to the survey, whilst the sums of money involved in investment make it more possible to justify funds for a thorough survey. Thoroughness tends to be viewed, however, as meaning survey on a large scale. What may be emphasized is that this is not the only requirement. The information necessary for a sound assessment of the land potential and hazards for irrigation requires to an equal degree a soil survey that is wide-ranging in its coverage of land evaluation and soil response.

SITE APPRAISAL

Government soil survey organizations receive frequent requests to carry out *ad hoc* surveys for local projects, e.g. farm plans or village land redevelopment schemes. Whilst not belittling the importance of such surveys – they start with the advantage of actually having been asked for by a potential user – they are undoubtedly an interruption to the main survey programme of resource inventory and major projects. In cases of small local surveys of this kind the information needed can often be supplied by a simplified form of soil survey known as *site appraisal*.

The information requested is usually a plan showing land suitable for cultivation and for pasture, swamps, hilly or rocky land, a provisional layout for roads and soil conservation works, and perhaps a central farm or services site. To produce such a plan does not require a soil survey in the formal sense of the term; it is not necessary, for example, to identify soil series, but merely to map 'arable land'.

In making a site appraisal, valley floors, swamps, hills and watersheds are first delineated on air photographs; approximate slope angle classes may also be mapped. A short field visit is then made, examining soils on what appears to be the arable land. A division into two or more classes (e.g. 'red loams, sandy soils') may or may not be necessary. It will often be possible to indicate which is the agricultural experimental site possessing the most similar soil and climate. Returning to the office, a brief re-appraisal is made and a farm plan (or other information as requested) drawn on the air photographs. A map is drawn by simple tracing, without photogrammetric control. A report of a few pages is produced, which will probably remain as an unpublished typescript. (A short note of the existence and location of the survey should be inserted in the published Department Annual Report.) It will sometimes be possible to identify the type of land present on the site as belonging to one described in an existing reconnaissance or other soil map. A farm plan, fully as satisfactory to the user as a map showing soil series, can sometimes be produced in this way in a week: 1 day's air photograph interpretation, 2 days of field survey, $1-1\frac{1}{2}$ days of map preparation and report writing, and a day spent in travelling.

It should perhaps be added that site appraisal is a skilled task, possibly more so than any other form of survey since it requires rapid identification both of development potential and, more important, potential hazards. It is not a job to be left to a revenue officer with an arts training and an urban bias.

SELECTED REFERENCES

Remote sensing. Soil Conservation Service (1966), Vink (1968), Carroll (1973).

Field survey. Beckett (1971), Beckett *et al.* (1971), Young (1973*a*), FAO (*Soil survey in irrigation investigations*, in preparation).

Map and report preparation. Smyth (1970).

CHAPTER 20

LAND EVALUATION

Land evaluation is the process of estimating the potential of land for one use or several alternative uses. The data employed come from three main sources: natural resource survey, the technology of resource use, and economics. The results are expressed as land evaluation maps with supporting statistical and other information, showing the suitability of various parts of the survey area for different kinds of land use. These include directly productive use, principally arable farming, pastoral use and forestry, together with other forms of use such as water catchments and tourism.

PLANNING ENVIRONMENTS

As with soil survey, there is no distinction in principle between land evaluation in developed and less developed countries, but there are differences in practice. These arise from differences between planning environments and purposes (Young, 1973*b*).

The major differences are between regions which are densely or sparsely settled, and countries with high or low standards of living. In densely settled developed countries, such as those of Western Europe, both land and labour are highly priced; most land has a potential for several alternative uses, and is already occupied by one of these. Planning is therefore primarily concerned with decisions between competing demands from different uses. In sparsely settled regions with high living standards, such as much of Canada and Australia, land development may involve the bringing of new land into productive use; labour is expensive and land development must compete for capital with other investment opportunities. In both these types of planning environment, development for productive use is usually privately financed; there are also substantial and growing demands, mainly from government sources, for land for non-productive uses such as recreation and nature conservation.

In developing countries, land and labour are less costly but capital is scarce. The critical feature of planning is therefore the allocation of investment to enterprises in which it will be most beneficial. Productive forms of use take clear precedence, the only substantial exception being land for game parks, which are non-productive only in a restricted sense. In sparsely settled regions, new land settlement is a major form of development. Such regions are, for example, extensive in Brazil and common (although shrinking) in tropical Africa. In densely settled regions, exemplified by India, most land is under some form of use and, except for the irrigation of arid areas, reorganization of land use on areas already occupied must be the main form of development. Features of land development planning in all less developed countries are the dominant, sometimes almost exclusive, role of government agencies, and the use of capital derived in part from foreign aid.

Many of the approaches and methods employed in land evaluation originated in developed countries. The earliest methods were developed in the United States, and recent advances have been made particularly in the Netherlands, where land is unusually scarce and highly-valued, and Canada. These methods have been modified for use in the planning environments of developing countries. They are now widely employed in land development at the project level, and to a lesser but growing extent in national and regional planning. The Food and Agriculture Organization of the United Nations is the international agency responsible for inter-governmental liason in the field of land development, as well as for conducting surveys itself. It has taken an important role in co-ordinating land evaluation methods, particularly terminology. An FAO 'Framework for land suitability evaluation' is in preparation, a draft version of which has been published (Brinkman and Smyth, 1973, pp. 55–101).

SOME DEFINITIONS

It is useful to preface discussion with some distinctions and definitions of terms referring to land itself, land use and the process of land evaluation. The formal definitions given in this chapter are for the most part those adopted by the FAO, in some cases modified.

There is a distinction between land and soil. *Land* comprises all the conditions of the physical environment, of which the *soil* is only one. The productivity of soils is sometimes assessed in isolation, but this can only be done by making tacit assumptions about climate and by allowing that slope angle is a property of soil. Evaluation is properly applied only to land. The term *terrain*, employed particularly in evaluation for engineering and military purposes, is sometimes used synonymously with land and sometimes to refer more particularly to landform and regolith properties.

Any specified area of the earth's land surface, at whatever level of generalization or category in a classification system, is termed a land mapping unit. The formal definition is as follows:

A *land mapping unit* is a specific area of the earth's surface; its characteristics embrace all reasonably stable, or predictably cyclic, attributes
of the biosphere vertically above and below this area including those
of the atmosphere, the soil and underlying geology, the hydrology,
the plant and animal populations and the results of past and present
human activity, to the extent that these attributes exert a significant
influence on present and future uses of the land by man.

Land units are defined and delineated by natural resource survey, particularly soil survey, and serve as the basis for land evaluation.

Evaluation is carried out with respect to specified forms of land use. The
expression *land use* may refer either to present or to possible future use;
where it is not clear from the context, *present land use* may be employed
with the former meaning. Other terms employed are:

Major kind of land use: one of the few different major alternatives of land
use, such as rainfed cultivation of annual crops, irrigated agriculture,
natural grassland, forestry plantations, or recreation.
Land utilization type: any lower categorical level in a classification or
specification of forms of land use.
Combined use refers to situations where two or more major kinds of land
use are involved in a single farming or other organizational system, as in
the inclusion of arable cropping and improved pastures in a single farm.
Multiple use refers to situations where two or more major kinds of land
use occupy the same land, as in the case of the use of forests not only for
timber but for tourism and wildlife conservation.

In combined use the two kinds of use occupy discrete and separate land
areas but are organizationally associated; in multiple use the various uses
simultaneously take place on the same piece of land.

Land classification is sometimes used as a synonym for land evaluation, but
more properly includes any method of grouping land, or elements of land,
into classes. The land systems method of resource survey is a form of land
classification but not of land evaluation (although evaluation may be applied
to the mapping units which it yields). *Land evaluation* is the process of
estimating the potential of land for one use or several alternative uses. Land
evaluation is therefore one branch of land classification, in which the basis
for classification is suitability for land use.

Land suitability is the fitness of a given tract of land for a defined land use.
Land capability is sometimes used with a closely similar meaning; however,
this term is so strongly associated with one evaluation system, that of the US
Soil Conservation Service, that it is better to employ it only with reference
to that system. *Land suitability classification* is the process of appraisal and

grouping of specific land mapping units according to their suitability for defined forms of use.

TYPES OF LAND EVALUATION

Land evaluation systems may be single-, multiple- or general-purpose; they may refer to current or potential suitability; and they may be qualitative or quantitative, physical or economic.

A *single-purpose classification* evaluates land for one specified purpose. This may be a major kind of land use, a particular crop or tree species, e.g. maize, *Pinus patula*, or some more complex farming or other land use system. A *multiple-purpose classification* is one in which a number of single-purpose classifications are combined according to stated principles. The term *general-purpose classification* is reserved for systems which directly compare capability for various land use alternatives without being constructed from single purpose systems.

Current land suitability is the potential of land in its present condition, with recurrent inputs (e.g. fertilizers) and minor land improvements (e.g. stone clearance) but without major improvements. *Potential land suitability* is the potential at some future date after major improvements have been carried out. *Major land improvements* are those which require a substantial non-recurrent input of capital, which can rarely be financed or executed by the individual farmer, and which will cause a significant and reasonably permanent change to the land characteristics. Examples are drainage, reclamation and, the most common case, irrigation. *Minor land improvements* can be financed by the individual farmer from his own resources or short-term loans and which cause no substantial permanent change. Examples are bush clearance, the hoeing of bunds for swamp rice, or simple soil conservation works.

Qualitative evaluation systems give the suitability of land in general physical terms. *Quantitative physical* systems specify the inputs and the production from the forms of land use under consideration. They involve relating environmental characteristics to the technology of land use; economic considerations are necessarily taken into account but as a general background only. *Economic* evaluation systems assess land suitability in terms of costs of inputs and value of production.

Qualitative classifications are always in physical terms only, and are usually employed in reconnaissance surveys. Economic classifications are necessarily quantitative and are only practicable in moderately detailed surveys. However, quantitative classifications need not necessarily be economic. One of the most useful types, especially in feasibility surveys, is quantitative physical evaluation, in which inputs are specified as, e.g. seed variety, amounts of fertilizer and man-days of labour, and output as crop yields or other measures of production.

THE ROLE OF LAND EVALUATION IN DEVELOPMENT PLANNING

The three-phase approach

In the standard approach to land development planning it is regarded as a process consisting of three phases: description, evaluation and development. The phase of *description* includes natural resource surveys, of which soil survey forms one, along with the collection of information on agriculture, economics and other aspects of the existing situation. In the *evaluation* or appraisal phase information from resource survey is combined with that from technology, such as crop requirements and agricultural methods, and expressed as productive potential. The *development* phase is concerned with the physical planning necessary to convert this potential into production.

These phases are to some extent successive in time. Information collected during one phase is passed on to, and serves as a basis for, the next. The completion of each phase is marked *inter alia* by the production of maps: soil, landform and other maps showing the physical environment at the conclusion of resource survey, land suitability maps following evaluation, and development plans at the conclusion of the project survey.

Where this three-phase approach is combined with different scales and intensities of survey a cyclic pattern can be envisaged, illustrated in an idealized form in table 36. A country or region conducts a survey at the reconnaissance scale, perhaps a land systems survey, for resource inventory purposes. The evaluation map derived from this survey is qualitative and shows suitability for major kinds of land use. Let it be assumed that such a survey is sufficient, with supplementary investigations, to identify one or more possible development projects. A feasibility survey of at least one such project is now set in hand. The resource surveys for this include a soil survey at the semi-detailed scale and, if it is an irrigation project, also thorough studies of hydrology. The land evaluation map from such a survey is produced initially in quantitative physical land suitability terms, since the timing of development is unknown and there is hence a substantial degree of uncertainty over costs and market prices; subsequently, but still within the feasibility survey, the physical suitability specifications are translated into current economic terms. Assuming that the project proves to be technically feasible and economically competitive, a decision to proceed with it may be made. In the final development survey the timing of implementation and production is known, thus removing some measure of uncertainty from costs and prices; the same quantitative physical evaluation that was produced

386

TABLE 36 *The place of land evaluation in an idealized sequence of development planning. Based in part on Brinkman and Smyth (1973)*

Purpose of survey	Phase
Resource inventory/Project location	1. Resource surveys, reconnaissance scale; including soil–landform or land system map
	2. Land evaluation: qualitative suitability for major kinds of land use
	3. Preparation of regional plan; identification of possible projects
Feasibility survey of possible project(s)	1. Resource surveys, semi-detailed scale; including soil map
	2. Land evaluation: (i) quantitative physical suitability for land utilization types; (ii) economic analysis
	3. Selection of project; funding and decision to proceed
Development survey of project	1. Resource surveys, detailed scale, or supplementary to those of feasibility survey.
	2. Land evaluation: suitability for defined land utilization types in economic terms
	3. Preparation of project plan

during the feasibility survey can be used as the basis for a revised economic evaluation using the most recent cost data.

Some alternative approaches

The three-phase approach, to the extent that it involves the separation of resource survey, evaluation and development planning into separate and largely successive phases, can be criticised on several grounds. One of these, the practice of treating evaluation as an office exercise commenced only after the completion of field soil survey, has been discussed in Chapter 19. Three other criticisms, and alternative approaches, may be considered; the first of these is principally concerned with methods of resource survey and the second with the place of resource survey in the development process, but inasmuch as both have a bearing upon land evaluation they are conveniently discussed in this context.

A. Contemporary functional relationships. One criticism of conventional methods is an argument that methods of resource survey based upon geomorphology lead to a static approach to environmental resources, and that this needs to be replaced by a dynamic approach based upon ecology. This criticism was made by Moss (1968, 1969), and his alternative proposal is termed that of contemporary functional relationships, or the biocenological approach. He starts with the proposition that in densely settled tropical areas the distinction between vegetation and land use is not meaningful. All

387

parts of the land are used for productive purposes of some kind, there is a large element of rotation between so-called 'natural vegetation' and crops, and this has everywhere caused greater or lesser modifications to both vegetation and soils. Every area of land consists of a geo-ecosystem, components of which include the landforms, soils, soil moisture, vegetation, and the cropping pattern over a period of years. Each of these components affects the other; for example there are multiple and reciprocal relations between vegetation/land use, soil moisture, soil organic matter and nutrient levels. Many components of the geo-ecosystem are in a state of change, cyclic or permanent. For example soil organic matter and nutrient levels vary cyclically between fallow and cropping periods; there may be insufficient replanting of a perennial crop, so that the tree population is ageing. Some of the changes involve interactions between economic and ecological systems; failure to replant a tree crop may be caused either by attack from insect parasites or by a fall in the market price.

Moss argues that by taking geomorphology as the basis for natural resource survey, as is done in the land systems method and in much soil mapping, a static approach is engendered since landforms change only very slowly over time. The result is a picture of the environment which sees it is offering a given, fixed quantity of resources, which may be made use of by man. In reality there is a continuous interaction between land use and resources. Hence a change in the type or intensity of land use, which is invariably a consequence of development projects, will lead to changes in the resources themselves. The way to take account of this interaction is to base resource survey on the approach of ecology, viewing land/man relations as functioning systems.

This criticism draws attention to some limitations of treating resource potential as static. The three-phase approach tends to encourage such a treatment, since if carried to its logical extreme it requires the 'handing over' of a fixed quantity of data on the environment, for example soil and vegetation maps, for persons other than natural scientists to do with what they wish in development planning.

Moss (1968) presents a convincing case for the area of south-west Nigeria on the basis of which his arguments were developed. What is not yet clear is how the approach of contemporary functional relationships can be translated into practical survey procedures, capable of relatively routine application to a variety of regions. The most practical step is that resource surveys should seek to predict the effects of various alternative land use changes upon resources; in the case of soils this requires prediction of moisture regime, organic matter status, nutrient levels and erosion hazard. What is made clear is that the resource survey phase cannot be regarded as complete when the primary sets of maps and descriptive environmental information are produced; the ultimate planning decisions will have consequences for the environment, assessment of the effects of which should be one of the considerations taken into account in making such decisions. This necessarily means that the role of resource survey, including soil survey, continues in some measure into the planning phase of development.

B. Economics first. The second challenge to conventional methods is in effect a suggestion that the sequence description – evaluation – development should be reversed. This was made by Davidson (1965), specifically as a criticism of land systems surveys carried out in northern and central Australia. The information presented by these surveys (CSIRO 1953–75) consists largely of descriptive accounts of the physical environment, with only relatively brief evaluation sections and no economic analysis. Davidson argued that most of the lands surveyed could be shown to have little or no potential for development on an economic basis; any money invested in land development would be more profitably spent on the already settled areas in the southern and eastern parts of the continent. The surveys were therefor collecting information which was of no use as a guide to investment, and were therefore a waste of government money.

The alternative approach which Davidson suggested is essentially as follows. Those crops or other forms of production which gave a sufficiently high return from capital to compete with other forms of investment would first be determined, by considering market prices of crops in relation to costs of production. This analysis would lead to the conclusion that certain types of land development would repay investment, given the present state of technology and market prices, provided that yields were above a certain figure whilst production costs remained below a corresponding level. The environmental conditions requisite for such levels of output and costs could then be determined; location and transport costs would play an important part in such analysis. Finally the specific conditions required could be sought in the field, with a great saving of field survey costs. In theory, nothing but 'useful' information would be collected.

Taken literally, this 'backwards' or economics-first approach would have disadvantages at least equal to those of the conventional approach carried to a similar extreme. Natural scientists would find themselves in a position of servitude, with no constructive contribution to make as to how the complex and varied resource potential of the environment might be put to use. Much resource data of relevance to development would be missed through the economic blinkers worn. What can be learnt from consideration of the economics-first approach, however, are some disadvantages of proceeding in an equally rigid way with the conventional, 'forwards', approach, without any feedback from economics to resource description. Natural resource surveys are prone to collect data which is of little use in evaluation and almost none in development, such as the detailed differentiation of soil

389

types on land which are clearly non-arable. The resolution of the disadvantages of either of these procedures when followed rigidly is that there should be greater communication between natural scientists and economists at all stages of a project survey.

C. Matching land use types with environment. This approach, proposed by Beek and Bennema (1974; also Beek, 1974) has been advanced not as a criticism of existing practices but as a positive contribution to methods of land evaluation. It is based upon a view of land use types which regards these as something much more than simply a matter of what is done with the land itself, e.g. maize cultivation. A land utilization type is regarded as 'a technical organization unit in a specific socio-institutional setting'. The key attributes by which land utilization types can be identified are:

A. *Biological*	1. Produce (e.g. crops, timber, beef)
B. *Socio-economic*	2. Land tenure system (legal status)
	3. Size of farms
	4. Labour intensity
	5. Capital intensity
	6. Level of technical knowledge
	7. Income level
C. *Technical*	8. Farm power (power source, implements)
	9. Technology

Given this view of land utilization types as organizational units defined in some detail, the process of land evaluation is regarded as one of 'matching' environmental resources with land use types. Matching is an iterative process. The results of resource surveys lead to a first provisional description of land utilization types which might be suited to them. The land utilization types so defined are compared in suitability terms with the land mapping units identified from resource survey. The provisional description of land utilization types is then modified in the light of this suitability, in such a way that it becomes more closely matched with the environmental potential. The end result of this matching process should be a series of land utilization types, defined in some detail, which are the most highly suited to each part of the survey area.

There are thus two inputs to the process of land evaluation: the results of natural resource surveys and the specifications of relevant land utilization types. The matching of the two may be systematized by means of a conversion table. From the land mapping units defined by resource survey, land qualities are obtained (e.g. moisture availability, rooting conditions); conversely, from the specifications of the land utilization types, the land qualities required for their successful functioning are derived. The one set of land qualities is compared with the other, resulting not only in land suitability classes but also management specifications for each land mapping unit. The improvement capacities of the land and their practicability can be incorporated into the system (Beek and Bennema, 1974, fig. 5).

The basic concept behind this approach is that land can only be related according to its suitability for a specified purpose; and therefore the purpose, namely the land utilization type, must be specified just as closely as are the environmental conditions of the land.

Requirements of a land evaluation system

There is something to be learnt from each of the above approaches. Resource survey should be concerned not only with relatively stable and static aspects of land but with processes of interaction, both within the environment and between environment and land use; the information which it provides should include 'what is happening' as well as 'what is there'. Development planning should consider the possible effects of proposed changes in land use upon the environment. Land evaluation should be carried out with respect to forms of production that are economically realistic, if not necessarily completely viable under present circumstances. Resource surveys should not spend undue time in describing land for which no form of productive use can be foreseen. Land suitability can only be evaluated if the intended forms of land use are specified.

A further aspect worthy of emphasis is that land suitability for a given form of use is as much a matter of inputs (or costs) as of production. Land alone has no productive potential: even a hunting and gathering economy requires a labour input. Moreover, the degree to which land suitability varies as between different areas is as much a matter of costs as of output. This point has been made above with respect to the soil requirements for individual crops. A given crop, e.g. bananas, can be grown under a wide range of environmental conditions. Under fairly unfavourable circumstances, say on steeply sloping land with sandy soil, it may still be possible to obtain yields one half or more of those reached under optimal conditions, but the costs of inputs, such a terracing and heavy fertilizer applications, could well be five times as high. Land becomes marginal for a particular crop in part because the limits of tolerance are exceeded, but in part also because of the rise in the farm costs necessary to obtain satisfactory yields. The same reasoning can be applied to other major kinds of land use.

Putting together some positive aspects of the above approaches, and adding consideration of the role of evaluation in development planning, a number of requirements or *desiderata* for a land evaluation system may be suggested. These are as follows (based in part on Brinkman and Smyth, 1973):

1. The system should evaluate land for specified forms of use, defined as closely as the intensity of the study requires.

391

2. The land use alternatives considered should be those which are not only physically possible but also economically and socially relevant.

3. Evaluation should take into account both the production, or other benefits, from each land use alternative, and the inputs or costs necessary to achieve this production.

4. The effects of the land use alternatives on the environment, particularly possible adverse effects (hazards), should be considered.

5. The evaluation system should permit interpretation in stages, according to different purposes and intensities of survey.

6. The system should be versatile, capable of adaptation to a wide variety of circumstances, both environmental and economic. For example, it should permit adaptation both to smallholder farming and to estate or other large-scale forms of agriculture.

7. The results of the evaluation should have a degree of permanence appropriate to its expected application; this means not only the period over which it will be consulted for planning purposes, but the anticipated duration of the planned changes in land use to which it refers. The results should not be unduly sensitive to short-term economic fluctuations (p. 415).

8. At the same time the system itself should be flexible, permitting periodic revision, for example with changing technology or with substantial and reasonably permanent changes in economic conditions.

9. The final results of the evaluation are going to be read by economists, planners and those responsible for administering foreign aid. They should therefore be presented in terms which are simple, capable of being understood by the non-specialist, whatever may have been the complexity of the processes which led up to them. The presentation should be in terms which inspire confidence in government agencies and investment institutions.

SOME SYSTEMS OF LAND EVALUATION

The USDA Land-Capability Classification

The most widely used evaluation system is the Land-Capability Classification of the Soil Conservation Service of the US Department of Agriculture (Klingebiel and Montgomery, 1961). Although constructed for the United States it has been widely adapted for use in less developed countries. Maps based on this system can be recognized by the use of Roman numerals for land classes.

The system is intended as a means of grouping soil mapping units. Other features of land, namely slope angle, climate and frequency of flooding, are taken into account. The main concept used is that of *limitations*, land characteristics which adversely affect land use. *Permanent limitations* are those which cannot be changed, at least by minor land improvements; these include slope angle, soil depth and climate. *Temporary limitations* are those which can be ameliorated by land management; examples are soil nutrient content and a minor degree of drainage impedance. Land is classified mainly on the basis of permanent limitations. The level of land management assumed is that 'within the ability of a majority of farmers', a pragmatic definition that enables the system to be modified according to farming standards in different countries.

There is a three-category classification:

(i) A *capability class* is a grouping of capability sub-classes that have the same relative degree of limitation or hazard. Classes are indicated by Roman numerals, the limitations to type of land use and risks of damage to the environment increasing from Class I to Class VIII. The following are abbreviated definitions of the capability classes:

Class I   Soils with few limitations that restrict their use.

Class II   Soils with some limitations that reduce the choice of plants or require moderate conservation practices.

Class III   Soils with severe limitations that reduce the choice of plants or require special conservation practices, or both.

Class IV   Soils with very severe limitations that restrict the choice of plants, require very careful management, or both.

Class V   Soils with little or no erosion hazard but with other limitations impractical to remove that limit their use largely to pasture, range, woodland, or wildlife food and cover. (In practice this class is mainly used for level valley-floor lands that are swampy or subject to frequent flooding.)

Class VI   Soils with very severe limitations that make them generally un-suited to cultivation and limit their use largely to pasture or range, woodland, or wildlife.

Class VII   Soils with very severe limitations that make them unsuited to cultivation and restrict their use largely to grazing, woodland, or wildlife.

Class VIII Soils and landforms with limitations that preclude their use for commercial plant production and restrict it to recreation, wild-life, water supply or esthetic purposes.

(ii) A *capability sub-class* is a grouping of capability units that have the *same kinds* of limitation or hazard. These kinds are indicated by lower-case letter subscripts, of which the original system gives four: erosion hazard (e), excess water (w), soil root-zone limitations (s) and climatic limitations (c). Later adaptations of the system employ additional kinds of limitations, e.g. stoniness, salinity. The meaning of sub-class letters changes according to the class to which they are attached; thus the 'e' in Sub-class IIIe indicates a more severe erosion hazard than that in IIe.

(iii) A *capability unit* is a grouping of soil mapping units that have the same potential, limitations and management responses. Units are shown by Arabic numbers, as IIIe-1, IIIe-2. All soils within a capability unit can be used for similar crops, require similar management practices and soil conservation measures, and have a comparable productive potential. With equal management, the yield range within a unit is not expected to exceed 25 percent.

The usual manner of applying the system to a local region is to draw up a conversion table that shows kinds of limitation down the left-hand column and capability classes across the top. For each kind of limitation, the value of the land characteristic judged appropriate to each degree of severity of limitation is filled in. Thus in table 37 the maximum slope angle permitted for Class I land is 1°, for Class II 3°, for Class III 5° and so on. Only in the case of soil depth does the original system suggest limits to be used. Land belongs to the lowest class to which it is allocated by this procedure; thus level and freely-drained land, which would fall into Class I on the basis of slope angle and wetness, would be put into Class IV, Sub-class IVs, if the soil depth were less than 30 cm. The original system does not consider interaction between different kinds of limitation, but that can be allowed for to a limited extent by subdividing the class limits in the conversion table; it is frequently necessary to give. different slope angle limits for sandy and heavy-textured soils, e.g. the limit for Class III is put at 5° for heavy-textured soils but 10 ° for sandy soils. Features of location, such as distance to markets, are explicitly left out of consideration. No account is taken of the scarcity value of a particular type of land in a given location.

This is a qualitative, general-purpose land evaluation system for current suitability. At the class level it results in a single ordering of the relative value of land, with a major distinction between cultivable and non-cultivable land between Classes IV and V. It assumes a decreasing order of value from arable use through grazing and forestry to recreation, wildlife conservation and water-catchment uses. Referring to the desiderata listed on pp. 391–2, the USDA system is flexible, extremely versatile, places stress on possible adverse effects to the environment, permits interpretation in stages, and has the merit of simplicity. It is perhaps this last feature above all others that accounts for its popularity in less developed countries; to have an (apparently) clear separation of cultivable from non-cultivable land, and to be able to make such statements as 'Project X contains 35000 hectares of Class I and II land' are features of much value in communicating the results of a survey to administrators, politicians and aid-giving agencies. The versatility of the system lies in the fact that the limiting values for sub-classes can (and must) be specified differently for each survey area. Thus, in a region where the rainfall is marginal for rainfed agriculture, a soil moisture-holding capacity limitation can be set up, with sub-class limits given either as

TABLE 37 A simplified example of a conversion table for use with the USDA Land-Capability Classification. To apply the table to a land unit, examine the columns successively from left to right until a column is found in which the values given are not exceeded for any limitation. Otherwise favourable land with a severe wetness limitation is placed in Class Vw

Limitation	Arable classes				Non-arable classes			Special class
	I	II	III	IV	VI	VII	VIII	Vw
Slope angle, degrees	1	3	5	10	18	35	Any	2
Outcrops and boulders, percent surface occupied	0	1	2	5	10	25	Any	
Wetness, class	Nil	Nil	Slight	Slight	Moderate	Moderate	Severe	Severe
Soil effective depth, cm	150	100	60	30	20	20	0	30
Soil texture	SCL-C	SL-C	SL-C	LS-C	LS-Heavy C	LS-Heavy C	Any	LS-Heavy C
Soil permeability	Moderate	Rapid-Slow	Rapid-Slow	Rapid-Slow	Any	Any	Any	Rapid-Slow
Available water capacity, cm	25	20	15	10	5	2	0	10
Cation exchange capacity, subsoil, m.e./100 g	20	15	10	5	5	2	0	5
Total soluble salts, percent	0.2	0.2	0.4	0.4	0.8	1.0	Any	0.4

available water capacity or indirectly as soil texture. There is a further advantage in the situation of less developed countries in that once a national system (that is, conversion table) has been set up by skilled staff, the table can be used in a relatively routine manner by less highly trained local staff.

Being a general-purpose system, it does not take adequate account of the requirements of major kinds of land use other than arable; thus land which may be very highly suited to grazing could fall into Class VI. There are both objections in principle and difficulties in practice to the use of limiting values determined from a single land characteristic, in that the effects of an individual characteristic vary according to its interaction with others. Thus it is impossible to say that in all environments, or for all crops, sandy textured soils are less valuable than clays or *vice versa*. The interpretation of the climatic limitation is difficult, since crop requirements vary so widely. The allocation of wetlands to Class V frequently causes problems in the tropics, since such lands may be excellently suited to padi cultivation. The system is not explicit on the extent to which economic considerations, especially costs, are to be taken into account.

Perhaps the major objections to the system are its negative nature, being based on limitations rather than positive potential; that it is considerably biased towards emphasis on the soil erosion hazard; and the fact that it does not take sufficient account of the differing requirements of different types of land use, nor is its structure such that it can easily be adapted to do so. Set against these drawbacks are the advantages of its simplicity and of the prestige that it already holds, which taken together will make it difficult to dislodge from the leading position it holds among systems for the evaluation of land for purposes other than irrigation.

The US Bureau of Reclamation Land Classification for Irrigation

The most widely used system where irrigation projects are being assessed is that of the US Bureau of Reclamation (1953). This classifies land specifically in terms of its suitability for irrigation. There are six *land classes*, abbreviated definitions of which are as follows (note that 'arable' is used with the meaning 'irrigable'):

Class 1. Arable. Lands that are highly suitable for irrigation, being capable of producing sustained and relatively high yields of a wide range of crops at reasonable cost. They are smooth lying with deep soils, open soil structure allowing easy penetration of roots yet good available moisture capacity, and free from harmful quantities of salts. These lands potentially have a relatively high payment capacity.

Class 2. Arable. Lands of moderate suitability for irrigation, being lower than Class 1 in productive capacity, adapted to a somewhat narrower range of crops, more expensive to prepare for irrigation or more costly to farm. These lands have an intermediate payment capacity.

Class 3. Arable. Lands that are suitable but approaching marginality for irrigation. They have substantial soil, topographic or drainage limitations. A greater risk is involved in farming these lands than Classes 1 and 2, but under proper management they are expected to have adequate payment capacity.

Class 4. Limited arable or special use. These lands may either have excessive deficiencies susceptible of correction at high cost, but are suitable for irrigation of high-value crops such as vegetables or fruits; or they may have excessive non-correctible deficiencies precluding arable use but permitting use as irrigated pasture or orchard. They are capable of supporting a farm family if operated in farms of sufficient size or in association with better lands. Class 4 lands may have a range of payment capacity, under intensive use, greater than that of arable classes.

Class 5. Non-arable. These lands are nonarable under existing conditions, but have a potential value sufficient to warrant segregation for special study; or their arability (*sic*) is dependent upon additional project construction. The designation of Class 5 is tentative and must be changed to the proper arable class or to Class 6 prior to completion of the land classification.

Class 6. Non-arable. These lands do not have sufficient payment capacity to warrant consideration for irrigation.

The manner of applying the system initially involves a conversion table similar to that of the USDA system, in which the permitted values of various land characteristics within each irrigation suitability class are set down. Being specifically for the one purpose, the problems of deciding upon such limits are simpler. The maps produced show each area rated according to a standard mapping symbol, e.g.

$$\frac{3std}{C22BX} u_2 f_2$$

in which 3 is the land class, s, t and d are soils, topographic and drainage deficiences, C is a type of land use (cultivated), the remaining lower symbols refer to productivity, land development costs, water requirement and drainability, and u and f refer to need for levelling and flooding hazard. This initially complex map is simplified by a prominent colouring scheme which refers to the land classes only and which leaves the non-arable Class 6 uncoloured.

The final allocation of land to a particular class is based on economic considerations, specifically payment capacity. This means the capacity of the land to provide an acceptable income for the farmer and at the same time to pay water charges sufficient to amortize the capital costs of development. This is thus a single-purpose economic evaluation system. The handbook in which the system is described sets out in some detail the survey procedures to be followed. It lays stress on the need to determine quantitatively the measures necessary for land development, e.g. bush clearance, levelling, ditching, drains, desalinization, and requires that each of these operations be costed. The initial and recurrent costs are set against the value of expected production, and the payment capacity calculated, necessarily making assumptions about discount rates. Classes 1 to 3 and 6 have progressively less payment capacity, falling below the acceptable level in the case of Class 6.

Considered with reference to the requirements listed above, the USBR system is specific, referring only to potential for irrigation and not intended for other forms of land use. It lays much stress on inputs and their costs, and it includes assessment of environmental hazards. It can be adapted to tropical as well as temperate regions, and to low-income countries as well as high. The system lacks flexibility; the specified procedures, in particular scale and intensity of survey, are rather rigid and often inappropriate for the lower intensity of much irrigation development in the tropics.

The survey procedures are also complex, although made to appear still more so by the characteristic American verbosity of the Manual. However, the system produces final results of simplicity, namely the areas of land in Classes 1–3 (irrigable), 4 (special use) and 6 (non-irrigable). The interpretation of Class 4 causes considerable difficulties in practice and it is sometimes omitted. The system is not designed for interpretation in stages, but only for detailed and intensive surveys. As with any economic system, the validity of the results can be altered by fluctuations in production costs or market prices.

The fact that survey procedures are specified has favoured use of the system when drawing up contracts for irrigation surveys. This feature, coupled with the prestige of the US Bureau of Reclamation, has led to its frequent adoption in less developed countries. The stress which it lays upon costs and payment capacity makes it well suited to irrigation projects, in which recovery of the considerable capital investment is a major consideration. It requires modification when applied to irrigation schemes which are of large size but relatively low value of output per unit area. Some of the economic principles of this system, particularly that of payment capacity, can be applied to methods for the economic evaluation of major kinds of land use other than irrigation.

Productivity indices

There have been several attempts at devising systems which provide a productivity index, or rating, by means of parametric methods. In these methods, the effects of individual land or soil characteristics are assessed individually and then arithmetically combined. Examples are the Storie Index developed in California (Edwards, 1970), a system devised for Hawaii (Nelson, 1963), methods for soil productivity rating proposed by Riquier *et al.* (1970) and Sys and Frankart (1971), both developed with particular reference to tropical conditions, and a soil irrigation rating for Antigua, West Indies, by Borden and Warkentin (1974). Most of these refer to soil rather than land, although some incorporate factors of slope and rainfall.

The general structure of these parametric systems is the same. The combined effect of individual factors is more nearly multiplicative than additive. In particular, a soil with several excellent qualities, say a deep and freely-drained profile with high nutrient content, can be rendered unproductive by one adverse property, e.g. the presence of soluble salts. The usual practice is therefore to estimate percentage ratings for each characteristic on a scale in which 100 percent represents optimal conditions, for example an effective profile depth over 150 cm, and 0 percent represents conditions sufficiently unfavourable to render the soil almost totally unproductive, say an effective depth of less than 20 cm. These individual ratings are then multiplied together. If all are 100 percent then the productivity index will be 100 percent. If any one individual rating is zero the index will also be zero. This latter effect is akin to the use of the limitation in the USDA system; the operation of the parametric method differs from the use of the limitation, however, in that the combined effect of several moderately adverse factors can produce an index lower than any single individual rating.

The productivity index of Riquier *et al.* (1970) may be taken as an example. The basic formula is

$$\text{Productivity Index} = H \times D \times P \times T \times N \quad or \quad S \times O \times A \times M$$

where the nine factors are respectively moisture, drainage, effective depth, texture/structure, base saturation *or* soluble salt concentration, organic matter content, cation exchange capacity/nature of clay, and mineral reserves. The conversion of soil characteristics to productivity rating is a two-stage process. First, each of the characteristics is classed on ordinal scales with a varying numbers of classes; for example effective depth is classed as P_1 = no or very thin soil cover, P_2 = soil < 30 cm deep, P_3 = 30–60 cm, P_4 = 60–90 cm, P_5 = 90–120 cm and P_6 = > 120 cm. Each of these classes is allocated a percentage rating, these ratings differing according to the type of land use under consideration. Thus, for crop growing, P_1 is 5 percent, P_2 20 percent, P_3 50 percent, P_4 80 percent, and both P_5 and P_6 100 percent, whereas for pasture P_1 is 20 percent and P_2 60 percent. Certain complications and interactions between factors are catered for by subsidiary tables. Methods are given for obtaining both a Productivity Index of the soil

399

in its present condition and a Potentiality Index indicating its potential productivity if improvements were carried out (e.g. irrigation, drainage, deep ploughing).

Sys and Frankart (1971), using a similarly-based system but with different factors and ratings, give examples of their index applied to 10 representative soils of the Congo, ranging from 97 percent on a 'hygroferrisol intergrade to brown tropical soil on basic rocks' (ferrisol/eutrophic brown soil intergrade?) to 11 percent on an 'arenoferralsol on Kalahari Sand' (arenosol or sandy ferrallitic soil?).

Disadvantages of systems of this type are their rigidity and restricted applicability. All of those cited are able to demonstrate good correlations between the Productivity Index and yields of a particular crop within the area for which the index was developed. As soon as one system is transferred to a substantially different climate the ratings require substantial alteration if they are not to give results greatly at variance with farming experience. All of them become complex as attempts are made to allow for interaction between factors, which the nature of a multiplicative system is unfitted to take into account. Indices of this type are more suited to assessing the productivity of soil than of land, and any one index can only do so within a restricted range of environments. They may have uses for research and possibly fiscal purposes, but are not well fitted to evaluation for the purpose of land development planning.

Land suitability evaluation

Land suitability evaluation is a type of land evaluation in which separate assessments are made of the suitability of land units for each of a number of different, defined forms of use.

An early example of this type is a land classification for use in New Guinea developed by Haantjens (1965c). This provides suitability ratings for four types of use: annual crops, tree crops, improved pastures and padi cultivation. The method of deriving these ratings is intermediate between the 'limitations' method and the parametric approach. Each of 14 environmental factors is given a rating from 0 to 6. Every rating is assessed according to its 'individual suitability' for each of the four types of land use. The 'overall suitability' is derived from the individual suitabilities by a modified summation process, in which a single very low rating or a combination of moderately low ratings can have similar effects on the overall suitability.

The outline to be described here is based on the *Framework for land suitability evaluation* of the FAO, as given in Brinkman and Smyth (1973) with subsequent modifications. It is not a complete evaluation system, but a framework around which national or local systems may be constructed.

Some of the definitions employed in the framework have been given above (pp. 383–4). The forms of land use which serve as the subject of evaluation are defined as closely as the purposes and intensity of the survey require. In reconnaissance surveys they may be major kinds of land use, e.g. perennial cropping, forestry. In more detailed surveys they may be individual crops or plant species, e.g. oil palm, *Pinus patula*; farming systems, e.g. smallholder mixed arable and dairy farming; or other land utilization types, e.g. national parks. It is essential that the forms of land use should be properly defined.

There is a classification structure with four categories:

Land suitability orders divide land according to whether it is *Suitable* or *Not suitable* for the defined use. They are identified by the capital letters S and N. Under certain restricted circumstances a *Conditionally suitable* phase of the Suitable order may be employed.

Land suitability classes indicate degrees of suitability within orders. They are identified by arabic numbers. Within the Suitable order any desired number of classes may be distinguished, for example:

S1 *Highly suitable.* Land having no significant limitations to the sustained application of the defined use.

S2 *Moderately suitable.* Land having limitations which will reduce production levels and/or increase costs, but which is physically and economically suitable for the defined use.

S3 *Marginally suitable.* Land having limitations which will reduce production levels and/or increase costs such that it is economically marginal for the defined use.

It is open to individual evaluation systems to have less or more classes with the Suitable order. Within the Not suitable order two classes are defined:

N1 *Currently not suitable.* Land having limitations which may be surmountable in time but which cannot be corrected with existing knowledge at presently acceptable costs, and which preclude successful sustained use in the defined manner.

N2 *Permanently not suitable.* Land having limitations so severe as to prevent any possibility of successful sustained use in the defined manner.

Land suitability sub-classes are divisions of classes distinguished by the nature of the limitations that have determined their classification. They are identified by lower-case letters with mnemonic significance, e.g. w = wetness limitation, t = topographic limitation. Subclasses are written thus: S2w, S2t, S2wt.

Land suitability units are divisions of sub-classes which differ in minor aspects of the management requirements. They are identified by arabic numbers in brackets, e.g. S2w(1), S2w(2).

The presentation is initially in the form of a table, in which the columns are forms of land use, the rows land units, and the boxes in the table indicate the suitability of each unit for each use (table 38, p. 411). Maps showing suitability for any selected use can be compiled from the table. The question of comparing suitability for different types of use is regarded as lying outside the terms of reference of a land evaluation, the function of which is to present physically possible and economically promising forms of land use and to provide information about their probable consequences, both beneficial and detrimental.

The FAO framework borrows certain features from both the USDA and USBR systems. The terms (capability/suitability) class, sub-class and unit have much the same meanings as in the USDA system, and in the two latter cases the same symbols. The framework does not specify how suitability is to be determined, but in practice systems based upon it are likely to use conversion tables which employ, in whole or in part, the concept of limitations. It differs radically from the USDA system in three respects: first, that suitability is assessed separately for each form of use; secondly, in that it makes use of land qualities (p. 414) in place of individual land characteristics; and thirdly, that it places emphasis on economic aspects. In this last respect it has borrowed the concept of payment capacity from the USBR system. The framework may be applied to qualitative or quantitative physical suitability evaluation or to economic evaluation, and to the assessment of either current or potential land suitability. When applied to the economic evaluation of a proposed irrigation scheme its approaches and methods become very similar to those of the USBR system, FAO Classes S1, S2, S3 and N2 corresponding to USBR Classes 1, 2, 3 and 6. Even when applied to qualitative evaluation, the framework clearly implies that land shall be classed as suitable only if the defined form of use is not only physically possible but also economically viable, in the circumstances prevailing in the country concerned. The USBR system is indeed fully compatible with the framework, whereas the USDA system, owing to its general-purpose nature, is not.

The framework was specifically designed to meet eight of the nine requirements on pp. 391–2. The only one of these in which it might be held wanting (and which is not a declared basis for it) is with respect to the degree of permanence of the results. By giving a clear priority to economic viability over physical land characteristics, both the class and order to which a given land unit belongs may change with movements of market prices. The most important contribution is the emphasis placed on the fact that different forms of land use have different requirements, and that consequently these uses must be defined and land suitability assessed separately for each. It remains to be seen whether this approach will succeed in gaining adoption in competition with the two main established systems.

REQUIREMENTS OF MAJOR KINDS OF LAND USE

Major kinds of land use

The concept of major kinds of land use is a loosely defined one, and the kinds of use distinguished must be partly a matter of convenience. In some circumstances it is useful to consider the rainfed cultivation of annual crops and of perennial crops as separate uses, in others to group them together as rainfed arable cultivation. Although rice is technically an annual crop its environmental requirements are so distinctive that in any evaluation system it is better treated separately. Irrigated agriculture may include the growing of annuals, perennials, and rice as well as fodder crops for grazing, but is best treated as a separate kind of use. Forms of land use based on grazing may be divided into those utilizing natural pastures (ranching) and improved pastures. Similarly there are two main forms of forestry, the logging of natural forests and forestry plantations; each may be either for softwoods or hardwoods, but this latter distinction does not warrant separation as a major form of use. Omitting land for urban expansion, mining, military purposes and specialized uses such as dam construction, the major kinds of rural land use are shown in the accompanying list.

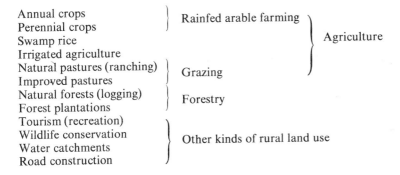

Annual crops	Rainfed arable farming	Agriculture
Perennial crops		
Swamp rice		
Irrigated agriculture		
Natural pastures (ranching)	Grazing	
Improved pastures		
Natural forests (logging)	Forestry	
Forest plantations		
Tourism (recreation)	Other kinds of rural land use	
Wildlife conservation		
Water catchments		
Road construction		

The apportionment of land between different kinds of use is a primary task of development planning, and the one for which land evaluation is mainly called upon to supply information. The basic decision is usually the allocation of land to arable farming, grazing and forestry. In proposed irrigation projects the decision is to determine the extent and location of land to be irrigated.

It is current orthodoxy (and also practical politics) to maintain that the role of land evaluation is to provide the information necessary to make such decisions; to take them is the responsibility of planners, administrators and, ultimately, politicians. Whilst there is a measure of truth in this attitude, it should not be interpreted to mean that the results of evaluation should be

403

'handed over', with no further concern over the decisions subsequently taken. As argued elsewhere with respect to soil survey, the relations between environment and proposed land use are not wholly one-way, and land evaluation has the potentiality for playing a continuing part in the planning and implementation processes.

Land requirements for arable farming

It is broadly true to say that landforms have a greater effect than soil characteristics in determining the suitability of land for arable use; both present land use patterns and the land use plans of most development projects bear more resemblance to maps of landforms than of soils. The main determinants of suitability are slope angle and site drainage. Slopes exceeding some particular angle, which varies with local circumstances from 6° to 18° or occasionally more (e.g. for tea growing), are unsuitable for cultivation in the tropics owing to the risk of erosion or the high cost of the conservation works necessary to prevent it. Valley floors and other poorly-drained sites are unsuitable for general arable use, although they may be employed for padi or for specialized cropping such as vegetables. Frequency of flooding is an additional hazard of such sites. A threefold division of land into steep slopes with thin soils, poorly-drained land, and 'arable land' is the starting-point of land use planning.

The soil requirements for arable farming have been partly considered above under the heading 'Conditions for high soil fertility' (p. 300). When deciding whether land is to be cultivated or not it is the presence of conditions causing low fertility which is important. The adverse conditions, or soil limitations, fall broadly into two types. The first comprises limitations which, if any single one of them is present in severe form, render the soil non-arable; these are shallowness, stoniness, soluble salts and sandiness. The second type comprises limitations which individually do not suffice to render a soil non-arable but which may do so by acting cumulatively; these include heavy intractable clays, massive or very coarse structure, low cation exchange capacity, low organic matter content and low nutrient content.

The climatic factor is usually crop-specific rather than applying to arable use as such. An important exception occurs near the drier limit for rainfed annual cropping, around 500–600 mm mean annual rainfall, where the combined effects of rainfall and available water capacity (the latter determined mainly by texture and depth) can tip the balance of physiological moisture availability between arable and non-arable land.

It may be added that the climatic factor is difficult to take account of in all evaluation systems. This is because climatic differences of sufficient magnitude to affect land potential occur over much larger distances than those of landform and soil variation. Climatic differentiation becomes important only in evaluation of large areas at small scales; in many semi-detailed and detailed surveys the climatic conditions show no variation of significance to land use within the project area, and are treated as a uniform background.

Land requirements for different kinds of arable farming have been discussed above as follows: annual crops, p. 306; perennial crops, p. 319; swamp rice, p. 230, irrigated agriculture, pp. 200, 378.

Land requirements for grazing

Three of the main restrictions on arable farming, steep slopes, shallow soils and poor drainage, have only minor adverse effects on use for pasture. Permanent grassland is satisfactory from a soil conservation viewpoint on steep slopes, provided it is not a fire-induced sub-climax replacing natural woodland. Although many savanna grasses are deep-rooted there are species adapted to shallow soils, and natural ecological competition ensures that these will be dominant upon them. Thus, hillsides with outcrops or boulders, moderately stony soils and a profile depth fluctuating between 0 and 50 cm may still have a moderate potential for wet-season grazing.

Hydromorphic grasslands of valley floors are the richest natural grasslands in the tropics. The species within them are adapted to both poor drainage and clayey soils. In the savanna zone these are the only sites in which grass growth continues during the dry season, and such is their value for grazing at that time that they should be given a substantial period of rest during the wet season, a management practice which is unfortunately rare.

Ranching may be a means of making use of land which, whilst topographically suited to cultivation, has moderately shallow, sandy, massive soils of low fertility.

Good arable soils are also good soils for sown pastures. They should be shown as 'highly suitable' for this purpose on land evaluation maps, notwithstanding the prior claim of arable use. This evaluation might make some contribution towards encouraging the development of rotational systems incorporating grass leys, which are urgently needed if high fertility is to be maintained on such soils.

Land requirements for forestry

Forest plantations are an important form of land use in the savanna zone, whilst the main potential for the utilization of natural forests lies in the rainforest zone.

(a) Forest plantations. Forestry is apt to be treated as the Cinderella of land use allocation, being given what no one else wants. Owing to the long period of growth it usually shows up unfavourably in cost/benefit analysis. Like other plants, trees grow best on land with deep soils of high fertility, but there is little chance that such land will be permitted to remain as forest reserve. Small patches of bluegum (*Eucalyptus grandis*) or fast-growing trees such as *Gmeligna arborea* may be included on farms on good land, as sources of firewood and poles.

Slope angle has little or no effect on tree growth. It does, however, have a considerable effect on costs of forest road construction and maintenance costs. Steep, bouldery slopes, especially if landsliding is frequent, may be sufficient to render land unsuitable for forestry. The consequences of shallow soils are not as serious in the tropics as in temperate lands. Wind throw, a major hazard in the temperate zone, is relatively unimportant. Hill-slope soils that are regarded as shallow for agricultural purposes in the tropics generally contain zones of gravelly soil and weathered rock along joints down which tree roots find their way.

Poor drainage is a serious limitation for forestry. There are few good timber species that tolerate waterlogging, and artificial drainage is unlikely to be economic.

As with crops, there are trees covering a wide range of climatic requirements, and commercial species exist which grow satisfactorily in all but the arid zone. A feature of special interest is that some conifers, notably a species of Mexican origin, *Pinus patula*, grow excellently at high altitudes in the tropics. Forest plantations are a potentially valuable source of production on many high-altitude plateaux.

(b) Utilization of natural forests. The assessment of suitability for timber production from natural forests is based largely on forest inventory and only to a minor extent on landform and soil surveys. The main determinant of suitability is the volume of merchantable timber species present. Rain-forest trees grow indifferently across plains, hills and sometimes swamps, and quite refined statistical treatment may be necessary to differentiate species frequency differences between soil types. The second determinant of suitability is extraction costs, which are adversely affectly by dissected relief, steep slopes and swamps.

The use of natural woodland in the savanna zone is a low-intensity activity owing to the slow rates of growth. It is compatible with ranching as a form of multiple use.

Requirements of other kinds of rural land use

The forms of rural land use other than agricultural and forestry might collectively be called 'non-productive uses', but as this is only true in a narrow sense of the word, and would certainly lead to misunderstanding if employed in an evaluation system, these are better grouped by some non-committal term.

The use that is called recreation in the Western world is better styled tourism in less developed countries, since in the foreseeable future it is only foreign tourism that is likely to be given weight in allotting land use priorities. Suitability evaluation for tourism, wildlife conservation (flora and fauna) and water catchments requires different criteria and to some extent a different approach to that used for agriculture and forestry. Factors other than soils take a prominent role. Steep and rocky slopes, instead of acting as severe limitations may take on a positive character for their scenic value, and inaccessibility becomes a virtue for wildlife conservation. A more fundamental difference is that individual land units cannot be assessed; an area of land acquires suitability as a national park or conservation area by virtue of the overall assemblage of features. Land value cannot entirely be converted into economic terms. There are many possibilities of multiple use.

Road construction is of a different nature to the other uses listed. Whilst the relative suitability of the entire land surface for this purpose can be assessed, in both physical and economic terms, the location of land in relation to the desired route connections is a major consideration. Only a small proportion of the 'suitable' land is eventually taken up by road, together of necessity with short lengths of inferior land. By means of cost/benefit analysis, the concept of repayment capacity is applicable to road suitability evaluation. As well as construction costs, maintenance costs should be included; these should be estimated separately for different land units since they can vary widely. Criteria to be taken into consideration in land evaluation for roads have been listed above (p. 349).

AN EXAMPLE

The test of any evaluation system ultimately lies not in whether it appears sound from a theoretical or logical viewpoint but in what happens when it is applied in practice. The principal questions are firstly, can the information required to carry out the evaluation be obtained? Secondly, and more important, are the results produced in accord with judgement and experience? Thus if a system produces the result that a tract of land is 'highly suitable' for annual crops, this ought to be in accord with the opinion that an experienced farmer in the area would express.

The systems described may be applied to an idealized sample area with five land mapping units, based upon a part of central Malaŵi. Fig. 25 shows the units brought together in semi-diagrammatic form. The real conditions have been modified slightly, to illustrate the operation of the evaluation systems, by assuming that units 1 and 3 have identical relief but different soils whilst units 3 and 4 have identical soils but different relief. The environmental conditions assumed are as follows:

Climate (all land mapping units). Savanna type with a single wet season and 7 months <50 mm; Köppen Aw. Mean annual rainfall 900 mm, mean annual temperature 18 °C. Monthly potential evapotranspiration approximately 150 mm, with a moisture surplus for 4 months.

Altitude (all units). *c.* 1200 m.

Geology and soil parent materials. Basement Complex metamorphic rocks. Land units 1 and 2, hornblende–biotite gneiss, of intermediate composition. Land units 3–5, perthosite gneiss of felsic composition.

Unit 1. Gently undulating plain, slopes 0 °–3 °, relative relief up to 50 m, valley width 1–3 km, convex–concave slopes. Ferruginous soil, dark reddish brown sandy clay, moderately well developed structure, over 2 m deep, pH *c.* 5.5, base saturation *c.* 70 percent, good rooting conditions, moderate nutrient content, moderately high fertility. Free site and profile drainage. Probably former *Piliostigma–Acacia* savanna woodland, at present entirely under cultivation.

Unit 2. Broad, gently-concave valley floors amid unit 1 (*dambos*), slopes <1 °. Gley soils, black heavy clays with strongly developed but very coarse prismatic structure, over 2 m deep, pH *c.* 6.0, base saturation *c.* 85 percent, moderately high nutrient content, no soluble salts. Water table at surface in wet season with intermittent flooding, falling to about 1 m depth in dry season. Profile drainage slow. Hydromorphic grassland, at present grazed all the year.

Unit 3. Landforms as unit 1. Weathered ferrallitic soils, yellowish brown sandy loams over sandy clay loams, massive, depth 60–100 cm, pH *c.* 5.0, base saturation *c.* 50 percent, low nutrient content. Free site and profile drainage. *Brachystegia* savanna woodland and tree savanna (*miombo*), about 50 percent under cultivation.

Unit 4. Lower hill slopes, up to 15 °. Soils as unit 3, but with 5 percent boulders and rock outcrops. Drainage and vegetation as unit 3. Mainly uncultivated.

Unit 5. Steep slopes, 25°–35°, many rock outcrops and boulders. Soils very thin. Vegetation as unit 3. High scenic value.

Economic and social conditions. The region is characterized by smallholder farming, mainly for subsistence with subsidiary cash crops. Population pressure is high, with a mean farm size of 2 ha. Traditional communal land tenure is giving place to *de facto* family freehold, and it is government policy to encourage this trend. National per capita income averages about £50 per annum but is lower in this area.

Some evaluation systems applied to these five land units are given in table 38. The *USDA land-capability classes* show the dominant influence

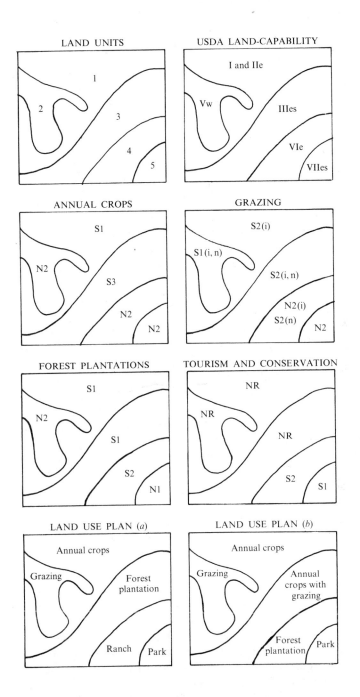

LAND UNITS

USDA LAND-CAPABILITY

1

2

3

4

5

I and IIe

Vw

IIIes

VIe

VIIes

ANNUAL CROPS

GRAZING

S1

N2

S3

N2

N2

S2(i)

S1(i, n)

S2(i, n)

N2(i)

S2(n)

N2

FOREST PLANTATIONS

TOURISM AND CONSERVATION

S1

N2

S1

S2

N1

NR

NR

NR

S2

S1

LAND USE PLAN (a)

LAND USE PLAN (b)

Annual crops

Grazing

Forest
plantation

Ranch

Park

Annual crops

Grazing

Annual
crops with
grazing

Forest
plantation

Park

Fig. 25. Land evaluation systems applied to a sample area. Cf. table 38. For details of land units see text, p. 408.

of soil conservation considerations in this system. The best land, unit 1, is Class II because conservation works are needed on 3° slopes; areas of < 1° on interfluve crests would be Class I. The very much poorer soils of unit 3 only lower it to Class III, whereas despite the similar soils of units 3 and 4, the steeper slopes of the latter put it into Class VI. The poorly-drained valley floors of unit 2 are put into the usual 'bottom lands' sub-class Vw, special use owing to a wetness limitation. Unit 5 is in Class VII rather than VIII as it is not totally barren.

Without fuller economic information the *USBR land classification for irrigation* cannot properly be applied. It is possible that, if a water supply were available, the valley floor land would be marginally suitable, Class 3, since it is not entirely flat and soil permeability is low. On the irrigation suitability rating of Borden and Warkentin (1974) these gley soils only reach 35 percent, 'fair', because of their poor profile drainage and coarse grade of structure. The soil itself of unit 1 reaches 90 percent, 'very good', but as there have been no investigations in the country on the economics of irrigating non-depositional relief this unit is put provisionally into Class 5, warranting further study.

The *soil productivity index of Riquier* et al. (1970) produces results quite at variance with farming experience. The excellent agricultural soil of unit 1, one of the best in the country, falls below 50 percent for crops, through the combined effect of 80 and 90 percent ratings on a series of properties. Unit 3 is lowered to 8 percent for crops, mostly because of its sandy texture and massive structure; these are the right reasons, but its actual crop yield is about half that of unit 1, not one sixth as suggested by the indices. The corresponding index of Sys and Frankart (1971) produces absolute and relative values for units 1 and 3 which are much more in accord with farming experience. Since they refer to soil only, both indices give identical values for units 3 and 4. Both correctly rate unit 5 very low, not because of steepness but soil depth. The Riquier index is correct in rating the valley floor land higher for pasture than for crops, but the Sys and Frankart index for this land is unacceptable. The experience of attempting to apply parametric indices to areas for which they were not developed clearly shows up the limitations of such indices; their values are carefully adjusted to give the 'right' results in one region, but become inapplicable in another region where background conditions and environmental interactions are different.

The remainder of table 38 shows the application of *land suitability evaluation*, based on the FAO framework, to various forms of land use. These are single-purpose qualitative evaluations of current suitability. The land suitability classes given are in this example based not on conversion tables but on subjective estimates of suitability. The fertility soils on gentle slopes of unit 1 are highly suitable for annual cropping and also for forestry planta-

TABLE 38. *Some land evaluation systems applied to a sample area. For environmental conditions of land units see text*

Land unit no.	USDA land-capability classification	USBR land-classification for irrigation	Soil productivity index (Riquier et al., 1970)		Soil productivity index (Sys and Frankart, 1971)	Irrigation suitability rating (Borden and Warkentin, 1974)	Qualitative land suitability evaluation based on the FAO framework							
			For crops	For pasture			Annual crops	Irrigated agriculture	Improved pasture	Natural pasture	Forest plantations	Tourism and conservation	Maize	Ground-nuts
1	IIe	5	47	46	82	90	S1	N1	S2	NR	S1	NR	S1	S1
2	Vw	3	12	23	56	35	N2	S3	S1	S1	N2	NR	N2	N2
3	IIIs	6	8	10	36		S3	N2	S2	S2	S1	NR	S3	S2
4	VIe	6	7	9	36		N2	N2	N2	S2	S2	S2	N2	N2
5	VIIes	6	1	3	12		N2	N2	N2	N2	N1	S1	N2	N2

411

tions, moderately suitable for planted pastures (the long dry season being a limitation) and currently unsuitable for irrigation. Other major kinds of land use are not assessed, since there is no question of their being applied to this unit and so they fall outside the requirement that forms of land use should be economically relevant. This land is highly suitable for both of the individual crops chosen as examples, maize and groundnuts. Unit 2 is suitable for both natural and improved pastures, but unsuitable for annual crops and forestry. The low fertility of the soils of unit 3 renders it only marginally suitable for annual cropping in general and for maize in particular, although satisfactory for groundnuts. Present maize yields are very low, and some farmers substitute cassava. It is moderately suitable for either natural or improved pastures. These soils are perfectly satisfactory for tree growth, which combined with the gentle topography makes the unit highly suitable for forest plantations. The moderate slopes of unit 4 render it unsuitable for cultivation and only moderately suitable for forestry, not because tree growth is any different from unit 3 but because of higher roading costs. The steep rocky slopes of unit 5 are unsuitable for agriculture, currently uneconomic for forestry, but could well be used as a tourism and conservation area; parts of unit 4 could be included in such an area without serious loss of production.

Two possible land use plans based on this evaluation are also shown in fig. 25. Land use plan (*a*) allocates each land unit to the kind of use for which it obtains the highest suitability class, taking relative economic productivity per unit area into consideration only where land is classed as highly suitable for two types of use. The other four units are allocated to the only form of use for which they receive an S1 evaluation. This gives a land use pattern of annual crops on the fertile soils, grazing in the valley floors within them, forestry plantations on the poorer soils of the plain but ranching, beneath natural woodland, on the foothills. The steep slopes, having no productive potential for agriculture or forestry, are placed as a national park or other recreation and conservation area.

Land use plan (*a*) is acceptable as regards units 1, 2 and 5, and corresponds with their present use (with recommendations for the introduction of leys into the farming system of unit 1 and rotational grazing into unit 2). For units 3 and 4, plan (*a*) is open to question on both economic and political grounds. There are presently a substantial number of farmers living in unit 3, and it would be socially and politically unacceptable to displace them. Moreover, the classification of this unit as marginally suitable for annual crops is in terms of present farming standards; if improved methods could be introduced, the economic return of this land from a farming system of maize and groundnuts integrated with beef cattle would substantially exceed that obtainable from even the highest standards of forestry. So the recommendation, shown as Land use plan (*b*), is that such a farming system be developed for unit 4. There is a considerable demand for firewood and timber from the fertile plain, so unit 4 is allocated to forest plantations despite its higher suitability for ranching.

SOME PROBLEMS

The parametric versus the synthetic approach

A frequent practice when applying evaluation systems is to draw up a conversion table of the type illustrated by table 37 (p. 395). The land characteristics, regarded as limitations, are listed separately, and each land unit is classified on the basis of the most severe limitation which it possesses. Alternatively the effects of each land characteristic on the use under consideration are assessed separately and combined by some arithmetic procedure, as in most productivity indices. The method of considering the effects of individual land characteristics, whether as limitations or in positive terms, and then combining these effects to obtain suitability is the *parametric approach*.

The opposite extreme is to consider the land as a whole, judge whether it is suitable for the use in question, and classify it accordingly. This is the *synthetic approach*, and is the one applied by farmers to their own land.

The main advantage of the parametric approach is that once the conversion table has been drawn up, this being a subjective and skilled task, it can be applied objectively so that different observers will obtain consistent results. Its drawback is that by its nature it fails to allow for interaction between factors; for example good permeability becomes of greater importance in soils subject to intermittent flooding. Where parametric systems attempt to allow for interactions by means of subdivisions within their conversion tables (e.g. separate ratings for permeability according to flooding frequency) they become progressively more complex.

The interaction problem can be held within bounds so long as the whole of the area under study has a climate which for practical purposes is uniform, as is the case in many project surveys. It becomes much more serious once there is substantial climatic differentiation, as in systems intended for application throughout a country. In the latter case it is probably better to draw up a separate conversion table, based on soils and landforms, for use within each of the major climatic regions of the country.

The problem of interactions can be overcome in part through the use of compound land characteristics. Available water capacity of the soil profile is an example of such, being compounded of the capacity of individual horizons, their thicknesses, and the effective depth; it may also be estimated indirectly from combining texture and depth.

413

Land evaluation

A further step in this direction is to employ the *land quality*, defined as a complex attribute of land which acts in a manner distinct from the actions of most other land qualities in its influence on the suitability of land for a defined purpose (Brinkman and Smyth, 1973). Examples of land qualities which affect arable farming are availability of water, availability of nutrients, and erosion hazard. Thus, availability of water to crops involves not only available water capacity (and hence texture and depth) but also rainfall amount and distribution. Erosion hazard is compounded of slope angle, soil texture, topsoil structure and rainfall characteristics.

In order to employ land qualities in an objective manner, scales are devised for each quality, e.g. erosion hazard as nil, light, moderate, severe and very severe. The main conversion table relates land suitability to land qualities. Secondary tables are drawn up which convert combinations of land characteristics into land quality classes. To take an extreme case, erosion hazard may be 'severe' on normal soils on slopes above 10°, on ferrisols not until 18°, but on vertisols on as little as 2°. The approach of land qualities has not yet been tested as extensively as that of limitations.

The synthetic approach depends on the experience and judgement of the person making the assessment. This is at the same time its weakness and its strength: a weakness inasmuch as two observers may reach different conclusions about similar land; a source of strength in that the classifications made will always be moderated by common sense.

Most surveys will probably continue to employ a parametric approach, partly to obtain consistency and partly for the mundane reason that it gives clients or employers the impression that 'scientific' methods are being used. The two approaches are not wholly incompatible nor even exclusive, since the values shown in conversion tables are drawn up on the basis of subjective judgement. The way to obtain the advantages of both approaches is to follow a conversion table procedure initially, but then to take the classifications that it produces and set these against what is known of the potentialities of the land, from present experience or analogies with similar areas elsewhere. If the objectively obtained class is not the same as that subjectively judged to be correct, then either the conversion table should be modified or else a qualification added that some additional factor, or particular combination of factors, occurs on the land unit concerned. In either event the overall judgement, in the hands of an experienced observer, should prevail.

Location and accessibility

Set in the middle of a mountainous region lies an isolated and uninhabited valley. After an arduous journey this is found to have fertile soils and to be capable of high productivity, were it not for the fact that there is no means of transporting the produce to a market. Should such land be classed as highly

414

suitable, on the basis of its on-site characteristics, or as unsuitable on grounds of inaccessibility?

Similar problems arise where it is distance and transport costs rather than physical inaccessibility which affects potential land use. The main markets in less developed countries are one or more nodes of development, containing the main urban centres, together with ports or land export outlets. Land situated several hundred miles from these markets has to bear substantial additional transport costs, both of bringing fertilizers in and of removing produce. Should such land be downgraded in a suitability classification for that reason?

There is no real conflict over these questions. Whether location and accessibility should be considered depends on the purposes for which an evaluation is to be used. If it is for an economic evaluation of current suitability, for the purpose of a proposed development project, then costs of transport and, where necessary, road construction and maintenance should be included in arriving at the repayment capacity or other economic measure of land suitability. On a regional planning scale, however, it may prejudge issues to include locational aspects. For example, the question may be asked whether the improvement of transport links to a particular area can be justified. Part of the data needed to reach such a decision is the potential production, or acreage, within the area, assessed independently of transport; this data can be combined subsequently with estimates of road costs in a cost/benefit or other economic analysis.

A safe rule in every case is that evaluation should be made in the first instance on the basis of site characteristics alone, independent of location. It can be modified subsequently to include transport costs when the purposes require. Governments may wish to take decisions on other than economic grounds, e.g. to spread development benefits to relatively poor regions. All that is essential is to state on any map whether location, accessibility and transport costs have been taken into consideration in arriving at suitability classes.

A related problem is whether existing man-made land improvement, such as drainage dykes on coastal lowlands, should be included as properties of land. In general they should be, otherwise, for example, the whole of the rainforest zone would be assessed as if it were covered with jungle (and much of the Netherlands as if it were under the sea). Hence the results of past and present human activity are included in the definition of land. Where existing improvements involve operation and maintenance costs these are included as costs in the evaluation.

The place of economics

The part played by economics in land evaluation systems ranges from that of a shadowy background presence, as in the USDA system and qualitative suitability evaluation, to that of a major control, as in the USBR system and economic suitability evaluation. It is impossible to ignore economic considerations entirely; to adapt a well-known geographical example, it is on grounds of economics as well as environment and technology that the suitability of the North Pole for banana-growing is classified as 'very low'.

The recent tendency has been to incorporate economics more fully into the evaluation process. The argument for so doing is that land evaluation is required primarily for the purpose of making decisions about investment, and the decisions will be made largely on the basis of return on capital invested; it is therefore unrealistic not to translate production and inputs into income and costs. Moreover the practicability of land improvements or cultivation practices is dependent on labour costs; for example, construction of soil conservation bunds by hand labour is possible and cheap in countries with low living standards and small farms but out of the question in Queensland.

The drawback with making evaluation too heavily dependent on economics is that its validity becomes short-lived. Relative market prices, both for products and inputs, can change substantially over five years or less. At the time of writing, sugar and fertilizer prices have both increased out of proportion to general inflation. The case of tung, a tree crop which produces a vegetable oil used in paint, illustrates the changes that can occur. Tung grows well on land with deep, acid soils of heavy texture and a well-distributed rainfall of about 1250 mm, with a short dry season. In the early 1950s such land was economically profitable for tung, and a number of estates were established in Central Africa using investment from aid sources. Towards the end of that decade the price dropped considerably, as a result of production from China, and it became uneconomic to harvest these estates. At the time of writing it is again a profitable crop. Throughout this period the environment has not changed and the tung trees have continued to grow and yield at the same levels. The economic suitability of the same land for the same crop has, however, twice changed. In the case of evaluation for major kinds of land use, such price changes can be partially allowed for by crop substitution.

It is not only changes in market prices that cause difficulties. The discounting rates employed in cost/benefit analysis prior to the 1970s are far lower than those in use today. This has caused an apparent increase in the economic desirability of forms of land use with immediate returns, and a relative decline in the repayment capacity of long-term forms of production such as forestry plantations. Moreover, in cost/benefit analysis alone there are at least four parameters by which desirability can be compared (cost/benefit ratio, excess of benefits over costs, internal rate of return, and repayment period), and no agreement about which is to be used. Economists change their assumptions more often than most other people.

The trend towards translating land evaluation into economic terms is currently in danger of being carried too far. The results of quantitative physical suitability evaluation have a longer life, and more real meaning, than those of economic analysis. For every form of land use there exists, at a given management level (p. 370), an accepted range of inputs. Thus arable cultivation at the improved management level includes ploughing, purchase of improved seed and fertilizers, and amounts of labour which can be estimated. Similarly in the establishment of forestry plantations on a given type of land the mileages of road construction required can be estimated, as can man-days of labour and other inputs.

Levels of production, e.g. of crops of timber, can also be estimated. For any form of land use there are production levels which are generally regarded as satisfactory, e.g. maize yields of 3000 kg/ha (cf. table 34, p. 310). If land is capable of producing these yields from the accepted range of inputs, then in terms of physical evaluation it is suitable for that crop, even though the market price may make its cultivation uneconomic at present. In the case of tung referred to above, the physical suitability of the land, as determined by the climatic and soil requirements of the crop, remained unchanged throughout the fluctuations in world prices.

Physical and economic suitability are both needed for particular purposes. Physical evaluation is appropriate for the purpose of resource inventory. Thus a country may wish to know that if a pulp mill were to be constructed, there would be so many thousand hectares of land which could be planted to softwoods but were not suitable for the priority use of agriculture. For project investment decisions, the results of physical evaluation must be translated into economic terms. However, quantitative estimates of inputs (material and labour) and production in physical terms are in any case a necessary prerequisite to economic analysis. It is therefore desirable that in presenting the results of land evaluation they should be retained in this form, and not subsumed as costs and revenue, when proceeding to the subsequent stage of economic analysis.

Comparing different kinds of land use

An example of a multiple-purpose evaluation is the land inventory programme of Malaysia (Panton, 1970). Productive potential for minerals (mainly tin), agriculture, utilization of natural forests, and conservational uses is first assessed separately. These assessments are then combined in a simple ordering of five land classes of decreasing value:

Class I   Land with high potential for mineral development
Class II   Land with high potential for agriculture
Class III   Land with moderate potential for agriculture
Class IV   Land with high potential for forestry [but not suitable for agriculture]

Class V   Land with little mineral, agricultural or productive forestry potential, but suitable for protective forest reserves, water catchments, game reserves or recreation.

This ordering is indicative of the priorities in less developed countries. Mineral exploitation has absolute priority over other uses, even at the expense of destroying good agricultural land, because of its much higher economic returns. Agriculture takes precedence over forestry on similar grounds, and conservational use comes at the bottom of the order.

A more refined system of comparing uses is the land capability analysis of the Canada Land Inventory (1970); it employs an overlay procedure, with certain fixed priorities and the resolution of conflicts in committee.

The main criteria to be considered when deciding between different kinds of land use are:

1. Economic return on capital invested.
2. Production per unit area.
3. Income per capita.
4. Environmental effects.

Return on capital invested usually takes priority. Production per unit area, which may also be regarded as the number of people who can be supported at a given standard of living, is important where land is scarce and there is pressure to 'settle' as many landless as possible. Maximization of income per capita takes a low importance in development financed by government or foreign aid; intensive forms of land use are preferred to extensive even where per capita income is higher from the latter. Environmental effects refer to such considerations as avoiding soil erosion and maintaining river flow by keeping headwater catchments under forest.

It would be possible to devise working rules for combining land suitabilities, e.g. 'agriculture takes precedence over forestry unless the latter is two suitability classes higher'. The four considerations listed above could be accorded weightings and a goals achievement matrix employed. But such quasi-quantitative procedures will always produce results that are at variance with common sense, judgement, and an awareness of political realities. The more complex quantitative techniques of land evaluation can be applied successfully to assessing suitability for specified uses. In comparing different uses, human factors become important and the matter is best left to judgement (cf. Gibbons *et al.*, 1968; Robertson, 1970).

THE PRODUCTIVE POTENTIAL OF LAND

The assessment of the productive potential of land is not identical with land evaluation. It is a matter of estimating the amount which can be produced from cultivable land at a given level of technology, but not necessarily with inputs which are currently economic. The question most frequently asked is what is the capacity of land resources to supply the minimum food requirements for good health.

The 'ultimate' productive potential of the earth's surface is determined by solar energy and, at present, photosynthetic efficiency, and when estimated on this basis is considerably above present production levels (Black, 1965; Monteith, 1972). This theoretical energy-based approach is, however, unrealistic for practical planning purposes. Productive potential can only be realistically estimated in terms of level of technology up to those of the most skilled practices presently known.

A useful concept is the *carrying capacity* of land, defined as the number of people that can be supported per unit area, with a given level of technology and at a specified standard of living. It is convenient to consider carrying capacity initially in terms of the ability of land to produce basic dietary requirements. These may be taken as 2500 kcal of energy, from carbohydrates, per day, together with 50 g of protein per day of which at least 10 g should be animal protein (Carpenter, 1969). After allowing for small storage and conversion losses the calorific requirement is equivalent to 1000 Mcal per adult per year, the latter quantity being known as a Standard Nutrition Unit (SNU). A procedure that is not too far from realities in areas of subsistence farming is to estimate carrying capacity solely in terms of the capacity of arable land to produce the basic energy requirement, and assume that protein requirements, together with fuel, timber and other needs, will come largely from non-arable land.

One approach is to base assessment on existing farming practices. This was employed by Allan (1965) in the context of shifting cultivation systems in Zambia. Let C = cultivated area per capita, in a given year, L = length of cultivation period in the cultivation–fallow cycle, F = length of fallow in the cultivation–fallow cycle, P = percentage of total area that is cultivable. If there is no unutilized land, $L/(L + F)$ is equivalent in spatial terms to the proportion of cultivable land that is cultivated in any one year; hence by the reverse procedure, the relative lengths of cultivation and fallow can be estimated from air photographs. The total area need per capita, A, is given by

$$A = C \frac{(L + F)}{L} \frac{100}{P}$$

For example, if $C = 0.25$ ha, $L = 3$ years, $F = 15$ years and $P = 50$ percent, then $A = 0.25 \times 18/3 \times 100/50 = 3$ ha per capita. The constants are determined separately for each of the types of land within a region, the area needed per capita calculated for each, and what Allan terms the 'critical population density' of the region is found. This latter may alternatively be called the land carrying capacity under existing levels of technology.

The existing ratio between population and land resources may be compared with such calculations of critical population density. The *agricultural density* has been defined as the agriculturally dependent population per unit area of cultivated land (Nand, 1966; Chakravarti 1970). The ratio of agricultural density to critical population density could be used as a measure of population pressure.

The advantage of this approach is its realism. It is based on existing cultivation practices and assumes there will be no improvements. Allan's method was devised for a time and place (Zambia) where land surplus to farmers' requirements was available, and it could therefore be assumed that the agriculture–soil system was in a steady state. It is frequently the case nowadays that bush fallowing systems are no longer in this state and are drawing upon soil reserves of organic matter and nutrients. In India it is likely that a condition of *low-level equilibrium* (p. 121), in which crop yields become stabilized at low levels on degraded soils, has been reached over wide areas. The method gives no indication of land potential under improved technology.

The alternative and more widely applicable approach is based on establishing the productivity of land under specified management practices, using the same methods as those employed in soil survey (Young, 1974*a*). As an example consider the land system represented by land units 1 and 2 in fig. 25. Cultivable land (unit 1) occupies 80 percent of the area of this land system, and pasture (unit 2) 20 percent. It has been found empirically that a rotation of four years' cropping and four years' fallow (or ley) maintains soil fertility and organic matter levels. A suitable rotation for this relatively fertile soil is groundnuts–tobacco–maize–maize, which includes a legume and gives both food and cash crops. Let it be assumed that the productivity under improved technology is to be estimated. This involves the use of purchased hybrid seed and fertilizers (100 kg/ha sulphate of ammonia for maize and tobacco, plus 100 kg/ha single superphosphate for tobacco). Given these inputs, the areas cultivated, estimated yields and consequent food production and cash income are given in table 39.

These assumptions give a production of 1400 Mcal plus £11.80 per hectare. Ten hectares would therefore provide subsistence for 14 adults, or 3 farm households each of 3 adults and 3 children. If this were in fact the population, there would be a cash income of £6.56 per capita, or £39.33 per farm, plus subsistence food. Farm size would be 3.3 ha, and agricultural density 1.8 people per hectare or 180 per square kilometre of productive land.

This example is for a soil type of above average fertility. On poorer soils, such as the weathered ferrallitic soils of land unit 3 in fig. 25, yields would be about 33 percent lower for similar inputs or 20 percent lower with

TABLE 39 *Estimated productivity under improved management of the land system represented by land units 1 and 2 in fig. 25. Prices are for 1974/75. Based on Young (1974a)*

Crop	Area (ha)	Yield (kg/ha)	Production (kg)	Calorific value (kcal/kg)	Subsistence food production (Mcal)	Price (£/kg)	Income (£)
Maize	2	2000 (grain)	4000	3500	14000		
Groundnuts	1	1000 (nuts in shell)	1000			0.08	80
Tobacco	1	600 (cured leaf)	600			0.15	90
Fallow	4						
Pasture	2	(milk, meat)			(milk, meat)		
Total	10				14000		170

Costs:	400 kg fertilizer	£32		Gross cash income	£170
	Seed, implements, etc.	£20		Costs	£52
	Total	£52		Net cash income	£118

Notes. It is assumed that all maize is consumed as food and all other crops sold. Supplementary food supplies come from milk and meat production from pasture and from patches of vegetables and fruit. The fallow land is assumed to be non-productive.

421

higher rates of fertilizer application. Hence an agricultural density of about 1.0–1.2 per hectare is the greatest that could be supported.

In the moist savanna zone, agricultural population densities are commonly of the order of two people per hectare, whilst in the dry savannas one person per hectare is typical. Densities are considerably lower in the semi-arid zone, of the order of 0.2 people per hectare. On soils derived from basic rocks in the savanna zone, densities up to five people per hectare are common. Agricultural densities of five people per hectare are typical of the rain-forest zone, rising to 15 per hectare or more in areas with double cropping including rice. Such figures are necessarily highly generalized, but for any particular combination of climatic, landform and soil conditions it is possible to estimate carrying capacity, provided that management techniques are specified.

Estimates of the productive potential of land under given levels of technology could be applied to regional planning. Assume that for a particular region, population estimates for the next 25 years have been prepared. It is then necessary to estimate the proportion of this population that is likely to be supported by urban activities and by income generated outside the region, and also the probable magnitude of food imports into the region. The product of this analysis will be the number of people whom it is anticipated will have to be supported, or at least their basic food requirements met, from land production within the region itself at various dates. From production estimates at two management levels, traditional farming and improved technology, the land productivity if given proportions of farmers follow improved methods can be obtained. Comparing productivity with projected demand, it would be possible to state that if demands were to be met, and if the land were not to degenerate through over-intensive use, then it will be necessary that given proportions of farmers adopt improved methods by certain dates. This leaves the most difficult problem, that of how to achieve such adoption, unanswered, but it does provide a scientific basis to demands for the allocation of financial resources to agricultural education and extension.

The same approach can potentially be applied on a national scale. It is open to those few countries that have completed resource inventory surveys to translate these into productive potential. Such estimates are already used in national planning, but they are at present nearly all based on present levels of productivity coupled with estimates of 'cultivable land' of very dubious origin and accuracy.

On a continental and world scale, the concern of the early post-war years for the adequacy of food production to meet demands, which abated during the technology-based hopes of the succeeding period, has again come to the forefront of international discussion. Whatever may be the more distant prospects for limiting population growth it is clear that world population will continue to increase at over two percent at least until the end of the century. The first substantial review of world food production in relation to population was made by Russell in 1954. Among more recent discussions are those of Hutchinson (1969), Thomas (1973) and the FAO *Provisional indicative world plan for agricultural development* (1970). The last took 1961–3 base data and produced forecasts of demand and production for 1975 and 1985; actual production is almost certain to fall short of the 1975 requirement.

Leaving aside the extremist fringe, as represented by systems-based 'futurology' predictions, it is undoubtedly true that the impact of population pressure on land resources is going to become increasingly felt over the next quarter of a century. The average world cultivated area per capita is declining steadily and will continue to do so. Notwithstanding international trade and aid shipments, the impact of this increasing pressure will be mainly at the national level. Bangladesh is already close to a condition in which its population will be limited by the Malthusian control of death through a combination of starvation and disease aggravated by malnutrition. There is a need for scientifically-based estimates of potential land productivity. Resolution VI of the United Nations World Food Conference in 1974 is entitled 'World Soil Charter and Land-Capability Assessment'. It calls for 'an assessment of the lands that can still be brought into cultivation, taking proper account of . . . land required for alternative uses', and for the establishment of a World Soil Charter, which would be the basis for international co-operation towards the most rational use of the world's land resources. The speculative estimates that are the only ones at present available could be replaced by estimates based on maps of landforms, climate and soils, to which scientific methods of evaluation are applied. The FAO–Unesco *Soil map of the world* provides for the first time a soils basis for such estimates. With continued increases in soil survey coverage, coupled with the use of remote sensing methods at the small scales at which it has its greatest potential, it should become possible to obtain increasingly realistic estimates of the productive potential of the world's land resources.

SELECTED REFERENCES

Steele (1967), Stewart (1968), Robertson *et al.* (1968), Moss (1968, 1969), Robertson (1970), Farmer (1971), Brinkman and Smyth (1973), Young (1973*b*, 1974*a*), Vink (1975).

Summaries of the main evaluation systems are given in FAO (1974) and Olson (1974).

SOIL PROFILE ANALYSES

Below are given examples of four main types of latosols together with a vertisol and an alluvial soil. For each profile, information is given on the site, profile morphology and laboratory data, followed by an analysis of the characteristics and origin of the soil and its agricultural potential. Space prevents the inclusion of other soil types but examples, classified according to the FAO system, are becoming available in the explanatory volumes of the *Soil map of the world* (FAO–Unesco, 1971; other volumes in preparation).

PROFILE 1 FERRUGINOUS SOIL

CCTA: Ferruginous soil
FAO: Ferric luvisol

Location: Chitedze Research Station, Lilongwe, Malaŵi; 14° S, 34° E.
Source: Original.
Altitude: 1150 m.   *Drainage:* Free.
Parent material: Basement Complex gneisses, probably of intermediate composition.
Climate: Moist to dry savanna with one wet season (Köppen Aw); rainfall 900 mm, 7 months < 60 mm.
Vegetation and land use: Former broad-leaved deciduous woodland entirely removed for cultivation. Continuous cropping of annuals.

Profile description

0–13 cm	Ap	Dark reddish brown (5YR 3/3), sandy clay; hoed into cultivation ridges; clear boundary.
13–30 cm	AB	Dark reddish brown (5YR 3/4), sandy clay; moderate fine to medium blocky structure with weak clay skins; moderately stiff when wet; merging boundary.
30–70 cm	Bt	Dark reddish brown (2.5YR 3/4), clay; moderate medium blocky structure with moderate clay skins; moderately stiff when wet; merging boundary.
70–160 cm	B2	Dark red (2.5YR–10R 3/6), sandy clay; weak blocky structure; very friable, much less firm than above, only slightly stiff when wet; few weatherable minerals, no weathered rock.

Analysis

This profile belongs to the more clayey and fertile type of ferruginous soil. The site is a gently undulating plain ('African surface'), and this is the uppermost soil in a five-member catena (p. 277, type 1). The differences between this ferruginous soil and weathered ferrallitic soils developed under similar climate and relief (Profile 3, below) is probably accounted for by parent material composition, with ferruginous soils on the less strongly felsic rocks.

Horizon boundaries are gradual. The dark reddish brown colour gradually becomes redder with depth, the red colour showing prominently in termite mounds. Lower down the catena the predominant colour is yellowish red and the structural grade weaker. In some profiles in this series the topsoil is a sandy clay loam. The Bt horizon with clay skins qualifies as an argillic horizon (FAO), which coupled with the high base saturation puts the profile into the FAO class of ferric luvisols. Silt content is moderate, clay skins are present, and weatherable minerals are visible in depth, all indications of only a moderate degree of weathering. The way in which the consistence 'goes soft' below the Bt horizon is characteristic of the series.

Reaction is weakly to moderately acid, and base saturation above 50 percent, indicative of leaching of moderate intensity. The cation exchange capacity is lower than might be expected, since the consistence suggests that 2:1 lattice clay minerals are present in small amounts. In view of the fact that this site has been cultivated for many years the organic matter level is quite high, and the carbon:nitrogen ratio very satisfactory.

This soil possesses most of the conditions favouring high fertility, with the exception that nutrient levels are only moderate. Physically it is extremely good, with free drainage, high available water capacity and excellent rooting conditions. There is sometimes a stone line at about 100 cm, but it is not sufficiently thick or stony to prevent some roots reaching beneath. The entire area of this soil series is under permanent annual cropping, either maize with groundnuts and fire-cured tobacco or maize monoculture, and it is rarely fallowed. Both monoculture and absence of rest periods are undesirable, and there should be between 33 and 50 per-

Depth (cm)	Stones (%)	Particle size analysis				Organic fraction				pH (H_2O)
		Coarse sand (%)	Fine sand (%)	Silt (%)	Clay (%)	Organic matter (%)	Carbon (%)	Nitrogen (%)	C/N ratio	
0–13	0	29	23	12	36	2.2	1.3	0.10	12	6.0
13–30	0	22	24	14	40	1.2	0.7	0.07	9	5.7
30–70	0	22	20	12	46					5.7
70–160	0	23	25	16	36					6.0

Depth (cm)	Cation exchange (m.e./100 g)						Apparent base saturation (%)	Approx. CEC of clay (m.e./100 g)	Available P (NH_4F) (p.p.m.)
	Ca	Mg	K	Na	TEB	CEC			
0–13	15.3	1.5	1.2		18.0	18.8	96		25
13–30	6.3	1.4	0.5		8.2	11.3	73	28	14
30–70	3.3	0.3	0.3		3.9	7.1	55	15	
70–160	2.0	0.3	0.2		2.5	4.3	58	12	

cent fallows or leys in the rotation. Under improved management, maize yields of some 5000 kg/ha can be obtained, and under advanced technology 10000 kg/ha is occasionally reached.

PROFILE 2 LEACHED FERRALLITIC SOIL

CCTA: Ferrallitic soil
FAO: Orthic ferralsol

Location: Jengka Triangle, Pahang State, West Malaysia; 4° N, 102° E.
Source: Original.
Altitude: 100 m.   *Drainage:* Free.
Parent material: Mesozoic sandstone.
Relief: Undulating, dissected relief; site 5°, mid-slope.
Climate: Rainforest type with no dry season (Köppen Af); rainfall 2300 mm, no month < 60 mm.
Vegetation and land use: Lowland evergreen rainforest.

Profile description

0–8 cm	Ah	Yellowish brown (10YR 5/4), sandy loam; moderate medium crumb structure; almost loose consistence; many roots; clear boundary.
8–45 cm	A2	Brownish yellow (10YR 6/6), sandy clay loam; weak medium blocky structure, no clay skins; very friable; roots common; merging boundary.
45–110 cm	Bt	Brownish yellow (10YR 6/6), sandy clay loam; moderate medium blocky structure, weakly developed clay skins; friable; few roots; merging boundary.
110–280 cm	BC	Strong brown (7.5YR 5/6), sandy clay loam; moderate medium angular blocky structure; firm; few roots; mottle from weathered rock increasing towards base.

Analysis

Although the climate is one of intense weathering the profile is only of moderate depth, owing to the quartz-rich sedimentary parent material; by comparison, granites in the same climate are weathered to 5 m and more. Other properties caused by the parent material are the brownish yellow colour, resulting from its low iron content, and the proportions of coarse and fine sand. Some clay translocation has occurred to give a textural B horizon with appreciable structure, but horizon boundaries are merging. As is normal in evergreen rainforest, roots are markedly concentrated in the topsoil. The organic matter content is rather low for natural forest as a result of the sandy texture.

The high rainfall and permeable rock give very intense leaching; over 1000 mm of water annually passes downwards through the profile. This causes the strongly acid reaction and very low apparent base saturation, although the latter is probably in part a consequence of the pH effect on cation exchange capacity (p. 95). These low values of pH and saturation are common to all parent materials in this climate.

Depth (cm)	Stones (%)	Particle size analysis				Organic fraction				
		Coarse sand (%)	Fine sand (%)	Silt (%)	Clay (%)	Organic matter (%)	Carbon (%)	Nitrogen (%)	C/N ratio	pH (H$_2$O)
0–8	0	18	52	12	20	2.7	1.6	0.08	19.4	3.8
8–45	0	14	52	12	26	0.8	0.5	0.06	8.2	4.3
45–110	0	15	47	12	30	0.4	0.2			4.5
110–280	0	15	47	12	32	0.1	<0.1			4.7

Depth (cm)	Cation exchange (m.e./100 g)						Apparent base saturation (%)	Approx. CEC of clay (m.e./100 g)	Available P (NaOH) (p.p.m.)
	Ca	Mg	K	Na	TEB	CEC			
0–8	0.5	0.3	0.1		0.9	10.7	8		28
0–45	0.2	0.2	<0.1		0.4	7.8	5	30	23
45–110	0.1	0.1	<0.1		0.2	8.3	3	24	21
110–280	0.2	0.1	0.0		0.3	6.4	5	20	21

The cation exchange capacity of the clay fraction just falls below the limit of 24 m.e./100 g commonly taken as the limit for intensely weathered soils; it would be lower if more sesquioxides were present.

There is adequate depth for crop growth, excellent root penetration, and although available water capacity is fairly low this is not a problem as dry spells rarely exceed one month. Plant nutrition problems are considerable, with low inherent fertility, lowish cation exchange capacity and strong leaching. Potassium in particular will undergo rapid leaching. Whilst climatically favourable for oil palm, its high nutrient demands would involve heavy and continuous fertilization, and a less demanding crop such as rubber is better suited to this soil.

PROFILE 3 WEATHERED FERRALLITIC SOIL

CCTA: Ferrallitic soil
FAO: Xanthic ferralsol

Location: East of Kapiri Mposhi, Central Province, Zambia; 14° S, 29° E.
Source: Clayton (1975).
Altitude: 1200 m.   *Drainage:* Free.
Parent material: Quartz-rich gneiss, Basement Complex.
Relief: Gently undulating plain, site on almost level crest, 1°.
Climate: Moist savanna with one wet season (Köppen Aw); rainfall 1000 mm, 7 months <60 mm.
Vegetation and land use: Savanna woodland, *Brachystegia-Julbernardia*.

Profile description

0–14 cm	Ah	Very dark greyish brown (10YR 3/2), loamy sand; massive; soft when dry; many roots; clear boundary.

14–23 cm	AB	Strong brown (7.5YR 4/6), sandy clay loam; massive; hard when dry; roots common; merging boundary.
23–110 cm	BT	Strong brown (7.5YR 4/6, becoming 5/8 below 80 cm), sandy clay; massive; hard to very hard when dry; few roots; clear, wavy boundary.
110–170 cm	BC	Highly weathered rock, with few iron and manganese concretions, a little soil.

Analysis

The site is typical for this soil type: a gently-undulating plain with a one-wet-season savanna climate and 'miombo' woodland. The metamorphic rocks of felsic composition are deeply weathered; although rock structure becomes visible at 2 m or less, weathering continues to well over 10 m. This profile is the upper member of a well-differentiated catena, which terminates in a sandy gley in broad valley floors.

Characteristic features are the shallow sandy topsoil passing into a deep, massive sandy clay, hard when dry and lacking clay skins and weatherable minerals; this is a textural B horizon but does not fulfil the requirements of an argillic horizon (FAO). The sandy topsoil results from clay eluviation but there are no signs of illuvial clay beneath, indicative of an advanced stage of soil evolution. The felsic parent material coupled with the advanced stage of weathering gives colours no redder than 7.5YR hue, reaching 5YR in some associated profiles. In many associated soils a stone line of quartz gravel, partly iron-stained, occurs at the upper margin of the BC horizon of weathered rock.

The reaction and base saturation in this profile are unusually high, values of pH 5.5–6.0 and saturation 50–80 percent being more typical. These values indicate

		Particle size analysis				Organic fraction				
Depth (cm)	Stones (%)	Coarse sand (%)	Fine sand (%)	Silt (%)	Clay (%)	Organic matter (%)	Carbon (%)	Nitrogen (%)	C/N ratio	pH (CaCl₂)
0–14		56	27	11	6	2.1	1.2	0.08	16	5.8
14–23		50	21	9	20	0.8	0.5	0.03	14	6.1
23–110		28	17	7	48					6.3
10–170		48	22	6	24					6.3

	Cation exchange, m.e./100 g						Apparent base saturation (%)	Approx. CEC of clay (m.e./100 g)	Available P (Bray) (p.p.m.)
Depth (cm)	Ca	Mg	K	Na	TEB	CEC			
0–14	2.9	1.0	0.6		4.6	5.2	88		44
14–23	2.2	0.8	0.5		3.5	3.8	94	19	32
23–110	2.6	1.5	0.7		4.8	5.8	82	13	
110–170	4.4	2.4	1.6		8.4	10.2	82	42	

429

that, despite the advanced stage of weathering, present-day leaching is of only moderate intensity. The low values of cation exchange capacity indicate entirely kaolinitic clay minerals in the Bt horizon. This profile is under natural woodland, hence the quite high organic matter content for so sandy a topsoil; once cultivated, this would rapidly fall to about one percent.

These 'plateau' or 'sandveldt' soils are of low fertility, with fairly poor physical conditions and, once the organic matter is lowered, a low nutrient content. Traditionally they are used for the *citemene* system of shifting cultivation, growing finger millet, which only supports a very low population density owing to the long fallow period needed. They can be used for permanent annual cropping, but it requires good management to do so. Groundnuts and flue-cured tobacco are suitable crops. To prevent the limited organic matter content declining too far, a fallow or ley of at least 50 percent, possibly 67 percent, of the rotation is desirable.

PROFILE 4 FERRISOL

CCTA: Ferrisol
FAO: Eutric nitosol

Location: Sao Paolo State, Brazil; 23° S, 50°W.
Source: FAO–Unesco (1971).
Altitude: 580 m.   *Drainage:* Free.
Parent material: Basalt.
Relief: Undulating; 3°, upper slope.

Climate: Humid subtropical (Köppen Cfa); rainfall 1750 mm, 2 months <60 mm.
Vegetation and land use: Coffee; formerly semi-deciduous forest.

Profile description

0–19 cm	Ap	Dark reddish brown (2.5YR 3/4), clay; strong fine and medium blocky structure; very firm, plastic; abundant roots; clear boundary.
19–80 cm	B1	Dark reddish brown (2.5YR 3/4), clay; strong fine and medium blocky structure, strongly developed clay skins; firm, plastic; clear boundary.
80–134 cm	B2	Dark reddish brown (2.5YR 3/4), clay; moderate fine blocky structure, strongly developed clay skins; soft, friable, slightly plastic; few roots; merging boundary.
134–224 cm	B3	Dark reddish brown (2.5YR 3/4), clay; massive, breaking into weak fine granular structure; soft, very friable; no roots; merging, wavy boundary.
224–250 cm	BC	Dark brown (7.5YR 5/6), clay; soft, very friable.

Analysis

The site has both of the features associated with ferrisols: basic parent material and a climate that is humid but with a short dry season. With a permanently-humid weathering front, the fine-grained rock gives a deep, stoneless profile with a high clay content. The fine sand is partly secondary in origin, consisting of iron concretions, magnetite and some quartz. The uniform colour is typical of ferrisols, and the main horizon differences are in structure and consistence: the strong blocky structure of the B1 horizon, with marked clay skins and hard peds, gives place to a soft, floury consistence in depth. The clay fraction is dominated by 1:1 minerals and iron oxides, accounting for the friability despite the high clay content.

The total element composition, approximately 29 percent silica, 24 percent iron and 23 percent alumina, remains similar at all depths, with desilication giving a low silica:sesquioxide ratio. The profile is moderately to strongly acid, with medium base saturation. For a cultivated soil the organic matter content is good, a consequence in part of the clay topsoil.

The main agricultural problems are the very firm consistence, hard when dry, and the fact that the soil commonly occurs on moderate and steep slopes. Peasant farmers dependent on hand implements sometimes leave such soils uncultivated, choosing in preference soils of lower fertility but sandier texture. For perennial crops, however, including coffee and sugar cane, this is an excellent soil. There is good root penetration, free drainage but good moisture retention, and an adequate nitrogen status. Potassium is low, and owing to the acidity and probable presence of exchangeable aluminium, phosphate fixation is likely to be a problem. With normal conservation measures, erosion resistance is good, although severe erosion has occured in Brazil where these soils have been misused. This is a valuable soil, capable under good management of continuous cropping with sustained high yields.

		Particle size analysis				Organic fraction				
Depth (cm)	Stones (%)	Coarse sand (%)	Fine sand (%)	Silt (%)	Clay (%)	Organic matter (%)	Carbon (%)	Nitrogen (%)	C/N ratio	pH (H$_2$)
0–19	0	1	20	15	64	2.6	1.5	0.18	8	6.2
19–80	0	1	5	12	82	1.0	0.6	0.07	9	5.8
80–134	0	1	12	14	73	0.7	0.4	0.07		4.8
134–224	0	1	18	20	61	0.3	0.2	0.03		5.0
224–250	0	6	26	25	43	0.3	0.2	0.05		4.9

Depth (cm)	Cation exchange (m.e./100 g)						Apparent base saturation (%)	Approx. CEC of clay (m.e./100 g)	Available P (Truog) (p.p.m.)
	Ca	Mg	K	Na	TEB	CEC			
0–19	9.7	3.7	1.1	0.1	14.5	16.9	86		8
19–80	4.8	3.4	0.6	0.1	8.9	11.3	79	14	9
80–134	2.0	3.9	0.1	0.1	6.0	9.4	64	13	7
134–224	1.0	3.4	0.1	0.1	4.5	8.6	52	14	8
224–250	0.8	2.8	0.1	0.1	3.8	9.5	40	22	8

PROFILE 5   VERTISOL

CCTA: Vertisol
FAO: Pellic vertisol

Location: Shemankar Valley, Benue Plateau State, Nigeria; 9° E, 9° N.
Source: Original.
Altitude: 400 m.   *Drainage:* Poor.
Parent material: Basalt.
Relief: Almost level swampy plain, site ½°; micro-relief of round gilgai.
Climate: Moist savanna (Köppen Aw); rainfall 1250 mm, 5 months <60 mm.
Vegetation and land use: Grassland. Mainly used for nomadic grazing; small patches cultivated successfully for rice and less so for sorghum.

Profile description

0–10 cm	Ah	Dark greyish brown (10YR 4/2), clay; strong fine blocky structure; plastic, very sticky; many grass roots; clear boundary.
10–45 cm	Bg1	Greyish brown (10YR 5/2) with ped surfaces grey, heavy clay; strong fine and medium angular blocky structure, with strong continuous clay skins; plastic, very stiff, very sticky; roots common; merging boundary.
45–150 cm	Bg2	Dark greyish brown (2.5Y 4/2) with ped surfaces slightly greyer, heavy clay; strong coarse prismatic plus moderate to strong medium angular blocky structure, with strong continuous clay skins; plastic, very stiff, very sticky; permeability very slow; occasional $CaCO_3$ concretions; very few roots.

Analysis

The site is a level plain formed by a flow of Tertiary basalt. Climatically a 1250 mm rainfall is toward the wetter limit for vertisols, and their formation here is explained by a high water table in conjunction with basic rock. Nearby basalt flows slightly above the water table have red ferrisols.

This is the dark coloured, poorly-drained type of vertisol, tending towards a gley. The description was taken during the wet season when cracking was not apparent but the vertisol nature is indicated by the strongly developed composite prismatic and blocky structure with polished ped surfaces, and the micro-relief of round gilgai with an unusually close spacing. There are sink-holes in the centres of the gilgai shelves, filled in the wet season with water or soft mud.

Apart from the slightly less dark, more finely structured and humic topsoil the profile is very uniform; horizon differences are small and the boundary at 45 cm is arbitrary. The uniformity is also shown in the notably small variations in analytical data with depth. It is consistent with a self-mixing process, which could also explain the presence of occasional basalt stones on the surface. Neither rock nor a transitional horizon was reached. The clay content is not exceptionally high but the presence of much montmorillonite, indicated by the high exchange capacity values of the clay fraction, account for the stiff, plastic and sticky consistency and the very low permeability. The reaction and saturation values indicate intermittent

Depth (cm)	Stones (%)	Particle size analysis				Organic fraction				
		Coarse sand (%)	Fine sand (%)	Silt (%)	Clay (%)	Organic matter (%)	Carbon (%)	Nitrogen (%)	C/N ratio	pH (H$_2$O)
0–10	0	6	16	30	48	2.1	1.2	0.08	15	6.2
25–40	0	5	12	27	56	0.9	0.5	0.03	15	6.1
00—125	1	4	11	27	58					8.0

Depth (cm)	Cation exchange (m.e./100 g)						Apparent base saturation (%)	Approx. CEC of clay (m.e./100 g)	Available P (p.p.m.)
	Ca	Mg	K	Na	TEB	CEC			
0–10	15.6	11.6	0.5	0.2	27.9	31.3	89		No data
25–40	15.6	11.3	0.5	0.2	27.6	30.6	90	55	
100–125	13.1	15.6	0.8	3.1	32.6	29.9	100	52	

leaching in the upper part, caused by the high rainfall early in the wet season when the water-table is low; in the lower part of the profile, infrequently above the water table, the reaction becomes alkaline. The rainfall also accounts for the rarity of calcium carbonate in comparison with most vertisols. Waterlogging causes inhibited nitrification, indicated by the high carbon:nitrogen ratio.

Under traditional technology these soils are largely ignored in favour of sandier, substantially less fertile but more easily cultivable soils. With the marked seasonal climate the soil alternates between a waterlogged state, with much standing water, and drying out hard. Peripheral areas are used for rice, and with bunding this could be extended. There are possibilities for perennial irrigation in the region and with winter irrigation, combined with maintenance of the water table at a more nearly constant depth by drainage, there are possibilities of double cropping, with rice in summer and a different crop in winter. The fertility level is moderate, although well above most other soils in the region. Moisture retained is 36 percent at field capacity but is also high, 23 percent, at wilting point (15 atmospheres), giving only a moderate available water capacity. The slow permeability gives difficulties to both drainage and water application. Correctly managed this could be a soil of quite high productivity, but it also possesses a high hazard potential.

PROFILE 6 ALLUVIAL SOIL

CCTA: Juvenile soils on riverine alluvium/Brown calcareous soil
FAO: Eutric fluvisol/Orthic luvisol

Location: Indian Agricultural Research Institute Farm, Delhi; 28° N, 77° E.
Source: Chibber *et al*. (1970).
Altitude: 250 m.   *Drainage*: Imperfect; water table at 175 cm.
Parent material: Older alluvium.
Relief: Almost level cover flood plain; no current deposition.

433

Appendix

Climate: Dry savanna type of monsoonal origin (Köppen Cwa); rainfall 680 mm, 8 months <60 mm.
Vegetational and land use: Irrigated double-cropping of annual crops.

Profile description

0–25 cm	Ap	Dark yellowish brown (10YR 4/4), loam; weak medium blocky structure, hard when dry; non-calcareous; many crop roots; clear boundary.
25–70 cm	AB	Dark yellowish brown (10YR 4/4), sandy loam; weak medium blocky structure; non-calcareous; few roots; clear, boundary.
70–130 cm	Bt	Dark brown (10YR 4/3), clay loam; weak medium blocky structure, friable, becoming plastic toward base; slightly calcareous; few roots; clear boundary
130–175 cm	Bca	Light brownish grey (10YR 6/2), clay loam; friable; highly calcareous, calcium carbonate concretions common.

Analysis

No alluvial soil can be called typical, since they may occur under any climate, drainage can vary from free to poor, and texture from sand to clay. This example has imperfect site drainage and is found near the drier margin of the humid tropics. It is derived from older alluvium, and two features of pedogenetic origin have begun to develop: a textural B horizon and calcium carbonate concretions in depth. The texture is moderately sandy, consequently there is a weak grade of structure and low available water capacity. The profile dries out to a light yellowish brown, becoming hard. Reaction is weakly alkaline, but carbonates are leached from horizons above the water table. The groundwater is slightly saline, and patches of saline soils occur on nearby low-lying sites.

On a narrow definition of an alluvial soil this profile would not be classed as such, since active deposition has ceased and textural boundaries of depositional origin are no longer sharply defined. On the FAO system it is probably no longer a fluvisol but an orthic luvisol, the depth to carbonates being just greater than that required for the calcic luvisol class. To classify such profiles together with soils derived from consolidated rocks is, however, misleading, and it is better considered as belonging to the broad group of alluvial soils.

On both physical and chemical grounds this soil has a low fertility. Irrigation

		Particle size analysis				Organic fraction			
Depth (cm)	Stones (%)	Coarse sand (%)	Fine sand (%)	Silt (%)	Clay (%)	Organic matter (%)	Carbon (%)	pH (H₂O)	Total soluble salts (%)
0–25	0	21	49	13	17	0.9	0.5	7.8	0.07
25–70	0	18	56	10	16	0.3	0.2	7.7	0.04
70–130	0	14	44	16	26	0.5	0.3	7.8	0.06
130–175	3[a]	15	43	16	26	0.4	0.2	7.7	0.12

[a] Calcium carbonate concretions.
Other data not available.

water is available, and it is intensively and continuously cropped. As a result, organic matter has fallen to a low level. Yields are low if unfertilized, rising to moderate levels with high and balanced fertilizer application. The soil would benefit from manuring. There is a salinization hazard, calling for careful water management. Set against these limitations is the principal merit of alluvial soils, that the ground is level, permitting irrigation by basin flooding.

BIBLIOGRAPHY

Abedi, M. J. and Talibudeen, O. (1974) The calcareous soils of Azerbaijan. I. Catena development related to the distribution and surface properties of soil carbonate. *J. Soil Sci.* **25**, 357–72

Abu-Zeid, M. O. (1973) Continuous cropping in areas of shifting cultivation in the southern Sudan. *Trop. Agric., Trin.*, **50**, 285–90

Ackermann, E. (1936) Dambos in Nordrhodesien. *Wiss. Veröff. Museums Länderkunde Leipzig* **4**, 147–57

Acland, J. D. (1971) *East African crops.* Longman, London

Ahmad, N. and Jones, R. L. (1969) Genesis, chemical properties and mineralogy of limestone-derived soils, Barbados, West Indies. *Trop. Agric., Trin.* **46**, 1–15

Ahmed, S., Swindale, L. D. and El-Swaify, S. A. (1969) Effects of adsorbed cations on physical properties of tropical red earths and tropical black earths. *J. Soil Sci.* **20**, 255–68

Ahn, P. M. (1970) *West African soils.* Oxford University Press, London

Aitchison, G. D. and Grant, K. (1968) Terrain evaluation for engineering. In *Land evaluation* (ed. G. A. Stewart), 125–46

Akehurst, B. C. (1968) *Tobacco.* Longman, London

Alexander, L. T. and Cady, J. G. (1962) Genesis and hardening of laterite in soils. *US Dept Agric. Soil Conserv. Serv. Tech. Bull.* **1282**

Alexander, L. T., Cady, J. G., Whittig, L. D. and Dever, R. F. (1956) Mineralogical and chemical changes in the hardening of laterite. *Trans. 6th Int. Congr. Soil Sci.* **E** 67–72.

Allan, A. Y. (1971) Fertilizer use on maize in Kenya. *Soils Bull., FAO* **14**, 10–25

Allan, W. (1965) *The African husbandman.* Oliver & Boyd, London

Allison, R. E. (1973) *Soil organic matter and its role in crop production.* Elsevier, Amsterdam

Allison, L. E. (1964) Salinity in relation to irrigation. *Adv. Agron.* **16**, 139–80

Anderson, B. (1957) *A survey of soils in the Kongwa and Nachingwea Districts of Tanganyika.* University of Reading

Andriesse, J. P. (1968/9) A study of the environment and characteristics of tropical podzols in Sarawak (East-Malaysia). *Geoderma* **2**, 201–28

(1969/70) The development of the podzol morphology in the tropical lowlands of of Sarawak (Malaysia). *Geoderma* **3**, 261–79

(1971) The influence of the nature of parent rock on soil formation under similar atmospheric climates. *Nat. Resour. Res., Unesco* **11**, 95–110

Arnon, D. I. (1972) *Crop production in dry regions.* Leonard Hill, Aylesbury

Askew, G. P. (1964) The mountain soils of the east ridge of Mt Kinabalu. *Proc. Roy. Soc. B* **161**, 65–74

Askew, G. P., Moffatt, D. J., Montgomery, R. F. and Searl, P. L. (1970) Soil landscapes in north eastern Mato Grosso. *Geogr. J.* **136**, 211–27

Atlas of Australian soils (1960–68) CSIRO, Melbourne

Aubert, G. (1962) Arid zone soils. *Arid Zone Res., Unesco* **18**, 115–37

(1963) Soils with ferruginous or ferrallitic crusts of tropical regions. *Soil Sci.* **95**, 235–42

(1964) The classification of soils as used by French pedologists in tropical or arid areas. *Sols Afr.* **9**, 97–115

(1965*a*) Classification des sols utilisée par la Section de la pedologie de l'ORSTOM. *Cah. ORSTOM Sér. Pédol.* **3**, 269–88

(1965*b*) La classification pédologique utilisée en France. *Pédologie, Ghent, numero spec.* **3**, 25–56

(1968) Classification des sols utilisée par les pédologues français. *World Soil Resour. Rep., FAO* **32**, 78–94

Aubert, G. and Tavernier, R. (1972) Soil survey. In *Soils of the humid tropics* (National Academy of Sciences, Wash. DC), 17–44

Baeyans, J. (1938) *Les sols de l'Afrique centrale, specialement de Congo belge.* Pubns INEAC hors Serie

(1949) Classifying banana soils in tropical West Africa. *Comm. Bur. Soil Sci. Tech. Commun.* **46**, 203–9

Baldwin, M., Kellogg, C. W. and Thorp, J. (1938) Soil classification. In *Soils and man. Yearbook of Agriculture 1938* (Wash. DC), 979–1001

Ball, D. F. (1964) Loss-on-ignition as an estimate of organic matter and organic carbon in non-calcareous soils. *J. Soil Sci.* **15**, 84–92

Barber, R. G. and Rowell, D. L. (1972) Charge distribution and the cation exchange capacity of an iron-rich kaolinitic soil. *J. Soil Sci.* **23**, 135–46

Barnes, A. C. (1974) *The sugar cane.* Leonard Hill, Aylesbury

Basinski, J. J. (1959) The Russian approach to soil classification and its recent development. *J. Soil Sci.* **10**, 14–26

Bawden, M. G. (1965) A reconnaissance of the land resources of eastern Bechuanaland. *J. Appl. Ecol.* **2**, 357–65

Bawden, M. G. and Carroll, D. M. (1967) *The land resources of Lesotho.* Directorate of Overseas Surveys, Tolworth

Bawden, M. G. and Tuley, P. (1967) *The land resources of southern Sardauna and southern Adamawa Provinces, northern Nigeria.* Directorate of Overseas Surveys, Tolworth

Bazilevich, N. I., Drozov, A. V. and Rodin, L. E. (1971) World forest productivity, its basic regularities and relationship with climatic factors. *Nat. Resour. Res., Unesco* **4**, 345–53

Beaven, P. J., Lawrance, C. J. and Newill, D. (1971) A study of terrain evaluation in West Malaysia for road location and design. *Fourth Asian Regional Conf. Soil Mechs Foundation Engng* **1**, 411–16

Beckett, P. H. T. (1971) The cost-effectiveness of soil survey. *Outlook on Agric.* **6**, 191–8

Beckett, P. H. T. and Webster, R. (1971) Soil variability – a review. *Soils Fertil.* **34**, 1–15

Beckett, P. H. T., Burrough, P. A. and Jarvis, M. G. (1971) The relation between cost and utility in soil survey. I–V. *J. Soil Sci.* **22**, 359–94 and 466–89

Beckett, P. H. T., Webster, R., McNeil, G. M. and Mitchell, C. W. (1972) Terrain evaluation by means of a data bank. *Geogr. J.* **138**, 430–56

Beckmann, G. G., Thompson, C. H. and Hubble, G. D. (1974) Genesis of red and black soils on basalt on the Darling Downs, Queensland, Australia. *J. Soil Sci.* **25**, 265–81

437

Bibliography

Beek, K. J. (1974) The concept of land utilization types. *Soils Bull.*, *FAO* **22**, 103–20

Beek, K. J. and Bennema, J. (1974) Land evaluation for agricultural land use planning – an ecological method. *Soils Bull.*, *FAO* **22**, 54–70

Bennema, J. (1963) The red and yellow soils of the tropical and subtropical uplands. *Soil Sci.* **95**, 250–7

Bennema, J., Jongerius, A. and Lemos, R. C. (1970) Micromorphology of some oxic and argillic horizons in south Brazil in relation to weathering sequences. *Geoderma* **4**, 333–55

Bennett, H. H. and Allison, R. V. (1928) *Soil map of Cuba 1 :800 000.* Tropical Plant Research Foundation, Wash. DC

Benninson, R. N. and Evans, D. D. (1968) Some effects of crop rotation on the productivity of crops on a red earth in a semi-arid tropical climate. *J. Agric. Sci., Camb.* **71**, 365–80

Berger, J. (1962) *Maize production and the manuring of maize.* Centre d'Étude de l'Azote, Geneva

Berry, L. and Ruxton, B. P. (1959) Notes on weathering zones and soils on granitic rocks in two tropical regions. *J. Soil Sci.* **10**, 54–63

Bidwell, O. W. and Hole, F. D. (1965) Man as a factor in soil formation. *Soil Sci.* **99**, 65–72

Bie, S. W. and Beckett, P. H. T. (1970) The costs of soil survey. *Soils Fertil.* **33**, 203–17

Bie, S. W. and Beckett, P. H. T. (1971) Quality control in soil survey, I and II. *J. Soil Sci.* **22**, 32–49 and 453–65

Bie, S. W., Ulph, A. and Beckett, P. H. T. (1973) Calculating the economic benefits of soil survey. *J. Soil Sci.* **24**, 429–35

Birch, H. F. (1958) The effect of soil drying on humus decomposition and nitrogen availability. *Plant Soil* **10**, 9–31

 (1960) Soil drying and soil fertility. *Trop. Agric., Trin.* **37**, 3–10

Birch, H. F. and Friend, M. T. (1956) The organic-matter and nitrogen status of East African soils. *J. Soil Sci.* **7**, 156–67

Birot, P. (1960/68) *Le cycle d'erosion sous les differents climats.* Univ. Brazil, Rio de Janiero (1960). Transl. as *The cycle of erosion in different climates.* (1968)

Black, J. N. (1965) The ultimate limits of crop production. *Proc. Nutr. Soc.* **24**, 2–8

Bleackley, D. and Khan, E. J. A. (1963) Observations on the white-sand areas of the Berbice Formation, British Guiana. *J. Soil Sci.* **14**, 44–51

Bloomfield, C., Coulter, J. K. and Kanaris-Sotiriou, R. (1968) Oil palms on acid sulphate soils in Malaya. *Trop. Agric., Trin.* **45**, 289–300

Bolton, J. L. (1962) *Alfalfa.* Leonard Hill, London

Borden, R. W. and Warkentin, B. P. (1974) An irrigation rating for some soils in Antigua W.I. *Trop. Agric., Trin.* **51**, 501–14

Botelho da Costa, J. V. and Cardoso Franco, E. P. (1965) Note on the concepts of ferrallitic soils and oxisols. *Pédologie, Ghent,* numero spec. **3**, 181–4

Boughey, A. S. (1957a) The vegetation types of the Federation. *Proc. Trans. Rhodesia Sci. Ass.* **45**, 73–91

 (1957b) Ecological studies of tropical coastlines. I. The Gold Coast, West Africa. *J. Ecol.* **45**, 665–87

Boyko, H. (ed.) (1966) *Salinity and aridity.* Junk, The Hague

Boyko, H. (ed.) (1968) *Saline irrigation for agriculture and forestry.* Junk, The Hague

Bradfield, R. (1974) Intensive multiple cropping. *Trop. Agric., Trin.* **51**, 91–3

Brammer, H. (1952) *Detailed soil survey of Kpong Pilot Irrigation Area.* Map 1: 5000. Dept. Soils, Aburi, Ghana

438

(1962) Soils. In *Agriculture and land use in Ghana* (ed. J. B. Wills, Oxford University Press, London), 88–126

(1968) Decalcification of soils developed in calcareous Gangetic alluvium in East Pakistan. *Pakist. J. Soil Sci.* **4**, 8–20

(1971) Coatings in seasonally-flooded soils. *Geoderma* **5**, 5–16

Brewer, R. (1968) Clay illuviation as a factor in particle-size differentiation in soil profiles. *Trans. 9th Int. Congr. Soil Sci.* **4**, 489–98

Breznak, J. A., Brill, W. J., Mertins, J. W. and Coppel, H. C. (1973) Nitrogen fixation in termites. *Nature* **244**, 577–9

Brink, A. B. A. and Williams, A. A. B. (1964) Soil engineering mapping for roads in South Africa. *CSIR Res. Rep.* **227**, Pretoria

Brink, A. B., Mabbutt, J. A., Webster, R. and Beckett, P. H. T. (1966) Report of the working group on land classification and data storage. *MEXE Rep.* **940**, Christchurch, Hants

Brinkman, R. and Pons, L. J. (1968) A pedo-geomorphological classification and map of the Holocene sediments in the coastal plain of the three Guianas. *Soil Surv. Paper Neth. Soil Surv. Inst.* **4**

Brinkman, R. and Smyth, A. J. (ed.) (1973) *Land evaluation for rural purposes.* International Institute for Land Reclamation and Improvement, Wageningen

Broadbent, F. E. (1953) The soil organic fraction. *Adv. Agron.* **5**, 153–83

Brosh, A. (1970) Observations on the geomorphic relationships of laterite in southeastern Ankole (Uganda). *Jerusalem Stud. Geogr.* **1**, 153–79

Buchanan, F. (1807) *A journey from Madras through the countries of Mysore, Canara and Malabar.* Vol. 2, 436–60. East India Co., London

Büdel, J. (1957) Die 'doppelten Einebnungsflächen' in den feuchten tropen. *Z. Geomorph.* **1**, 201–28

Bullock, J. A. (1967) The Arthropoda of tropical soils and leaf litter. *Trop. Ecol.* **8**, 74–87

Buol, S. W. (1965) Present soil-forming factors and processes in arid and semi-arid regions. *Soil Sci.* **99**, 45–9

Butler, B. E. (1955) A system for the description of soil structure and consistence in the field. *J. Aust. Inst. Agric. Sci.* **21**, 239–49

(1959) Periodic phenomena in landscapes as a basis for soil studies. *CSIRO Soil Pub.* **14**, Melbourne

(1967) Soil periodicity in relation to landform development in southeastern Australia. In *Landform studies from Australia and New Guinea* (ed. J. N. Jennings and J. A. Mabbutt, Cambridge University Press), 231–55

Campbell, J. M. (1917) Laterite: its origin, structure and minerals. *Mineral Mag.* **17**, 67–77, 120–8, 171–9 and 220–9

Canada Land Inventory (1970) The Canada land inventory: objectives, scope and organization. *CLI Rep.* **1**, 2nd edn, Ottawa

Carpenter, K. J. (1969) Man's dietary needs. In *Population and food supply* (ed. J. Hutchinson), 61–74

Carroll, D. M. (1973) Remote sensing techniques and their application to soil science. 1–2. *Soils Fertil.* **36**, 259–66 and 313–20

Carroll, D. M. and Bascomb, C. L. (1967) Notes on the soils of Lesotho. *Tech. Bull. Land Res. Div.* **1**

Carter, G. F. and Pendleton, R. L. (1956) The humid soil: process and time. *Geogr. Rev.* **46**, 488–507

Chakravarti, A. K. (1970) Foodgrain sufficiency patterns in India. *Geogr. Rev.* **60**, 208–28

Bibliography

Chang, J.-H. (1957) World patterns of monthly soil temperature distribution. *Ann. Ass. Am. Geogr.* **47**, 241–9

Charter, C. F. (1957*a*) The aims and objects of tropical soil surveys. *Soils Fertil.* **20**, 127–8

(1957*b*) Suggestions for a classification of tropical soils. *Div. Soil Land Use Survey Dept Agric. Ghana Misc. Paper* **4**

(1958) *Report on the environmental conditions prevailing in Block 'A', Southern Province, Tanganyika Territory.* Dept. of Agriculture, Accra

Chibber, R. K., Das, B. and Verma, H. K. G. (1970) *Soil and land use survey of the IARI Farm, New Delhi.* Indian Agriculture Research Institute, New Delhi

Child, R. (1953) The selection of soils suitable for tea. *Tea Res. Inst. E. Afr. Pamphlet* **5**

(1964) *Coconuts.* Longman, London

Childs, E. C. and Youngs, E. G. (1974) Soil physics: twenty-five years on. *J. Soil Sci.* **25**, 399–407

Christian, C. S. and Stewart, G. A. (1953) Survey of Katherine–Darwin region 1946. *CSIRO Land Res. Ser.* **1**, Melbourne

Christian, C. S. and Stewart, G. A. (1968) Methodology of integrated surveys. *Nat. Resour. Res. Unesco* **6**, 233–80

Clare, K. E. (1967) *Engineering soils on part of the coastal plain of Guyana.* Map 1:250000, Directorate of Overseas Surveys, Tolworth

Clare, K. E. and Beaven, P. J. (1962) *Soils and roadmaking materials in Nigeria.* HMSO

(1965) Roadmaking materials in northern Borneo. *DSIR Road Res. Lab. Road Res. Tech. Paper* **68**

Clayton, D. B. (1975) The sandveldt soils of Central Province. *Mt Makulu Res. Stn, Chilanga, Zambia, Soil Surv. Rep.* **32**

Cole, M. M. (1963*a*) Vegetation and geomorphology in Northern Rhodesia. *Geogr. J.* **129**, 290–310

Cole, M. M. (1963*b*) Vegetation nomenclature and classification with particular reference to the savannas. *S. Afr. Geogr. J.* **45**, 3–14

Colmet-Daage, F. (1967) Caractéristiques de quelques sols d'Equateur dérivés de cendres volcaniques. *Cah. ORSTOM Sér. Pédol.* **5**, 353–92

Commission de Pédologie et de Cartographie des Sols (1967) *Classification des sols.* Laboratoire de Géologie, Commission de Pédologie de l'Ecole Nationale Supérieure Agronomique de Grignon, France

Cook, A. (1974) The use of photo-interpretation in the assessment of physical and biological resources in Tanzania. *BRALUP Res. Paper* **31**, Dar es Salaam

Cooke, G. W. (1962) Chemical aspects of soil fertility. *Soils Fertil.* **25**, 417–20

Cornforth, I. S. (1970) Reafforestation and nutrient reserves in the humid tropics. *J. Appl. Ecol.* **7**, 609–15

Cortes, A. and Franzmeier, D. P. (1972) Climosequence of ash-derived soils in the Central Cordillera of Colombia. *Proc. Soil Sci. Soc. Am.* **36**, 653–9

Coulter, J. K. (1964) Soil surveys and their application in tropical agriculture. *Trop. Agric., Trin.* **41**, 185–96

(1972) Soils of Malaysia. A review of investigations on their fertility and management. *Soils Fertil.* **35**, 475–98

Coursey, D. G. (1967) *Yams.* Longman, London

Cox, R. A. J., Spurling, A. T. and Spurling, D. (1972) Unit farms in Malawi. *Min. Agric. Nat. Resour. Res. Circular* **4/72**, Zomba

Crocker, R. L. (1952) Soil genesis and pedogenic factors. *Q. Rev. Biol.* **27**, 139–68

Crompton, E. (1960) The significance of the weathering/leaching ratio in the differentiation of major soil groups. *Trans 7th Int. Cong. Soil Sci.* **4**, 406–12.

—— (1962) Soil formation. *Outlook on Agric.* **3**, 209–18

CSIRO (1953–75) *Land Research Series*, Nos 1–35, Commonw. Sci. Indust. Res. Organization, Melbourne.

Cunningham, R. K. (1963) The effect of clearing a tropical forest soil. *J. Soil Sci.* **14**, 334–45

Dan, J. (1973) Arid-zone soils. In *Arid zone irrigation* (ed. B. Yaron, E. Danfors and Y. Vaadia), 11–28.

Dancette, C. and Poulain, J. F. (1969) Influence de l'*Acacia albida* sur les facteurs pédoclimatiques et les rondements des cultures. *Sols Afr.* **13**, 197–239

Davidson, B. R. (1965) *The northern myth.* Melbourne University

Dawson, J. A. and Doornkamp, J. C. (1973) *Evaluating the human environment.* Arnold, London

D'Costa, V. and Ominde, S. H. (1973) Soil and land-use survey of the Kano Plain—Nyanza Provinze—Kenya. *Dept. Geogr. Univ. Nairobi Occas. Mem.* **2**

De Alwis, K. D. and Panabokke, C. R. (1972–3) Handbook of the soils of Sri Lanka (Ceylon). *J. Soil Sci. Soc. Ceylon* **2**, 1–97

De Datta, S. K. and Magnaye, C. P. (1969) A survey of the forms and sources of fertilizer nitrogen for flooded rice. *Soils Fertil.* **32**, 103–9

Delvigne, J. (1965) Pédogenèse en zone tropicale. La formation des minéraux secondaires en milieu ferrallitique. *Mém. ORSTOM* **13**

Denisoff, I. (1959) Le concept de la zonalité verticale appliqué a quelques sols caracteristiques du Ruanda-Urundi. *Proc. 3rd Interafr. Soils Conf.* 313–6

De Swart, A. M. J. (1964) Lateritisation and landscape development in parts of equatorial Africa. *Z. Geomorph.* **8**, 313–33

De Vos t.N.C., J. H. and Virgo, K. J. (1969) Soil structure in vertisols of the Blue Nile clay plains, Sudan. *J. Soil Sci.* **20**, 189–206

D'Hoore, J. (1954*a*) L'accumulation des sesquioxides libres dans les sols tropicaux. *Publ. INEAC Ser. Scientif.* **62**

—— (1954*b*) Proposed classification of the accumulation zones of free sesquioxides on a genetic basis. *Sols Afr.* **3**, 66–81

—— (1955) The description and classification of free sesquioxide accumulation zones. *Trans. 5th Int. Congr. Soil Sci.* **4**, 39–44

—— (1964) *Soil map of Africa scale 1 to 5000000.* Explanatory monograph. *CCTA Publ.* **93**, Lagos

—— (1968*a*) Influence de la mise en culture sur l'évolution des sols dans zone de forêt dense de basse et moyen altitude. *Sols Afr.* **13**, 155–68

—— (1968*b*) The classification of tropical soils. In *The soil resources of tropical Africa* (ed. R. P. Moss, Cambridge University Press), 7–28

Doggett, H. (1970) *Sorghum,* Longman, London

Douglass, J. F., Austin, M. E. and Smith, G. D. (1969) General soil map of the United States. *Proc. Soil Sci. Soc. Am.* **33**, 746–9

Dowling, J. W. F. (1968) Land evaluation for engineering purposes in northern Nigeria. In *Land evaluation* (ed. G. A. Stewart), 147–59

Dowling, J. W. F. and Beaven, P. J. (1969) Terrain evaluation for road engineers in developing countries. *J. Inst. Highway Engineers* **14**, 5–22

Dowling, J. W. F. and Williams, F. H. P. (1964) *The classification of landforms in northern Nigeria and their use with aerial photographs in engineering soil surveys.* Road Res Lab.

Bibliography

Du Bois, C. G. B. and Jeffery, P. G. (1955) The composition and origin of the laterites of the Entebbe Peninsula, Uganda. *Col. Geol. Min. Resour.* **5**, 387–408

Duchaufour, P. (1963) Soil classification: a comparison of the American and the French systems. *J. Soil Sci.* **14**, 149–55

(1960/70) Précis de pédologie. 1st edn (1960), 3rd edn (1970) Masson, Paris

Dudal, R. (1963) Dark clay soils of tropical and subtropical regions. *Soil Sci.* **95**, 264–70

(ed.) (1965) Dark clay soils of tropical and subtropical regions. *FAO Agric. Dev. Paper* **83**, Rome

(1966) Soil resources for rice production. In *Mechanization and the world's rice* (publ. Massey-Ferguson), Blackwell, Oxford

(1968*a*) Definitions of soil units for the Soil Map of the World. *FAO World Soil Resour. Rep.* **33**

(1968*b*) Genesis and classification of paddy soils. In *Geography and classification of soils of Asia* (ed. V. A. Kovda and E. V. Lobova, Jerusalem), 194–7

Dudal, R. and Soepraptohardjo, M. (1960) Some considerations on the genetic relationship between latosols and andosols in Java. *Trans 7th Int. Congr. Soil Sci.* **4**, 229–37

Dudal, R., Moorman, F. and Riquier, J. (1974) Soils of humid tropical Asia. *Nat. Resour. Res. Unesco* **12**, 159–78

Dury, G. H. (1969) Rational descriptive classification of duricrusts. *Earth Sci. J.* **3**, 77–86

(1971) Relict deep weathering and duricrusting in relation to the palaeoenvironments of middle latitudes. *Geogr. J.* **137**, 511–22

Edelman, C. H. (1963) Applications of soil survey in land development in Europe; with special reference to experiences in the Netherlands. *ILRI Pub.* **12**, Wageningen

Edelman, C. H. and Brinkman, R. (1962) Physiography of gilgai soils. *Soil Sci.* **94**, 366–70

Edelman, C. H. and van der Voorde, P. K. J. (1963) Important characteristics of alluvial soils in the tropics. *Soil Sci.* **95**, 258–63

Eden, T. (1965) *Tea.* 2nd edn, Longman, London

Edwards, R. D. (1970) *Soil survey of Ventura area, California.* Wash. DC

Ellis, B. S. (1950) A guide to some Rhodesian soils. II. A note on mopani soils. *Rhod. Agr. J.* **47**, 49–61

(1952) Genesis of a tropical red soil. *J. Soil Sci.* **3**, 52–62

(1958) Soil genesis and classification. *Soils Fertil.* **21**, 145–7

Eswaran, H. (1970) Micromorphological indicators of pedogenesis in some tropical soils derived from basalts from Nicaragua. *Geoderma* **7**, 15–31

Eyles, R. J. (1970) Physiographic implications of laterite in West Malaysia. *Bull. Geol. Soc. Malaysia* **3**, 1–7

Fadl, A. (1971) A mineralogical characterization of some vertisols in the Gezira and the Kenana clay plains of the Sudan. *J. Soil. Sci.* **22**, 129–35

FAO (1965*a*) Meeting on the classification and correlation of soils from volcanic ash. *World Soil Resour. Rep.* **14**, Rome

(1965*b*) Meeting on soil correlation and soil resources appraisal in India. *World Soil Resour. Rep.* **26**, Rome

(1965*c*) Soil erosion by water. Some measures for its control on cultivated lands. *Agric. Dev. Paper* **81**, Rome

(1965/73) *Catalogue of maps. Soil map of the world.* 3rd edn (1965), 4th edn (1973), Rome

(1968) Approaches to soil classification. *World Soil Resour. Rep.* **32**, Rome.

(1970) *Provisional indicative world plan for agricultural development. Vols 1 and 2.* Rome

(1973*a*) *Irrigation, drainage and salinity. An international source book.* Rome

(1973*b*) Calcareous soils. *Soils Bull.* **21**, Rome

(1974) Approaches to land classification. *Soils Bull.* **22**, Rome

(n.d.) *Guidelines for soil description.* Rome

FAO–Unesco (1971) *FAO–Unesco soil map of the world,* 1:5000000. *Vol. IV, South America.* (Map and memoir.) Paris

(1974) *FAO–Unesco soil map of the world,* 1:5000000. *Vol. I, Legend.* (Legend sheet and memoir.) Paris

Farmer, B. H. (1971) The environmental sciences and economic development. *J. Dev. Stud.* **7**, 257–69

Fittkau, E. J. and Klinge, H. (1973) On biomass and trophic structure of the central Amazonian rain forest ecosystem. *Biotropica* **5**, 1.14

Flach, K. W., Cady, J. G. and Nettleton, W. D. (1968) Pedogenetic alteration of highly weathered parent materials. *Trans. 9th Int. Congr. Soil Sci.* **4**, 343–51

Fölster, H. (1969) Slope development in SW-Nigeria during late Pleistocene and Holocene. *Göttinger Bodenkundl. Berichte* **10**, 3–56

Fölster, H, Kalk, E. and Moshrefi, N. (1971). Complex pedogenesis of ferrallitic savanna soils in south Sudan. *Geoderma* **6**, 135–49

Fosberg, F. R. (1967) A classification of vegetation for general purposes. In *Guide to the check sheet for IBP areas* (ed. G. F. Peterken, Blackwell, Oxford), 75–120

Fosberg, F. R., Garnier, B. J. and Küchler, A. W. (1961) Delimitation of the humid tropics. *Geogr. J.* **51**, 333–47

Fox, J. E. D. and Hing, T. T. (1971) Soils and forest on an ultrabasic hill north-east of Ranau, Sabah. *J. Trop. Geogr.* **32**, 38–48

Fraser, I. S. *et al.* (1958) *Report on a reconnaissance survey of the landforms, soils and present land use of the Indus plains, West Pakistan.* With maps 1:253440. Publ. for Govt of Pakistan by Govt of Canada

Gerasimov, I. P. (1968) World soil maps compiled by Soviet soil scientists. *FAO World Soil Resour. Rep.* **32**, 25–36

Gibbons, F. R., Rowan, J. N. and Downes, R. G. (1968) The role of humans in land evaluation. In *Land evaluation* (ed. G. A. Stewart), 231–8

Gile, L. H., Peterson, F. F. and Grossman, R. B. (1966) Morphological and genetic sequences of carbonate accumulation in desert soils. *Soil Sci.* **101**, 347–60

Gokhale, N. G. (1959) Soil nitrogen status under continuous cropping and with manuring in the case of unshaded tea. *Soil Sci.* **87**, 331–3

Goudie, A. (1973*a*) *Duricrusts in tropical and subtropical landscapes.* Clarendon, Oxford

(1973*b*) The geomorphic and resource significance of calcrete. *Progress Geogr.* **5**, 77–118

Goudie, A. S., Allchin, B. and Hedge, K. T. M. (1973) The former extensions of the Great Indian sand desert. *Geogr. J.* **139**, 243–57

Grant, C. J. (1964) Soil characteristics associated with the wet cultivation of rice. In *The mineral nutrition of the rice plant* (Baltimore), 15–28

Greene, H. (1945) Classification and use of tropical soils. *Proc. Soil Sci. Soc. Am.* **10**, 392–6

(1947) Soil formation and water movement in the tropics. *Soils Fertil.* **10**, 253–6

(1960) Paddy soils and rice production. *Nature* **186**, 511–13

Bibliography

(1961) Some recent work on soils of the humid tropics. *Soils Fertil.* **24**, 325–7

(1963) Prospects in soil science. *J. Soil Sci.* **14**, 1–11

Greenland, D. J. and Nye, P. H. (1959) Increases in the carbon and nitrogen contents of tropical soils under natural fallows. *J. Soil Sci.* **10**, 284–99

Grist, D. H. (1965) *Rice.* 4th edn, Longman, London

Grove, A. T. (1958) The ancient erg of Hausaland, and similar formations on the south side of the Sahara. *Geogr. J.* **124**, 528–33

(1969) Landforms and climatic change in the Kalahari and Ngamiland. *Geogr. J.* **135**, 192–212

Grove, A. T. and Warren, A. (1968) Quaternary landforms and climate on the south side of the Sahara. *Geogr. J.* **134**, 194–208

Guha, M. M. (1969) Recent advances in fertilizer usage for rubber in Malaya. *J. Rubber Res. Inst. Malaya* **21**, 207–16.

Haantjens, H. A. (1965*a*) The classification of oxisols (latosols). *CSIRO Div. Land Res. Reg. Surv. Tech. Memor.* **65/5**, Canberra

(1965*b*) Practical aspects of land system surveys in New Guinea. *J. Trop. Geogr.* **21**, 12–20

(1965*c*) Agricultural land classification for New Guinea land resources surveys. *CSIRO Div. Land Res. Reg. Surv. Tech. Memor.* **65/8**, Canberra

(1967) Major soil groups of New Guinea and their distribution. *Dept. Agric. Res. Roy. Trop. Inst. Austr. Commun.* **55**

Hallsworth, E. G. and Costin, A. B. (1953) Studies in pedogenesis in New South Wales. IV. The ironstone soils. *J. Soil Sci.* **4**, 24–47

Hallsworth, E. G., Robertson, F. R. and Gibbons, F. R. (1955) Studies in pedogenesis in New South Wales. VII. The 'gilgai' soils. *J. Soil Sci.* **6**, 1–31

Hanna, L. W. (1974) Bioclimatology and land evaluation in Uganda. In *Spatial aspects of development* (ed. B. S. Hoyle), 75–94

Hardy, F. and Ahmad, N. (1974) Soil science at ICTA/UWI, 1922–1972. *Trop. Agric., Trin.* **51**, 468–76

Harris, S. A. (1958) The gilgaied and bad-structured soils of central Iraq. *J. Soil Sci.* **9**, 169–85

(1959) The classification of gilgaied soils: some evidence from northern Iraq. *J. Soil Sci.* **10**, 27–33

Hartley, C. W. S. (1967) *The oil palm.* Longman, London

(1968) The soil relations and fertilizer requirements of some permanent crops in West and Central Africa. In *The soil resources of tropical Africa* (ed. R. P. Moss, Cambridge University Press), 155–83

Herbillon, A. J. and Tran Vinh An, J. (1969) Heterogeneity in silicon iron mixed hydroxides. *J. Soil Sci.* **20**, 223–35

Hesse, P. R. (1955) A chemical and physical study of the soils of termite mounds in East Africa. *J. Ecol.* **43**, 449–61

Hill, I. D. (1969) An assessment of the possibilities of oil palm cultivation in Western Division, The Gambia. *Land Resour. Div., Land Resour. Stud.* **6**, Tolworth

(1970) Quantitative micromorphological evidence of clay movement. In *Micromorphological techniques and applications* (ed. D. A. Osmond and P. Bullock, Soil Surv. Tech. Monograph 2), 33–42

Hissink, D. J. (1901) *Groondsoortenkaart van een gedeete van Deli.* Map, 1:100000, with explanatory notes, 18 pp. Top. Bureau te Batavia, Buitenzorg

Holmes, D. A. and Western, S. (1969) Soil-texture patterns in the alluvium of the lower Indus plains. *J. Soil Sci.* **20**, 23–37

Hornby, A. J. W. (1938) *Soil map of central Nyasaland.* Zomba

Howard, J. A. (1971) *Aerial photo-ecology.* Faber, London

444

Hoyle, B. S. (ed.) (1974) *Spatial aspects of development.* Wiley, London

Huang, W. H. and Keller, D. (1972) Organic acids as agents of chemical weathering of silicate minerals. *Nature Phys. Sci.* **239**, 149–51

Hudson, J. C. (1965) Agronomic use of a soil survey in Barbados. *Expl. Agric.* **1**, 215–24

Hudson, N. (1971) *Soil conservation.* Batsford, London

Hunting Surveys Corporation (Canada) Ltd (1958) *Landforms, soils and land use of the Indus Plains, West Pakistan*

Hunting Technical Services Ltd and Sir M. Macdonald and Partners (1966) *Lower Indus Report.* Lahore

Hutchinson, J. (ed.) (1969) *Population and food supply.* Cambridge University Press

Ignatieff, V. and Lemos, P. (1963) Some management aspects of more important tropical soils. *Soil Sci.* **95**, 243–9

Ignatieff, V. and Page, H. J. (ed.) (1958) Efficient use of fertilizers. *FAO Agric. Stud.* **43**

International Institute for Land Reclamation and Improvement (1973) Acid sulphate soils. Proceedings of the International Symposium, Wageningen, 1972. Vols I and II. *ILRI Pub.* **18**, Wageningen

Ireland, H. A., Sharpe, C. F. S. and Eargle, D. H. (1939) Principles of Gully erosion in the Peidmont of South Carolina. *US Dept. Agric. Tech. Bull.* **633**

Irvine, F. R. (1969) *West African crops.* Oxford University Press, London

Irving, H. (1956) Fertilizer experiments on yams in eastern Nigeria, 1947–1951. *Trop. Agric., Trin.* **33**, 67–78

Islam, M. A. (1966) Soils of East Pakistan. In *Scientific problems of the humid tropical zone: deltas and their implications* (Unesco), 83–7

Ivanova, E. N. and Rozov, N. N. (1960) Classification of soils and the soil map of the USSR. *Trans. 7th Int. Congr. Soil Sci.* **4**, 77–87

Jacks, G. V. (1956) The influence of man on soil fertility. *Adv. Sci.* **13**, 137–45
(1963) The biological nature of soil productivity. *Soils Fertil.* **26**, 147–50

Jackson, M. L. (1968) Weathering of primary and secondary minerals in soils. *Trans. 9th Int. Congr. Soil Sci.* **4**, 281–92

Jackson, M. L. and Sherman, G. D. (1953) Chemical weathering of minerals in soils. *Adv. Agron.* **5**, 219–318

Jacob, A. and Uexküll, H. V. (1963) *Fertilizer use. Nutrition and manuring of tropical crops.* 3rd Eng. edn, Hannover

James, P. E. (1966) *A geography of man.* 3rd edn

Jeffery, J. W. O. (1960) Iron and the Eh of waterlogged soils with particular reference to paddy. *J. Soil Sci.* **11**, 140–8
(1961) Defining the state of reduction of a paddy soil. *J. Soil Sci.* **12**, 172–9

Jenkin, R. N. and Foale, M. A. (1968) An investigation of the coconut-growing potential of Christmas Island. Vols 1 and 2. *Land Resour. Div., Land Resour. Stud.* **4**, Tolworth

Jenny, H. (1941) *Factors of soil formation. A system of quantitative pedology.* McGraw-Hill, New York
(1946) Arrangement of soil series and types according to functions of soilforming factors. *Soil Sci.* **61**, 375–91

Jenny, H. and Raychaudhuri, S. (1960) *Effect of climate and cultivation on nitrogen and organic matter reserves in Indian soils.* Indian Council Agric. Res., New Delhi

Jenny, H., Bingham, F. and Padilla-Saravia, B. (1948) Nitrogen and organic matter contents of equatorial soils of Columbia, South America. *Soil Sci.* **66**, 173–86

Bibliography

Jenny, H., Gessel, S. P. and Bingham, F. T. (1949) Comparative study of decomposition rates of organic matter in temperate and tropical regions. *Soil Sci.* **68**, 419–32

Jewitt, T. N. (1955) Gezira soil. *Sudan Min. Agric. Bull.* **12**, Khartoum

Jolly, A. L. (1952) Unit farms. *Trop. Agric., Trin.* **29**, 172–9

(1955) Peasant experimental farms. *Trop. Agric., Trin.* **32**, 257–73

Jones, M. J. (1971) The maintenance of soil organic matter under continuous cultivation at Samaru, Nigeria. *J. Agric. Sci.* **77**, 473–82

(1973) The organic matter content of the savanna soils of West Africa. *J. Soil Sci.* **24**, 42–53

(1975) Leaching of nitrate under maize at Samaru, Nigeria. *Trop. Agric., Trin.* **52**, 1–10

Jones, M. J. and Bromfield, A. R. (1970) Nitrogen in the rainfall at Samaru, Nigeria. *Nature* **227**, 86

Jones, W. O. (1959) *Manioc in Africa.* Stanford University Press

Jordan, H. D. (1964) The relation of vegetation and soil to development of mangrove swamps for rice growing in Sierra Leone. *J. Appl. Ecol.* **1**, 209–12

Jordan, C. F. (1971) Productivity of a tropical forest and its relation to a world pattern of energy storage. *J. Ecol.* **59**, 127–42

Joseph, K. T. (1968) A toposequence of limestone parent material in north Kedah, Malaya. *J. Trop. Geogr.* **27**, 19–22

Jurion, F. and Henry, J. (1969) *Can primitive farming be modernized?* INEAC, Brussels

Kantey, B. A. and Williams, A. A. B. (1962) The use of soil engineering maps for road projects. *Civil Engineer in S. Afr.* **4**, 149–59

Kantor, W. and Schwertmann, U. (1974) Mineralogy and genesis of clays in red-black soil toposequences on basic igneous rocks in Kenya. *J. Soil Sci.* **25**, 67–78

Keay, R. W. J. (1959) *Vegetation map of Africa south of the Tropic of Cancer.* Map 1 : 10 000 000 with memoir. Oxford University Press, London

Kelley, W. P. (1963) Use of saline irrigation water. *Soil Sci.* **95**, 385–91

Kellman, M. C. (1969) Some environmental components of shifting cultivation in upland Mindanao. *J. Trop. Geogr.* **28**, 40–56

Kellogg, C. E. (1949) Preliminary suggestions for the classification and nomenclature of great soil groups in tropical and equatorial regions. *Commonw. Bur. Soil Sci. Tech. Commun.* **46**, 76–85

King, L. C. (1948) A theory of bornhardts. *Geogr. J.* **112**, 83–7

(1950) The study of the world's plainlands: a new approach. *Q. J. Geol. Soc. Lond.* **106**, 101–31

(1962) *Morphology of the earth.* Oliver & Boyd, Edinburgh

(1966) The origin of bornhardts. *Z. Geomorph.* **10**, 97–8

King, R. B. (1975) Geomorphic and soil correlation analysis of land systems in the Northern and Luapula Provinces of Zambia. *Trans. Inst. Brit. Geogr.* **64**, 67–76

Kira, T. and Ogawa, H. (1971) Assessment of primary production in tropical and equatorial forests. *Nat. Resour. Res. Unesco,* **4**, 309–21

Klinge, H. (1965) Podzol soils in the Amazon basin. *J. Soil Sci.* **16**, 95–103

(1966) Verbreitung tropischer Tieflandpodsole. *Naturwissenschaften* **17**, 442–3

(1969) Climatic conditions in lowland tropical podzol areas. *Trop. Ecol.* **10**, 222–39

Klingebiel, A. A. (1966) Costs and returns of soil surveys. *Soil Conserv.* **32**, 3–6

Klingebiel, A. A. and Montgomery, P. H. (1961) Land capability classification. *US Dept. Agric. Handbk* 210

Köppen, W. (1931) *Grundriss der Klimakunde.* Berlin
(1936) *Das geographische System der Klimate.* Berlin

Kubiena, W. L. (1953) *The soils of Europe.* Madrid
(1956) Rubefizierung und Laterisierung (zu ihrer Unterscheidung durch mikromorphologische Merkmale). *Trans. 6th Int. Cong. Soil Sci.* E, 247–9
(1958) The classification of soils. *J. Soil Sci.* **9**, 9–19

Küchler, A. W. (1949) A physiognomic classification of vegetation. *Ann. Ass. Am. Geogr.* **39**, 201–10
(1967) *Vegetation mapping.* Ronald, New York

Lake, P. (1890) The geology of south Malabar, between the Beypore and Ponnáni Rivers. *Mem. Geol. Surv. India* **24**(3)

Lee, K. E. and Wood, T. G. (1971) *Termites and soils.* Academic Press, London

Leneuf, N. and Aubert, G. (1960) Essai d'évaluation de la vitesse de ferallitisation. *Trans. 7th Int. Congr. Soil Sci.* **4**, 225–8

Lock, G. W. (1969) *Sisal.* 2nd edn, Longman, London

Loughnan, F. C. (1969) *Chemical weathering of the silicate minerals.* Elsevier, Amsterdam.

Louis, H. (1964) Über Rumpfflächen- und Talbildung in den wechselfeuchten Tropen besonders nach Studien in Tanganyika. *Z. Geomorph. Sonderheft* **8**, 43–70

Mabbutt, J. A. (1965) The weathered land surface in central Australia. *Z. Geomorph.* **9**, 82–114
(1968) Review of concepts of land classification. In *Land evaluation* (ed. G. A. Stewart) 11–28

Mabbutt, J. A. and Scott, R. M. (1966) Periodicity of morphogenesis and soil formation in a savannah landscape near Port Moresby, Papua. *Z. Geomorph.* **10**, 69–89

Mäckel, R. (1974) Dambos: a study in morphodynamic activity on the plateau regions of Zambia. *Catena* **1**, 327–66

Maignen, R. (1959) Soil cuirasses in tropical West Africa. *Sols Afr.* **4**, 4–41
(1961) The transition from ferruginous tropical soils to ferrallitic soils in the south-west of Senegal. *Sols Afr.* **6**, 171–228
(1966) Review of research on laterites. *Nat. Resour. Res., Unesco* **4**

Makin, J., Schilstra, J. and Theisen, A. A. (1969) The nature and genesis of certain aridisols in Kenya. *J. Soil Sci.* **20**, 111–25

Martini, J. (1970) Allocation of cation exchange capacity to soil fractions in seven surface soils from Panama and the application of a cation exchange factor as a weathering index. *Soil Sci.* **109**, 324–31

Maud, R. R. (1965) Laterite and lateritic soil in coastal Natal, South Africa. *J. Soil Sci.* **16**, 60–72

McCallien, W. J., Ruxton, B. P. and Walton, B. J. (1964) Mantle rock tectonics. A study in tropical weathering at Accra, Ghana. *Overs. Geol. Min. Resour.* **14**, 257–94

McFarlane, M. J. (1970) Laterization and landscape development in Kyagwe, Uganda. *Q. J. Geol. Soc. Lond.* **126**, 501–39

McGown, A. and Iley, P. (1973) A comparison of data from agricultural soil surveys with engineering investigations for roadworks in Ayrshire. *J. Soil Sci.* **24**, 145–56

McKeague, J. A. and Cline, M. G. (1963) Silica in soils. *Adv. Agron.* **15**, 339–96

Meiklejohn, J. (1955) Nitrogen problems in tropical soils. *Soils Fertil.* **18**, 459–63
    (1968) Numbers of nitrifying bacteria in some Rhodesian soils under natural grass and improved pastures. *J. Appl. Ecol.* **5**, 291–300

Milne, G. (1935) Composite units for the mapping of complex soil associations. *Trans. 3rd Int. Congr. Soil Sci.* **1**, 345–7
    (1935/6) *A provisional soil map of East Africa, with explanatory memoir.* Map 1:2000000 (1935), memoir (1936). London
    (1936) Normal erosion as a factor in soil profile development. *Nature*, **138**, 548–9
    (1937) Note on soil conditions and two East African vegetation types. *J. Ecol.* **25**, 254–8
    (1947) A soil reconnaissance journey through parts of Tanganyika Territory, December 1935 to February 1936. *J. Ecol.* **35**, 192–265

Mohr, E. C. J. and van Baren, F. A. (1954/72) *Tropical soils.* 1st edn (1954). 3rd edn (1972) with J. van Schuylenborgh, Mouton, The Hague

Monteith, J. L. (1972) Solar radiation and productivity in tropical ecosystems. *J. Appl. Ecol.* **9**, 747–66.

Moore, A. W. (1960) The influence of annual burning on a soil in the derived savanna zone of Nigeria. *Trans. 7th Int. Congr. Soil Sci.* **4**, 257–64
    (1963) Nitrogen fixation in latosolic soil under grass. *Plant Soil* **19**, 127–38

Moormann, F. R. (1963) Acid sulfate soils (cat-clays) of the tropics. *Soil Sci.* **95**, 271–80

Morin, W. J. and Parry, W. T. (1971) Geotechnical properties of Ethiopian volcanic soils. *Géotechnique* **21**, 223–32

Morison, C. G. T. (1949) The catena concept and the classification of tropical soils. *Comm. Bur. Soil Sci. Tech. Commun.* **46**, 124–8

Morison, C. G. T., Hoyle, A. C. and Hope-Simpson, J. F. (1948) Tropical soil-vegetation catenas and mosaics. A study in the south-western part of the Anglo-Egyptian Sudan. *J. Ecol.* **36**, 1–84

Moss, R. P. (1965) Slope development and soil morphology in a part of south-west Nigeria. *J. Soil Sci.* **16**, 192–209
    (1968) Land use, vegetation and soil factors in south west Nigeria, a new approach. *Pacific Viewpoint* **9**, 107–26
    (1969) The appraisal of land resources in tropical Africa: a critique of some concepts. *Pacific Viewpoint* **10**, 18–27

Moss, R. P. and Morgan, W. B. (1970) Soils, plants and farmers in West Africa. In *Human ecology in the tropics* (ed. J. P. Garlick and R. W. J. Keay, Pergamon, Oxford), 1–31

Muir, A. and Stephen, I. (1956) The superficial deposits of the lower Shire valley, Nyasaland. *Col. Geol. Min. Resour.* **6**, 391–406

Mulcahy, M. J. (1960) Laterites and lateritic soils in south-western Australia. *J. Soil Sci.* **11**, 206–25

Mulcahy, M. J. and Humphries, A. W. (1967) Soil classification, soil surveys and land use. *Soils Fertil.* **30**, 1–8

Murdoch, G., Webster, R. and Lawrance, C. J. (1971) *A land system atlas of Swaziland.* Christchurch, Hauts

Nand, N. (1966) Distribution and spatial arrangement of rural population in East Rajasthan, India. *Ann. Ass. Am. Geogr.* **56**, 205–19

National Academy of Sciences (1972) *Soils of the humid tropics.* Wash. DC

Nelson, L. A. (1963) Detailed land classification – Island of Oahu. *Univ. Hawaii Land Study Bur., Bull.* 3

Netting, R. McC. (1968) *Hill farmers of Nigeria.* University of Washington, Seattle

Nettleton, W. D., Flach, K. W. and Brasher, B. R. (1969) Argillic horizons without clay skins. *Proc. Soil Sci. Soc. Am.* **33**, 121–4

Ng, S. K. (1966) Soils. In *The oil palm in Malaya* (Min. Agric. Co-op., Kuala Lumpur), 24–47

Ng, S. K. and Law, W. M. (1971) Pedogenesis and soil fertility in West Malaysia. *Nat. Resour. Res., Unesco* **11**, 129–31

Nikiforoff, C. C. (1949) Weathering and soil evolution. *Soil Sci.* **67**, 219–30

Norman, A. G. (ed.) (1963) *The soybean*. Academic Press, New York

Norris, D. O. (1969) Observations on the nodulation status of rainforest leguminous species in Amazonia and Guyana. *Trop. Agric., Trin.* **46**, 145–51

Nortcliff, S. (1974) *Aspects of the spatial variability of soils in Norfolk*. Ph. D. thesis, Univ. E. Anglia

Northcote, K. H. (1971) *A factual key for the recognition of Australian soils*. 3rd Edn, Rellin, Adelaide

Northey, R. D. (1966) Correlation of engineering and pedological soil classification in New Zealand. *N. -Z. J. Sci.* **9**, 809–33

Nossin, J. J. (1962) Coastal sedimentation in northeastern Johore (Malaya). *Z. Geomorph.* **6**, 296–316

 (1964) Geomorphology of the surroundings of Kuantan (eastern Malaya). *Geol. en Mijnb.* **43**, 157–82

Nye, P. H. (1954/55) Some soil forming processes in the humid tropics. I–V. *J. Soil Sci.* **5**, 7–21 and **6**, 51–83

 (1961) Organic matter and nutrient cycles under moist tropical forest. *Plant Soil* **13**, 333–46

 (1963) Soil analysis and the assessment of fertility in tropical soils. *J. Sci. Food Agric.* **14**, 277–80

Nye, P. H. and Greenland, D. J. (1960) The soil under shifting cultivation. *Commonw. Bur. Soils Tech. Commun.* **51**

Oades, J. M. (1963) The nature and distribution of iron compounds in soils. *Soils Fertil.* **26**, 69–80

Oertel, A. C. (1968) Some observations incompatible with clay illuviation. *Trans. 9th Int. Congr. Soil Sci.* **4**, 481–8

Ojo-Atere, J. and Murdoch, G. (1971) Soil series straddling the savanna–forest boundary in the Western State of Nigeria. *Land Resour. Div. Misc. Rep.* **110**

Ollier, C. D. (1959) A two-cycle theory of tropical pedology. *J. Soil Sci.* **10**, 137–48

 (1967) Landform description without stage names. *Austr. Geogr. Stud.* **5**, 73–80

 (1969) *Weathering*. Oliver & Boyd, Edinburgh

Ollier, C. D. *et al.* (1969) *Land systems of Uganda: terrain classification and data storage*. MEXE Rep. **959**, Dept. Agric., Univ. Oxford

Olson, G. W. (1974) Land classifications. *Search, Agric.* **4**(7), 1–34

Opuwaribo, E. and Odu, C. T. I. (1974) Fixed ammonium in Nigerian soils. *J. Soil Sci.* **25**, 256–64

Orvedal, A. C. (1975) *Bibliography of soils of the tropics. I. Tropics in general and tropical Africa*. US Agency Internat. Dev., Wash DC

Pallister, J. W. (1956) Slope development in Buganda. *Geogr. J.* **122**, 80–7

Panabokke, C. R. (1959) A study of some soils in the dry zone of Ceylon. *Soil Sci.* **87**, 67–74

Panabokke, C. R. and Moorman, F. R. (1961) Soils of Ceylon. *Tropl Agriculturalist* **67**, 5–70

Panton, W. P. (1956) Types of Malayan laterite and factors affecting their distribution. *Trans. 6th Int. Congr. Soil Sci.* E, 419–23

 (1970) The application of land use and natural resource surveys to national planning: the Malaysian experience. In *New possibilities and techniques for land use and related surveys* (ed. I. H. Cox, Geographical Publications Ltd, Berkhamsted), 129–38

Papadakis, J. (1966) *Climates of the world and their agricultural potentialities.* Buenos Aires

Paton, T. R. (1953–59) *Soils of the Semporna Peninsula.* Map 1:50000, with report. HMSO

(1961) Soil genesis and classification in Central Africa. *Soils Fertil.* **24**, 249–51

(1974) Origin and terminology for gilgai in Australia. *Geoderma* **11**, 221–42

Patrick, W. H. and Mahapatra, I. C. (1968) Transformation and availability to rice of nitrogen and phosphorous in waterlogged soils. *Adv. Agron.* **20**, 323–60

Pédologie (1965) Soil classification. 3rd International Symposium. *Pédologie, Ghent,* no. spec. **3**

Pendleton, R. L. (1930) *Soil map of the La Carlota area, Occidental Negros, Phillipine Islands.* 1:30000

Penman, H. L. (1948) Natural evaporation from open water, bare soil and grass. *Proc. Roy. Soc. A* **193**, 120–45

Phillips, J. (1959) *Agriculture and ecology in Africa.* Faber, London

Pons, L. J. and Zonneveld, I. S. (1965) Soil ripening and soil classification. *Internat. Inst. Land Reclam. Improvement, Pub.* **13**

Prentice, A. N. (1972) *Cotton, with special reference to Africa.* Longman, London

Prescott, J. A. (1931) The soils of Australia in relation to vegetation and climate. *CSIR Bull.* **52**

Prescott, J. A. and Pendleton, R. L. (1952) *Laterite and lateritic soils.* Commonwealth Agric. Bureaux, Farnham Royal, Bucks

Pullan, R. A. (1967) A morphological classification of lateritic ironstones and ferruginized rocks in northern Nigeria. *Nigerian J. Sci.* **1**, 161–74

(1970) The soils, soil landscapes and geomorphological evolution of a metasedimentary area in northern Nigeria. *Dept. Geogr. Univ. Liverpool Res. Paper* **6**

(1974) Biogeographical studies and agricultural development in Zambia. *Geography* **59**, 309–21

Purseglove, J. W. (1968) *Tropical crops. Dicotyledons.* Vols 1 and 2. Longman, London

(1972) *Tropical crops. Monocotyledons.* Vols 1 and 2. Longman, London

Pushparajah, E. and Guha, M. M. (1968) Fertilizer response in *Hevea brasiliensis* in relation to soil type and leaf nutrient status. *Trans. 9th Int. Congr. Soil Sci.* **4**, 85–93

Radwanski, S. A. (1968) Field observations of some physical properties in alluvial soils of arid and semi-arid regions. *Soil Sci.* **106**, 314–6

(1969) Improvement of red acid sands by the neem tree (*Azadirachta indica*) in Sokoto, North-Western State of Nigeria. *J. Appl. Ecol.* **6**, 507–11

Radwanski, S. A. and Ollier, C. D. (1959) A study of an East African catena. *J. Soil Sci.* **10**, 149–68

Radwanski, S. A. and Wickens, G. E. (1967) The ecology of *Acacia albida* on mantle soils in Zalingei, Jebel Marra, Sudan. *J. Appl. Ecol.* **4**, 569–79

Ratter, J. A., Richards, P. W., Argent, G. and Gifford, D. R. (1973) Observations on the vegetation of northeastern Mato Grosso. *Phil. Trans. Roy. Soc. London B* **266**, 449–92

Rattray, J. M. (1960) *The grass cover of Africa.* Map 1:10000000 and memoir, Rome.

Rauschkolb, R. S. (1971) Land degradation. *Soils Bull., FAO* **13**

Raychaudhuri, S. P. (1962) Development of legends for classification and nomenclature of Indian soils. *J. Indian Soc. Soil Sci.* **10**, 1–18

(1963) *Land resources of India. Vol. I. Indian soils – their classification, occurrence and properties.* New Delhi

Raychaudhuri, S. P. and Govinda Rajan, S. V. (1971) Soil genesis and classification. In *Review of soil research in India* (ed. J. S. Kanwar and S. P. Raychaudhuri, New Delhi), 107–36

Reeve, R. C. and Bower, C. A. (1960) Use of high-salt waters as a flocculant and source of divalent cations for reclaiming sodic soils. *Soil Sci.* **90**, 139–44

Richards, B. N. (1974) *Introduction to the soil ecosystem.* Longman, Harlow

Richards, L. A. (1954) Diagnosis and improvement of saline and alkali soils. *US Dept. Agric. Handbk* **60**

Richards, P. W. (1941) Lowland tropical podzols and their vegetation. *Nature* **148**, 129–31

Richardson, H. L. (1963) The fertility potentialities and uses of tropical soils. *Trop. Sci.* **5**, 166–78

Riquier, J. (1969) Contribution a l'étude des 'stone lines' en régions tropicale et equatoriale. *Cah. ORSTOM Sér. Pédol.* **7**, 71–111

(1974) A summary of parametric methods of soil and land evaluation. *Soils Bull., FAO* **22**, 47–53

Riquier, J., Bramao, D. L. and Cornet, J. P. (1970) A new system of soil appraisal in terms of actual and potential productivity. *FAO Paper* AGL:TESR/70/6

Robertson, V. C. (1970) Land and water resource planning in developing countries. *Outlook on Agric.* **6**, 148–57

Robertson, V. C. and Stoner, R. F. (1970) Land use surveying: a case for reducing the costs. In *New possibilities and techniques for land use and related surveys* (ed. I. H. Cox), 3–16

Robertson, V. C., Jewitt, T. N., Forbes, A. P. S. and Law, R. (1968) The assessment of land quality for primary production. In *Land evaluation* (ed. G. A. Stewart, Melbourne), 88–103

Robinson, G. H. (1971) Exchangeable sodium and yields of cotton on certain clay soils of Sudan. *J. Soil Sci.* **22**, 328–35

Robinson, J. B. D. and Gacoka, P. (1956) Evidence of the upward movement of nitrate during the dry season in the Kikuyu red loam coffee soil. *J. Soil Sci.* **13**, 133–9

Rodin, L. E. and Basilevič (1968) World distribution of plant biomass. *Nat. Resour. Res. Unesco* **5**, 45–52

Rodríguez, F. A. (1968) Effect of soil acidity and liming on yields and composition of sugarcane growing on an ultisol. *J. Agric. Univ. P. Rico* **52**, 85–100

Rozanov, B. G. and Rozanova, I. M. (1968) Genesis of 'degraded' soils on paddies in the tropics. In *Geography and classification of soils of Asia* (ed. V. A. Kovda and E. V. Lobova, Jerusalem), 243–8

Rozov, N. N. and Ivanova, E. N. (1968) Soil classification and nomenclature used in Soviet pedology, agriculture and forestry. *World Soil Resour. Rep., FAO*, **53**, 53–77

Ruhe, R. V. (1959) Stone lines in soils. *Soil Sci.* **87**, 223–31

Ruhe, R. V. and Cady, J. G. (1954) Latosolic soils of central African interior high plateaus. *Trans 5th Int. Congr. Soil Sci.* **4**, 401–7

Russell, E. J. (1954) *World population and world food supplies.* Allen & Unwin, London

Russell, E. W. (1968) Climate and soils. *Nat. Resour. Res., Unesco* **7**, 193–200

Ruthenburg, H. (1971) *Farming systems in the tropics.* Oxford University Press, London

Ruxton, B. P. (1968) Measures of the degree of chemical weathering of rocks. *J. Geol.* **76**, 518–27

Ruxton, B. P. and Berry, L. (1957) Weathering of granite and associated erosional features in Hong Kong. *Bull. Geol. Soc. Am.* **68**, 1263–92

Sawney, B. L. and Norrish, K. (1971) pH dependent cation exchange capacity: minerals and soils of tropical regions. *Soil Sci.* **112**, 213–5

Schroth, C. L. (1971) Soil sequences of western Samoa. *Pacific Sci.* **25**, 291–300

Schwertmann, U. (1971) Transformation of haematite to goethite in soils. *Nature* **232**, 624–5

Scott, R. M., Webster, R. and Lawrence, C. J. (1971) *A land system atlas of western Kenya*. With map 1:500000. Christchurch, Hants

Segalen, P. (1969) Contribution a la connaissance de la couleur des sols a sesqui-oxydes de la zone intertropicale: sols jaunes et sols rouges. *Cah. ORSTOM Sér Pédol.* **7**, 225–36

(1971) Metallic oxides and hydroxides in soils of the warm and humid areas of the world: formation, identification, evolution. *Nat. Resour. Res. Unesco* **11**, 25–38

Sharpe, C. F. S. (1938) *Landslides and related phenomena*. Columbia University Press, New York

Shoji, S. and Masui, J.-I. (1971) Opaline silica of recent volcanic soils in Japan. *J. Soil Sci.* **22**, 101–8

Siband, P. (1972) Étude de l'évolution des dols sous culture traditionelle en Haute-Casamance. *Agron. Trop.* **27**, 674–91

Sieffermann, G. (1973) Les sols de quelques régions volcaniques du Cameroun. *Mém. ORSTOM* **66**

Sillanpää, M. (1972) Trace elements in soils and agriculture. *Soils Bull., FAO* **17**

Simakova, M. S. (1964) *Soil mapping by colour air photography*. Jerusalem

Simmonds, N. W. (1966) *Bananas*. 2nd edn, Longman, London

Simonson, R. W. (1954) Morphology and classification of the regur soils of India. *J. Soil Sci.* **5**, 275–88

Simonson, R. W. (ed.) (1974) *Non-agricultural applications of soil surveys*. Elsevier, Amsterdam

Sivarajasingham, S., Alexander, L. T., Cady, J. G. and Cline, M. G. (1962) Laterite. *Adv. Agron.* **14**, 1–60

Slager, S., Jongmans, A. G. and Pons, L. J. (1970) Micromorphology of some tropical alluvial clay soils. *J. Soil Sci.* **21**, 233–41

Smith, G. D. (1963) Objectives and basic assumptions of the new soil classification. *Soil Sci.* **96**, 6–16

(1965) Lectures on soil classification. *Pédologie, Ghent*, no. spec. **4**

(1968) Soil classification in the United States. *World Soil Resour. Rep., FAO* **32**, 6–24

Smith, R. and Robertson, V. C. (1962) Soil and irrigation classification of shallow soils overlying gypsum beds, northern Iraq. *J. Soil Sci.* **13**, 106–15

Smyth, A. J. (1966) The selection of soils for cocoa. *Soils Bull., FAO* **5**

(1970) The preparation of soil survey reports. *Soils Bull., FAO* **9**

Soil Conservation Service (1966) Aerial-photo interpretation in classifying and mapping soils. *US Dept. Agric Handbk* **294**

Soil Survey Staff (1951) *Soil survey manual*. Wash. DC

(1960) *Soil classification. A comprehensive system. 7th approximation*. Wash. DC

(1967) *Supplement to soil classification system (7th approximation)*. Wash. DC

Sombroek, W. G. (1971) Ancient levels of plinthisation in N. W. Nigeria. In *Paleo-*

pedology: origin, nature and dating of palaeosols (ed. D. H. Yaalon, Jerusalem), 329–37

Speight, J. G. (1967) Explanation of land system descriptions. In *Lands of Bougainville and Buka Islands, Territory of Papua and New Guinea* (ed. R. M. Scott *et al.*, Melbourne), 174–84

(1968) Parametric description of land form. In *Land evaluation* (ed. G. A. Stewart), 239–50

(1974) A parametric approach to landform regions. *Inst. Br. Geogr. Spec. Pub.* 7, 213–30

Steele, J. G. (1967) Soil survey interpretation and its use. *Soils Bull., FAO* 8

Stephen, I., Bellis, E. and Muir, A. (1956) Gilgai phenomena in tropical black clays of Kenya. *J. Soil Sci.* 7, 1–9

Stephens, C. G. (1947) Functional synthesis in pedogenesis. *Trans. Roy. Soc. Austr.* 71, 168–81

(1961) Laterite at the type locality, Angadipuram, Kerala, India. *J. Soil Sci.* 12, 214–7

(1962) *A manual of Australian soils.* 3rd edn, CSIRO, Melbourne

(1971) Laterite and silcrete in Australia. *Geoderma*, 5, 5–52

Stephens, D. (1962) Upward movement of nitrate in a bare soil in Uganda. *J. Soil Sci.* 13, 52–9

Stewart, G. A. (ed.) (1968) *Land evaluation.* Macmillan, Melbourne

Stobbs, A. R. (1970) Soil survey procedures for development purposes. In *New possibilities and techniques for land use and related surveys* (ed. I. H. Cox, Geographical Publications Ltd, Berhamsted), 41–64

Stobbs, A. R. (1971) *Soils of the Dwangwa lakeshore plain.* Map 1:2500. Land Resources Div., Tolworth

Stoops, G. (1968) Micromorphology of some characteristic soils of the lower Congo (Kinshasa). *Pédologie, Ghent* 18, 110–49

Sweeting, M. M. (1958) The karstlands of Jamaica. *Geogr. J.* 124, 184–99

Swindale, L. D. (1965) The properties of soils derived from volcanic ash. *World Soil Resour. Rep., FAO* 14, 82–6

Sys, C. (1955) The importance of termites in the formation of latosols in the region of Elizabethville. *Sols Afr.* 3, 393–5

(1960) Principles of soil classification in the Belgian Congo. *Trans. 7th Int. Congr. Soil Sci.* 4, 112–8

(1967) The concept of ferrallitic and fersiallitic soils in Central Africa. Their classification and their correlation with the 7th approximation. *Pédologie, Ghent* 17, 284–325

Sys, C. and Frankart, R. (1971) Land capability classification in the humid tropics. *Sols Afr.* 16, 153–75

Sys, C. *et al.* (1961) La cartographie des sols au Congo, ses principes et ses methodes. *INEAC Sér. Tech.* 66, Brussels

Tavernier, R. and Sys, C. (1965) Classification of soils of the Republic of Congo (Kinshasa). *Pedologie, Ghent,* no. spec. 3, 91–136

Thomas, D. (1973) Nitrogen from tropical pasture legumes on the African continent. *Herbage Abstr.* 43, 33–9

Thomas, M. F. (1965) Some aspects of the geomorphology of tors and domes in Nigeria. *Z. Geomorph.* 9, 63–81

(1966a) Some geomorphological implications of deep weathering patterns in crystalline rocks in Nigeria. *Trans. Inst. Br. Geogr.* 40, 173–93

(1966b) The origin of bornhardts. *Z. Geomorph.* 10, 478–80

(1969) Geomorphology and land classification in tropical Africa. In *Environment and land use in Africa* (ed. M. F. Thomas and G. W. Whittington, Methuen, London), 103–46

(1974) *Tropical geomorphology. A study of weathering and landform development in warm climates*. Macmillan, London

Thomas, W. J. (1973) Looking at the future of agriculture. *J. Agric. Econ.* **24**, 443–63

Thompson, J. G. (1960) A description of the growth habits of mopani in relation to soil and climatic conditions. *Proc. First Federal Sci. Congr., Salisbury* 181–6

Thorne, D. W. and Peterson, H. B. (1954) *Irrigated soils*. 2nd edn, Blakiston, New York

Thornthwaite, C. W. (1933) The climates of the earth. *Geogr. Rev.* **23**, 433–49

(1948) An approach towards a rational classification of climate. *Geogr. Rev.* **38**, 55–94

Thorp, J. (1936) *Geography and soils of China*. Nanking

Thorp, J. and Baldwin, M. (1938) New nomenclature of the higher categories of soil classification as used in the Department of Agriculture. *Proc. Soil Sci. Soc. Am.* **3**, 260–8

Tinkler, P. B. H. (1964) Studies on soil potassium. IV. Equilibrium cation activity ratios and responses to potassium fertilizer of Nigerian oil palms. *J. Soil Sci.* **15**, 35–41

Tiurin, I. V. (1965) The system of soil classification in the USSR. *Pédologie, Ghent*, no. spec. **3**, 7–24

Tomlinson, P. R. (1970) Variations in the usefulness of rapid soil mapping in the Nigerian savanna. *J. Soil Sci.* **21**, 162–72

Townsend, F. C. and Reed, L. W. (1971) Effects of amorphous constituents on some mineralogical and chemical properties of a Panamanian latosol. *Clays Clay Mins.* **19**, 303–10

Townshend, J. R. G. (1970) Geographical research on the Royal Society/Royal Geographical Society's Expedition to north-eastern Mato Grosso, Brazil: a symposium. V. Geology, slope form, and slope process, and their relation to the occurrence of laterite. *Geogr. J.* **136**, 392–9

Trapnell, C. G. (1943) *The soils, vegetation and agriculture of north-eastern Rhodesia*. Lusaka

Trapnell, C. G. and Clothier, J. N. (1937/57) *The soils, vegetation and agricultural systems of north western Rhodesia. Report of the ecological survey*. With map. 1st edn (1937), 2nd edn (1957), Lusaka.

Trapnell, C. G., Martin, J. D. and Allan, W. (1948/50) *Vegetation–soil map of Northern Rhodesia*. Map 1:1000000. With accompanying memorandum (by C. G. Trapnell). 1st edn (1948), 2nd edn (1950), Lusaka

Trendall, A. F. (1962) The formation of 'apparent peneplains' by a process of combined laterisation and surface wash. *Z. Geomorph.* **6**, 183–97

Trewartha, G. T. (1943/1968) *An introduction to weather and climate*. 2nd edn (1943), 4th edn (1968), McGraw-Hill, New York

Tricart, J. (1956) Aspects géomorphologiques du delta du Sénégal. *Rev. Géomorph. Dyn.* **7**, 65–86

Uganda Government (1958–61) *Soil maps* 1:250000. 12 sheets

Unesco (1966) *Scientific problems of the humid tropical zone: deltas and their implications*. Paris

(1968) Aerial surveys and integrated studies. *UNESCO Nat. Resour. Res.* **6**

(1973) International classification and mapping of vegetation. *UNESCO Ecology and Conservation* **6**

United States Bureau of Reclamation (1953) *Bureau of reclamation manual. Vol. V, Irrigated land use. Part 2, Land classification.* Wash. D.C.

Urquhart, D. H. (1961) *Cocoa.* 2nd edn, Longman, London

USSR Academy of Sciences (1964) *Physical geographic atlas of the world.* Moscow

Van Beers, W. F. J. (1962) Acid sulphate soils. *Inst. Land Reclamation Improvement Bull.* 3

Van Lopik, J. R. and Kolb, C. R. (1959) A technique for preparing desert terrain analogs. *Army Waterways Expt. Stn. Corps Engineers, Tech. Rep.* 3–506

Van Schuylenborgh, J. (1971) Weathering and soil-forming processes in the tropics. *Nat. Resour. Res., Unesco* 11, 39–50

Van Wambeke, A. R. (1962) Criteria for classifying tropical soils by age. *J. Soil Sci.* 13, 124–32

(1967) Recent developments in the classification of the soils of the tropics. *Soil Sci.* 104, 309–13

(1974) Management properties of ferralsols. *Soils Bull., FAO* 23

Vermeer, D. E. (1970) Population pressure and crop rotational changes among the Tiv of Nigeria. *Ann. Ass. Am. Geogr.* 60, 299–314

Vincent, V., Thomas, R. G. and Anderson, R. (1961) *An agricultural survey of Southern Rhodesia. Part I. The agro-ecological survey. Part II. The agro-economic survey.* Salisbury

Vink, A. P. A. (1968) Aerial photographs and the soil sciences. *Nat. Resour. Res., Unesco* 6, 81–141

(1975) *Land use in advancing agriculture.* Springer, Berlin

Walker, P. H. (1962) Soil layers on hillslopes: a study at Nowra, NSW, Australia. *J. Soil Sci.* 13, 167–77

Walther, J. (1915) Über den Laterit in Westaustralien. *Z. Dtsch. Geol. Ges.* 67B, 113–40

Watson, J. P. (1958) *Soil and land use surveys. No. 3. St Vincent.* Imperial College of Tropical Agriculture, Trinidad

(1962a) Leached, pallid soils of the African plateau. *Soils Fertil.* 25, 1–4

(1962b) Formation of gibbsite as a primary weathering product of acid igneous rocks. *Nature* 196, 1123–4

(1962c) The soil below a termite mound. *J. Soil Sci.* 13, 46–51

(1964/5) A soil catena on granite in Southern Rhodesia. *J. Soil Sci.* 15, 238–57

(1965) Soil catenas. *Soils Fertil.* 28, 307–10

(1967) A termite mound in an Iron Age burial ground in Rhodesia. *J. Ecol.* 55, 63–9

(1969) Water movement in two termite mounds in Rhodesia. *J. Ecol.* 57, 441–51

(1974a) Termites in relation to soil formation, groundwater, and geochemical prospecting. *Soils Fertil.* 37, 111–4

(1974b) Calcium carbonate in termite mounds. *Nature* 247, 74

Webster, R. (1960) Soil genesis and classification in Central Africa. *Soils Fertil.* 23, 77–9

(1965) A catena of soils on the Northern Rhodesia plateau. *J. Soil Sci.* 16, 31–43

(1968a) Fundamental objections to the 7th approximation. *J. Soil Sci.* 19, 354–66

(1968b) Soil classification in the United States: a short review of the seventh approximation. *Geogr. J.* 134, 394–6

Webster, R. and Beckett, P. H. T. (1968) Quality and usefulness of soil maps. *Nature* 219, 680–2

West, G. and Dumbleton, M. J. (1970) The mineralogy of tropical weathering illustrated by some West Malaysian soils. *Q. J. Engng Geol.* 3, 25–40

Bibliography

Western, S. (1972) The classification of arid zone soils. I and II. *J. Soil Sci.* **23**, 266–97

White, L. P. (1967) Ash soils in western Sudan. *J. Soil Sci.* **18**, 309–17

(1971) The ancient erg of Hausaland in south-western Niger. *Geogr. J.* **137**, 69–73

Whiteside, E. P. (1959) A proposed system of genetic soil-horizon designations. *Soils Fertil.* **22**, 1–8

Wild, A. (1972) Nitrate leaching under bare fallow at a site in northern Nigeria. *J. Soil Sci.* **23**, 315–24

Williams, M. A. J. (1968*a*) Termites and soil development near Brocks Creek, Northern Territory. *Austr. J. Sci.* **31**, 153–4

(1968*b*) Soil salinity in the west central Gezira, Republic of the Sudan. *Soil Sci.* **105**, 451–64

(1968*c*) The influence of salinity, alkalinity and clay content on the hydraulic conductivity of soils in the west central Gezira. *Sols Afr.* **13**, 35–49

(1968*d*) A dune catena on the clay plains of the west central Gezira, Republic of the Sudan. *J. Soil Sci.* **19**, 367–78

Williams, W. A. (1967) The role of Leguminosae in pasture and soil improvement in the neotropics. *Trop. Agric., Trin.* **44**, 103–15

Wilson, A. T. (1956) *Report of a soil and land-use survey, Copperbelt, Northern Rhodesia*. Lusaka

Woolnough, W. G. (1918) The physiographic significance of laterite in Western Australia. *Geol. Mag.* **65**, 385–93

(1930) The influence of climate and topography on the formation and distribution of products of weathering. *Geol. Mag.* **67**, 123–32

Wright, A. C. S. (1963) Soils and land use of Western Samoa. *DSIR (New-Zealand) Soil Bur. Bull.* **22**

Wrigley, G. (1969) *Tropical agriculture*. Faber, London

Yaalon, D. H. and Yaron, B. (1966) Framework for man-made soil changes – an outline of metapedogenesis. *Soil Sci.* **102**, 272–7

Young, A. (1959/60) Soil survey. Chikwawa-Ngabu. *Ann. Rep. Dept. Agric. Nyasaland, Part II*, 190–201

(1960/61) Ecology of experiment stations. *Ann. Rep. Dept. Agric. Nyasaland, Part II*, 7–27

(1968*a*) Natural resource surveys for land development in the tropics. *Geography* **53**, 229–48

(1968*b*) Mapping Africa's natural resources. *Geogr. J.* **134**, 236–41

(1968*c*) Slope form and the soil catena in savanna and rainforest environments. *Br. Geomorph. Res. Group, Occas. Paper* **5**, 3–12

(1969*a*) Natural resource survey in Malawi: some considerations of the regional method in environmental description. In *Environment and land use in Africa* (ed. M. F. Thomas and G. W. Whittington, Methuen, London), 355–84

(1969*b*) Present rate of land erosion. *Nature* **224**, 851–2

(1972*a*) *Slopes.* Oliver & Boyd, Edinburgh

(1972*b*) The soil catena: a systematic approach. In *International geography 1972* (ed. W. P. Adams and F. M. Helleiner, Toronto), 287–9

(1973*a*) Soil survey procedures in land development planning. *Geogr. J.* **139**, 53–64

(1973*b*) Rural land evaluation. In *Evaluating the human environment* (ed. J. A. Dawson and J. C. Doornkamp, Arnold), 5–33

(1974*a*) The appraisal of land resources. In *Spatial aspects of development* (ed. B. S. Hoyle, Wiley), 29–50

(1974*b*) Some aspects of tropical soils. *Geography* **59**, 233–9

(1974*c*) The rate of slope retreat. *Inst. Br. Geogr. Spec. Pub.* **7**, 65–78

(1975) Crop/land relationships and the nature of decision-making on land use. *FAO World Soil Resour. Rep.* **45**, 85–7

Young, A. and Brown, P. (1962) *The physical environment of northern Nyasaland with special reference to soils and agriculture.* Zomba

(1965) *The physical environment of central Malawi with special reference to soils and agriculture.* Zomba

Young, A. and Stephen, I. (1965) Rock weathering and soil formation on high-altitude plateaux of Malawi. *J. Soil Sci.* **16**, 322–33

Young, A. and Stobbs, A. R. (1965/71) *Malawi. Natural regions and areas. Environmental conditions and agriculture.* Map 1: 500 000. Dir. of Overseas Surveys, Tolworth

INDEX

Index

Index

relative accumulation 160
relative nutrient status 285
relative organic matter status 105, 285
relative relief 34–7
relative variance 347–8
relict features 269–71
relief 22–37, 59–60
relief unit 23, 37
remote sensing 357–8, 361–2
rendzina 190, 244–6
report, soil survey 376–8
representative profile description 365–6
requirements
    land evaluation 391–2
    major kinds of land use 403–7
reserve nutrients 292
reserve phosphorous 297
resilication 77
resistance to erosion 289
resource inventory 334, 339
response, soil 373
Rhodesia 62, 119, 208, 265–6, 270–1, 281, 330
rice 230, 312
Richards, B. N. 107, 294
Richards, L. A. 206, 209, 211, 309, 379
richness of weathering 74–6
ridge-and-ravine 29
rift valley 26, 62
ripening 220
Riquier, J. 302, 399–400, 4–9–11
road engineering 349–51, 407
Robertson, V. C. 369, 418
Robinson, G. H. 209
Robinson, J. B. D. 295
rock composition 17–9, 76
rock debris 177–9
Rodin, L. E. 43
root carbon 117
root crops 313–5
Rozanov, B. G. 230
Rozov, N. N. 254
rubber 324–5
rubefication 88
rubrisol 252–3
Ruhe, R. V. 151
Russell, E. J. 423
Russell, E. W. 70
Ruthenburg, H. 121
Ruxton, B. P. 261–2

Saharan zone 12
Sahel zone 12
saline irrigation 211
saline soil 206–9
salinity 202–4
Salinity Laboratory, US 205–7, 209

salinization 84, 204, 210–1
salt tolerance 209
sampling 366
sand fraction 97
sandy ferruginous soil 135
sandy leached ferrallitic soil 139
satellite imagery 362
saturation, base 95, 299
Samoa 55
savanna 51–2
savanna climate 12–13
savanna landforms 31
Sawney, B. L. 95
scale, air photograph 358
scale, soil survey and map 333–5, 366–8, 374–5, 379
scarpland 31
Schroth, C. L. 53, 55
Schwertmann, U. 87
Scott, R. M. 332
secondary nutrients 290–1
sector 276
sedimentary landforms 26–7
sedimentary rock 20–2
sedosol 197
Segalen, P. 88
self-sealing 220
semi-arid climate 14
semi-detailed survey 334–5
Senegal 115
series identification 365
serozem 196
seventh approximation 235, 250–1
shallow phase 178
Sharpe, C. F. S. 168
shelf 187
shield 20, 25
shifting cultivation 109, 419–20
Shoji, S. 174
Siband, P. 115
side-looking airborne radar 361
Sieffermann, G. 174
sierozem 196
Silanpää, M. 298
silcrete 167
silica 71–2
    silica: alumina ratio 262
    silica potential 76–7
    silica: sesquioxide ratio 97
silt 90
Simakova, M. S. 361
Simmonds, N. W. 306
Simonson, R. W. 182, 349
simple catena 272
simple mapping unit 345
single-purpose classification 385
sink-hole 187